AF810037

Compendium of Plant Genomes

Series Editor

Chittaranjan Kole, Raja Ramanna Fellow, Government of India, ICAR-National Research Center on Plant Biotechnology, Pusa, New Delhi, India

Whole-genome sequencing is at the cutting edge of life sciences in the new millennium. Since the first genome sequencing of the model plant Arabidopsis thaliana in 2000, whole genomes of about 100 plant species have been sequenced and genome sequences of several other plants are in the pipeline. Research publications on these genome initiatives are scattered on dedicated web sites and in journals with all too brief descriptions. The individual volumes elucidate the background history of the national and international genome initiatives; public and private partners involved; strategies and genomic resources and tools utilized; enumeration on the sequences and their assembly; repetitive sequences; gene annotation and genome duplication. In addition, synteny with other sequences, comparison of gene families and most importantly potential of the genome sequence information for gene pool characterization and genetic improvement of crop plants are described.

Interested in editing a volume on a crop or model plant?
Please contact Prof. C. Kole, Series Editor, at ckoleorg@gmail.com

More information about this series at http://www.springer.com/series/11805

Domenico Carputo •
Riccardo Aversano •
Maria Raffaella Ercolano
Editors

The Wild Solanums Genomes

Springer

Editors
Domenico Carputo
Department of Agricultural Sciences
University of Naples Federico II
Naples, Italy

Riccardo Aversano
Department of Agricultural Sciences
University of Naples Federico II
Naples, Italy

Maria Raffaella Ercolano
Department of Agricultural Sciences
University of Naples Federico II
Naples, Italy

ISSN 2199-4781 ISSN 2199-479X (electronic)
Compendium of Plant Genomes
ISBN 978-3-030-30342-6 ISBN 978-3-030-30343-3 (eBook)
https://doi.org/10.1007/978-3-030-30343-3

© Springer Nature Switzerland AG 2021
This work is subject to copyright. All rights are reserved by the Publisher, whether the whole or part of the material is concerned, specifically the rights of translation, reprinting, reuse of illustrations, recitation, broadcasting, reproduction on microfilms or in any other physical way, and transmission or information storage and retrieval, electronic adaptation, computer software, or by similar or dissimilar methodology now known or hereafter developed.
The use of general descriptive names, registered names, trademarks, service marks, etc. in this publication does not imply, even in the absence of a specific statement, that such names are exempt from the relevant protective laws and regulations and therefore free for general use.
The publisher, the authors and the editors are safe to assume that the advice and information in this book are believed to be true and accurate at the date of publication. Neither the publisher nor the authors or the editors give a warranty, expressed or implied, with respect to the material contained herein or for any errors or omissions that may have been made. The publisher remains neutral with regard to jurisdictional claims in published maps and institutional affiliations.

This Springer imprint is published by the registered company Springer Nature Switzerland AG
The registered company address is: Gewerbestrasse 11, 6330 Cham, Switzerland

This book series is dedicated to my wife Phullara
and our children Sourav and Devleena

Chittaranjan Kole.

This book is dedicated to Emeritus Professor Luigi Frusciante on the occasion of his 70th birthday.

He continues to be a true mentor of professional behavior for us.

Preface to the Series

Genome sequencing has emerged as the leading discipline in the plant sciences coinciding with the start of the new century. For much of the twentieth century, plant geneticists were only successful in delineating putative chromosomal location, function, and changes in genes indirectly through the use of a number of "markers" physically linked to them. These included visible or morphological, cytological, protein, and molecular or DNA markers. Among them, the first DNA marker, the RFLPs, introduced a revolutionary change in plant genetics and breeding in the mid-1980s, mainly because of their infinite number and thus potential to cover maximum chromosomal regions, phenotypic neutrality, absence of epistasis, and codominant nature. An array of other hybridization-based markers, PCR-based markers, and markers based on both facilitated the construction of genetic linkage maps, mapping of genes controlling simply inherited traits, and even gene clusters (QTLs) controlling polygenic traits in a large number of model and crop plants. During this period, a number of new mapping populations beyond F2 were utilized, and a number of computer programs were developed for map construction, mapping of genes, and for mapping of polygenic clusters or QTLs. Molecular markers were also used in the studies of evolution and phylogenetic relationship, genetic diversity, DNA fingerprinting, and map-based cloning. Markers tightly linked to the genes were used in crop improvement employing the so-called marker-assisted selection. These strategies of molecular genetic mapping and molecular breeding made a spectacular impact during the last one and a half decades of the twentieth century. But still they remained "indirect" approaches for elucidation and utilization of plant genomes since much of the chromosomes remained unknown, and the complete chemical depiction of them was yet to be unraveled.

Physical mapping of genomes was the obvious consequence that facilitated the development of the "genomic resources" including BAC and YAC libraries to develop physical maps in some plant genomes. Subsequently, integrated genetic–physical maps were also developed in many plants. This led to the concept of structural genomics. Later on, emphasis was laid on EST and transcriptome analysis to decipher the function of the active gene sequences leading to another concept defined as functional genomics. The advent of techniques of bacteriophage gene and DNA sequencing in the 1970s was extended to facilitate sequencing of these genomic resources in the last decade of the twentieth century.

As expected, sequencing of chromosomal regions would have led to too much data to store, characterize, and utilize with the-then available computer software could handle. But the development of information technology made the life of biologists easier by leading to a swift and sweet marriage of biology and informatics, and a new subject was born—bioinformatics.

Thus, the evolution of the concepts, strategies, and tools of sequencing and bioinformatics reinforced the subject of genomics—structural and functional. Today, genome sequencing has traveled much beyond biology and involves biophysics, biochemistry, and bioinformatics!

Thanks to the efforts of both public and private agencies, genome sequencing strategies are evolving very fast, leading to cheaper, quicker, and automated techniques right from clone-by-clone and whole-genome shotgun approaches to a succession of the second-generation sequencing methods. The development of software of different generations facilitated this genome sequencing. At the same time, newer concepts and strategies were emerging to handle sequencing of the complex genomes, particularly the polyploids.

It became a reality to chemically—and so directly—define plant genomes, popularly called whole-genome sequencing or simply genome sequencing.

The history of plant genome sequencing will always cite the sequencing of the genome of the model plant Arabidopsis thaliana in 2000 that was followed by sequencing the genome of the crop and model plant rice in 2002. Since then, the number of sequenced genomes of higher plants has been increasing exponentially, mainly due to the development of cheaper and quicker genomic techniques and, most importantly, the development of collaborative platforms such as national and international consortia involving partners from public and/or private agencies.

As I write this preface for the first volume of the new series "Compendium of Plant Genomes," a net search tells me that complete or nearly complete whole-genome sequencing of 45 crop plants, eight crop and model plants, eight model plants, 15 crop progenitors and relatives, and three basal plants is accomplished, the majority of which are in the public domain. This means that we, nowadays, know many of our model and crop plants chemically, i.e., directly, and we may depict them and utilize them precisely better than ever. Genome sequencing has covered all groups of crop plants. Hence, information on the precise depiction of plant genomes and the scope of their utilization are growing rapidly every day. However, the information is scattered in research articles and review papers in journals and dedicated Web pages of the consortia and databases. There is no compilation of plant genomes and the opportunity of using the information in sequence-assisted breeding or further genomic studies. This is the underlying rationale for starting this book series, with each volume dedicated to a particular plant.

Plant genome science has emerged as an important subject in academia, and the present compendium of plant genomes will be highly useful to both students and teaching faculties. Most importantly, research scientists involved in genomics research will have access to systematic deliberations on the plant genomes of their interest. Elucidation of plant genomes is of interest not only for the geneticists and breeders, but also for practitioners of an array of plant science disciplines, such as taxonomy, evolution, cytology, physiology, pathology, entomology, nematology, crop production, biochemistry, and obviously bioinformatics. It must be mentioned that information regarding each plant genome is ever-growing. The contents of the volumes of this compendium are, therefore, focusing on the basic aspects of the genomes and their utility. They include information on the academic and/or economic importance of the plants, description of their genomes from a molecular genetic and cytogenetic point of view, and the genomic resources developed. Detailed deliberations focus on the background history of the national and international genome initiatives, public and private partners involved, strategies and genomic resources and tools utilized, enumeration on the sequences and their assembly, repetitive sequences, gene annotation, and genome duplication. In addition, synteny with other sequences, comparison of gene families, and, most importantly, the potential of the genome sequence information for gene pool characterization through genotyping by sequencing (GBS) and genetic improvement of crop plants have been described. As expected, there is a lot of variation of these topics in the volumes based on the information available on the crop, model, or reference plants.

I must confess that as the series editor, it has been a daunting task for me to work on such a huge and broad knowledge base that spans so many diverse plant species. However, pioneering scientists with lifetime experience and expertise on the particular crops did excellent jobs editing the respective volumes. I myself have been a small science worker on plant genomes since the mid-1980s and that provided me the opportunity to personally know several stalwarts of plant genomics from all over the globe. Most, if not all, of the volume editors are my longtime friends and colleagues. It has been highly comfortable and enriching for me to work with them on this book series. To be honest, while working on this series I have been and will remain a student first, a science worker second, and a series editor last. And I must express my gratitude to the volume editors and the chapter authors for providing me the opportunity to work with them on this compendium.

I also wish to mention here my thanks and gratitude to the Springer staff, particularly Dr. Christina Eckey and Dr. Jutta Lindenborn, for the earlier set of volumes and presently Ing. Zuzana Bernhart for all their timely help and support.

I always had to set aside additional hours to edit books beside my professional and personal commitments—hours I could and should have given to my wife, Phullara, and our kids, Sourav and Devleena. I must mention that they not only allowed me the freedom to take away those hours from them but also offered their support in the editing job itself. I am really not sure whether my dedication of this compendium to them will suffice to do justice to their sacrifices for the interest of science and the science community.

New Delhi, India Chittaranjan Kole

Preface

Cultivated Solanums, such as potato, tomato, and pepper, possess many wild relatives of great importance for practical breeding and evolutionary studies. The diversity characterizing such germplasm represents a valuable reservoir of useful genes and alleles to improve the cultivated species. For many years, breeders have not fully exploited wild Solanums, mainly because of the lack of information regarding their genetics and genomics. However, the last twenty years were dominated by transformative technological developments in genomics and plummeting sequencing costs. Researchers have now deciphered the genome of important cultivated Solanaceae such as potato, tomato, eggplant, and pepper. On the heels of these recent developments, wild Solanum genomes are becoming available, opening an exciting new era for basic research, mapping, identification of genes, stress resilience and recovery mechanisms, and for downstream varietal development.

This book wants to take stoke of the current state of the play on the organization of genomes in wild Solanum species. It emphasizes how this information is yielding direct outcomes in molecular breeding and a better understanding of both the patterns and processes of evolution. Collectively, the unifying picture emerging is that genetics and genomics represent a quantum leap in describing wild Solanum genomes, digging out information on gene function and identifying agronomically superior alleles. However, genomic data for wild Solanums still lag well behind the massive amount of information available for their relative crops. This emphasizes the need to achieve more genomic resources for wild Solanum species for a deeper understanding of genomic variation and complex trait associations in wild species. The new knowledge acquired will help scientists to convert successful research into practical applications and face the critical challenges of this century: to meet the four Fs demands (food, feed, fiber, and fuel) for a growing human population living in a changing and unstable climate.

Chapters were written by 38 authors from 20 research institutes active in eight different countries. Their knowledge, vision, and dedication made this project possible and are all gratefully acknowledged. We are also thankful to Springer for promoting and supporting this endeavor.

Naples, Italy

Domenico Carputo
Maria Raffaella Ercolano
Riccardo Aversano

Contents

1 The Solanaceae Family: Botanical Features and Diversity

Riccardo Motti

Abstract

Among angiosperm families, the Solanaceae is one of the most important one to human beings and has extensive economic importance, mainly as a food source. It is distributed in all continents except Antarctica and the greatest concentration of diversity is in Central and South America. Many of the economically important crops in the world belong to the genus *Solanum, Capsicum, and Nicotiana* with about 28 million hectares cultivated globally. In addition to species that underwent domestication, this family encompasses several wild species used in traditional cuisine or as a font of useful genes/alleles for breeding efforts. Solanaceae are also known for possessing a diverse range of biologically active compounds that may be used to benefit human health and for crop protection. The Solanaceae family includes about 100 genera and approximately 2500 species; recent classification has identified nine clades at subfamily level and 14 at tribe level.

R. Motti (✉)
Department of Agricultural Sciences, University of Naples Federico II, Via Università 100, 80055 Portici, Italy
e-mail: motti@unina.it

1.1 Introduction

Solanaceae is a cosmopolitan family with mostly worldwide distribution on all continents except Antarctica. The greatest concentration of diversity is in Central and South America and it is believed that the family originated there (Hunziker 1979). A recent study by Palchetti et al. (2020) reported that Solanaceae comprises 1611 species in South America. Peru shows the greatest diversity, including genera, total species, and endemic species; in this country, *Solanum*, *Jaltomata* and *Nolana* are the most species-rich genera. Peru also holds the most significant number of potato and tomato species in the world (Peralta et al. 2008; Spooner et al. 2016). Members of the Solanaceae can be found growing in a wide variety of habitats, from deserts to tropical rainforests, from costs to mountain areas (Knapp 2020).

Among angiosperm families, the Solanaceae ranks one of the most important to human beings, containing a number of economically important crops (Table 1.1). Many of them belong to the genus *Solanum*, which includes *Solanum lycopersicum* L. (=*Lycopersicon esculentum* Miller) (tomato), *Solanum tuberosum* L. (potato), *Solanum melongena* L. (eggplant or aubergine). Several other species take on great importance such as those belonging to the genus *Nicotiana* (*N. tabacum* L.), *Capsicum annuum* L. (both chili peppers and bell peppers), *Physalis* spp.(tomatillo, Cape

© Springer Nature Switzerland AG 2021

D. Carputo et al. (eds.), *The Wild Solanums Genomes*, Compendium of Plant Genomes,
https://doi.org/10.1007/978-3-030-30343-3_1

Table 1.1 Worldwide production of the main Solanaceae crop species (Fao 2018)

	Hectares	Tonnes	Yield (T/ha)
Potato	17.578.672	368.168.914	20.9
Tomato	4.762.457	182.256.458	38.3
Tobacco	3.368.929	6.094.875	1.fv8
Chilli and pepper	1.990.423	36.771.482	18.5
Eggplant	1.864.556	54.077.210	29.0

gooseberry, and Chinese lantern), and *Lycium* (*L. barbarum* L. and *L. chinense* L.).

Many genera of Solanaceae provide ornamental plants widely used in gardens, including *Brunfelsia* (Lady of the night), *Cestrum* (Jessamine), the closely related *Brugmansia* and *Datura* (Angel's trumpet) and *Petunia*.

1.2 Morphological Characteristics

The Solanaceae family shows a great range of habitus; it generally comprises herbs and shrubs, but also trees, woody vines, lianas, and sometimes epiphytic plants (Fig. 1.1).

They can be annuals, biennials, or perennials; subterranean tubers are sometimes present. This family does not have either laticifers or latex and prickles, and stellate trichomes can be present in some species.

Leaves present a suite of characteristics useful for plant identification. They are entire to deeply lobed, never compound, with pinnate or reticulate venation, exstipulate, generally simple to pinnatifid or ternate, alternately arranged or opposite (Fig. 1.2). In some species (*Physalis*), leaves can be alternate at the base of the plant and opposed towards the apex. Petiole is generally present, although leaves can be also subsessile or rarely sessile. This family does not have laticifers nor latex.

Flowers are solitary or bore in cymose, racemose, paniculate or compound-corymbose inflorescences, terminal, lateral or axillary. Pollination is generally entomophilous, but some species can be bat-dependent (Sazima et al. 2003). Flowers themselves are dichlamydeous, having both a calyx and a corolla, bisexual or rarely unisexual (e.g., *Symonanthus,* some *Solanum* species). The calyx is typically

Fig. 1.1 Some Solanaceae habitus: **a** herbaceous (*Petunia* × *hybrida* hort. ex Vilm.); **b** shrubby (*Solanum pseudocapsicum* L. 'Variegatum'); **c** arborescent (small tree) (*Brugmansia arborea* (L.) Steud.); **d** woody climbing (*Solanum laxum* Spreng.)

Fig. 1.2 Solanaceae leaf-shapes: entire **a** *Capsicum annuum* L.; lobed **b** *Solanum melongena* L.; deeply lobed **c** *Solanum lycopersicum* L.; lobed and spiny **d** *Solanum mammosum* L.

gamosepalous, from tubular to campanulate, with 5 (3–9) lobes, persistent or sometimes accrescent. In a few species (e.g., *P. alkekengi* L.), sepals resume growth after pollination to encapsulate the mature fruit, forming the "Chinese lantern," a trait also termed inflated-calyx syndrome (ICS) (Hu and Saedler 2007; Wilf et al. 2017).

The corolla is ebracteate, gamopetalous, actinomorphic (Figs. 1.3 and 1.4) or slightly zygomorphic (Fig. 1.5), rotate, campanulate, tubular, funnel-shaped, trumpet-shaped, or urn-shaped; corolla lobes are generally 5 (varying from 3 to 9), and usually display valvate or plicate arrangement within the flower bud.

Stamens are free, usually 5 or 4 (rarely 2, 6, 8), epipetalous and alternate with lobes, equal or unequal in length; anthers are either bilocular or unilocular, dorsifixed, or basifixed, often connivant, longitudinal, or poricidal in dehiscence. Staminodes (stamens with non-bearing anthers) occur in some species (e.g., *Schizanthus*, *Salpiglossis*). Gynoecium is syncarpous, with 2 oblique carpels (sometimes 4–5), whereas the ovary is superior and two-loculated; it may be secondarily divided by false septa (e.g., *Datura*). Style is simple, and stigma is capitate or rarely bilobate. Ovules are numerous, anatropous, and placentation is axile, rarely basal. In this family, fruit is predominantly a berry, a drupe, a capsule (septicidal or rarely loculicidal) (Fig. 1.6). In each fruit, seeds are numerous and hold curved, spiral, or straight embryos; cotyledons are two.

1.3 Taxonomy and Domestication

Assigned to Solanales order in the APG IV classification (Stevens 2015), along with Convolvulaceae, Hydroleaceae, Montiniaceae, and Sphenocleaceae, Solanaceae are considered a monophyletic group based on chloroplast DNA restriction (Olmstead and Palmer 1992). It is the largest plant family after Poaceae and Fabaceae and, as typical of large plant families, its classification is difficult and shows several challenges. Traditional classifications of the family typically recognized two subfamilies, Cestroideae and Solanoideae, the former characterized by curved embryos, discoid seeds, and berry-like fruits, the latter with straight embryos, contained in small, angular to subglobose seeds and capsular fruits (Olmstead et al. 2008, D'Arcy 1991). Molecular phylogeny based on two chloroplast DNA regions (*ndhF* and *trnLF*) allowed the

Fig. 1.3 Actinomorphic corollas. Trumpet-shaped: **a** *Lycianthes rantonnetii* (Carrière) Bitter; Rotate: **b** *Solanum eleagnifolium* Cav.; Tubulose: **c** *Iochroma cyaneum* (Lindl.) G.H.M. Lawr. & J.M. Tucker; Funnel-shaped: **d** *Datura stramonium* L.; Campanulate: **e** *Solandra maxima* (Sessé & Moc.) P.S. Green; Urn-shaped: **f** *Atropa belladonna* L.

Fig. 1.4 *Datura metel* L. 'Ballerina Purple' is a cultivar with a double corolla, resembling a dancer's costume

identification of four main clades at the subfamily level, 14 main clades at the tribe level, around 100 genera, and 2500 species (Olmstead et al. 2008). Recent APG classification (2016) identifies nine clades at subfamily level and 14 at tribe level (Table 1.2).

Solanaceae family began to diversify 50 to 65 million years ago (Magallòn et al. 2013). A Bayesian study by Dupin et al. (2017) on the historical biogeography of Solanaceae pointed out that South America is both the family distributional center as well as its ancestral range, and that dispersal events were the principal mode by which members of the family have spread. The authors hypothesized that short-range movements account for most of the spread beyond South America, although multiple long-distance dispersals to Africa, Australia, and Eurasia occurred.

The tomato was introduced from the Andean region to Europe by the Spanish in the early sixteenth century. Today, it represents the most economically important vegetable crop worldwide (Bergougnoux 2014) (see also Chap. 3). While the Spanish and Italians seem to have been the first Europeans to adopt it as a food, it was initially grown as an ornamental plant in France and Northern Europe because botanists

Fig. 1.5 Slightly zygomorphic corollas: **a** *Nicotiana tabacum* L.; **b** *Hyoscyamus albus* L.; **c** *Nicandra physalodes* (L.) Gaertn

Fig. 1.6 Berries types in the Solanaceae family: **a** *Solanum lycopersicum* L.; **b** *S. melongena* L.; **c** *S. chenopodioides* Lam.; **d** *S. tuberosum* L.; **e** *S. mammosum* L.; **f** *Caspicum annuum* L.

recognized it as a relative of the poisonous belladonna and deadly nightshade (Petruzello 2020a, b).

The cultivated potato traces its origin to Andean and Chilean landraces developed by pre-Columbian cultivators (Spooner et al. 2005) (see also Chap. 4). Remains of cultivated potatoes, identified based on their distinctive starch grains, have been found at coastal sites in Peru dating from 2000 B.C. (Prance and Nesbitt 2012). It would seem likely that potatoes were not seen by Europeans before 1532 when Pizarro first ascended the Andes of northern Peru and was mentioned for the first time in what is now Colombia in 1537, and in Perù by Lopez de Gomara's in 1552 and Cieza de Leon in 1553 (Hawkes and Francisco-Ortega 1993). The earliest known records of potato cultivation in Europe date at the slightly earlier date of about 1562 in the Canary Islands and from between 1573 and 1576 in Seville in Spain (Hawkes and Francisco-Ortega 1993; Prance and Nesbitt 2012).

Eggplant was domesticated from a group of spiny Solanum species in the Indo-Burma region and was early adopted in China as a vegetable

Table 1.2 Recent classifications of the Solanaceae family

APG IV (Stevens 2015)		Olmestead (2008)	
Subfamily	*Tribe*	Subfamily	*Tribe*
Schizanthoideae		Nicotianoideae	*Anthocercideae*
Goetzeoideae		Cestroideae	*Salpiglossideae*
Duckeodendroideae			*Browallieae*
Cestroideae			*Cestreae*
Salpiglossideae	*Browallieae*	Goetzeoideae	
	Cestreae	Solanoideae	*Solaneae*
	Schwenkieeae		*Capsiceae*
Petunioideae			*Physaleae*
Nicotianoideae	*Nicotianeae*		*Datureae*
Anthocercideae			*Juanulloeae*
Solanoideae	*Hyoscyameae*		*Hyoscyameae*
	Nolaneae		*Lycieae*
	Lycieae	Unresolved clades	*Schwenckieae*
	Mandragoreae		*Petunieae*
	Solandreae		*Benthamielleae*
	Nicandrea		
	Datureae		
	Solaneae		
	Capsiceae		
	Physalideae		

crop (discussed also in Chap. 6). Eggplant was unknown by the ancient Greeks and Romans but spread throughout the Mediterranean Basin due to Muslim expansion in the early Middle Ages (Daunay and Janick 2007; Yılmaz et al. 2013) and nowadays is widely grown in the Americas, Europe, and Asia. Its close relatives: *S. aethiopicum* L. (scarlet eggplant) and *S. macrocarpon* L. (African eggplant) are of African origin (Sękara et al. 2007).

The genus *Capsicum* (chilies and other peppers) comprises annual or perennial herbs and shrubs native to South and Central America. Chilies are believed as one of the earliest domesticated plants in the New World. The oldest evidence of domestication of these plants dates to 5000–6500 B.C. (Davenport 1970) (also reported in Chap. 6). The chili was brought to Europe by Colombus, and enthusiastically and rapidly was incorporated into many cultures. Within 50 years, this spice spread from Spain to England (Lippert et al. 1966). The genus *Capsicum* consists of 20–27 species (Walsh and Hoot 2001), five of which are widely cultivated (*C. annuum* L., *C. baccatum* L., *C. frutescens* L., *C. chinense* Jacq. and *C. pubescens* Ruiz & Pav.) (Heiser and Pickersgill 1969). *C. annuum* is the most common and extensively cultivated.

The tomatillo (*Physalis philadelphica* Lam.) is native to Mexico and is thought to have been first domesticated by the Aztecs around 800 BCE and was an important food crop to a number of pre-Columbian peoples in Mesoamerica, including the Mayans (Small 2011; Petruzzello 2020a, b).

Nicotiana tabacum and *Nicotiana rustica* are native plants of the Americas, having evolved in the Andes around Peru/Ecuador. Tobacco is thought to have been cultivated since about 5000–3000 BC, dispersed by Amerindias through both the southern and the northern American continent. Everywhere in the New World tobacco was employed in sacred magical and medicinal contexts and was undoubtedly more widely used for these purposes than any other plant (Schultes 1979). Tobacco was widely used by native people by the time that Columbus arrived in North America in 1492 (Musk and De Klerk 2003; Goodman 2005). It is likely that the first Europeans to smoke were Columbus' crew returning from the Americas in the late 15th and early sixteenth centuries (Musk and De Klerk 2003). During the sixteenth century, tobacco use spread throughout all of Europe (Davey 1999).

Recently the dried fruits of *L. barbarum* L. and *L. chinense* Mill., have become popular in Europe, where they are known as goji berries, and are considered functional food on account of their high vitamin and mineral content (Samuels 2015). Both species are native to Asia, and most commercial cultivation occurs in Northern China (Wang 2007).

1.4 Importance of *Solanaceae* Wild Species

In addition to species that underwent domestication, this family encompasses several wild species used in traditional cuisine, as a source of phytochemicals and as font of useful genes/alleles for breeding efforts.

Besides the well-known cultivated species, many wild species are used in the folk cuisine of some rural communities. They are utilized principally as leafy or fruit vegetables thanks to the richness of bioactive compounds contained in their edible parts. Cooked and consumed like spinach are, for example, the young leaves of *Jaltomata repandidentata* (Dunal) Hunz., (Samuels 2015), while *Cestrum nocturnum* L. leaves are eaten cooked with tortillas (Kermath et al. 2014). Eaten in the Americas, are the fruits of *Acnistus arborescens* (L.) Schltdl., *Chamaesarcha coronopus* (Dunal) A.Gray, *Jaltomata* spp. (Kermath et al. 2014), *Solanum stramonifolium* Jacq., *S. sessiliflorum* Dunal (Turner et al. 2011), *Lycianthes asarifolia* (Kunth and Bouché) Bitter and *L. lycioides* (L.) Hassl (Samuels 2015).

Solanaceae are also known for possessing a diverse range of biologically active compounds that may be used both for the benefit of human health and for crop protection. Based on their chemical structure, these metabolites can be classified into three main groups: terpenes (composed almost entirely of carbon and hydrogen and including plant volatiles, cardiac glycosides, carotenoids, and sterols), phenolics (possessing one or more phenol rings and comprising phenolic acids, coumarins, flavonoids, tannins, and lignin), and nitrogen-containing compounds (including alkaloids and glucosinolates) (Chowański et al. 2016). Of all the various classes of compounds, by far alkaloids have received the most attention due to their abundance and properties. They comprise nicotine and tropane alkaloids, and are localized in all plant parts, with the highest concentrations in roots and seeds in proportions that vary among species, time of year, location, and plant organs (Maga 1994). Tropane alkaloids constitute one of the distinctive groups of secondary metabolites of the Solanaceae and, despite their extreme toxicity, are important drugs when administered in appropriate dosages (Shah et al. 2013). Atropine is an anticholinergic drug occurring naturally in a number of species (e.g., *Atropa belladonna*, *Datura stramonium* L.) and is classified as muscarinic receptor antagonist (Brown and Taylor 2001). This alkaloid can affect the central as well as the peripheral nervous systems and, in overdoses, is poisonous. Atropine is used in the clinical practice in the management of myopia, as an antidote in Organophosphate poisoning and the treatment of bradycardia (Al 2014; Almubayedh et al. 2018). Hyoscyamine and scopolamine are secondary metabolites of some plants, particularly henbane (*Hyoscyamus niger, H. albus*). Both these tropane alkaloids are anticholinergics and, when used in small doses, they have medical uses such as treating gastrointestinal disorders (Kumar and Tewari 2018).

Wild Solanaceae are also important for crop improvement. They represent a precious reservoir of genes and alleles lacking in the cultivated species that can be moved into the cultivated gene pools through specific breeding approaches and technologies. Many cultivated varieties of the Solanaceae family have been developed through the exploitation of wild germplasm. A well-known and nice example of this is the contribution of Mexican species *Solanum demissum* to broaden the resistance to light blight in the common potato, *S. tuberosum*. Dominant *R* genes from *S. demissum* are currently present in most potato varieties, although new pathogen strains have now overcome many of these resistances. Finally, the contribution of wild species from the Solanaceae family is also crucial in genetic and genomic plant research. This part, as well as that related to the use of wild species in breeding, will be better expanded in the next chapters.

References

Al B (2014) The source-synthesis-history and use of atropine. Eurasian J Emerg Med 13(1):2

Almubayedh H, Albannay R, Alelq K, Ahmad R, Ahmad N, Naqvi AA (2018) Clinical uses and toxicity of *Atropa belladonna*; an evidence based comprehensive retrospective review. Biosci Biotech Res Comm 11:41–48

Bergougnoux V (2014) The history of tomato:from domestication to biopharming. Biotechnol Adv 32 (1):170–189

Brown JH, Taylor P (2001) Muscarinic receceplamine, and this treatment seemed to assist with tor agonists and antagonists. In:Goodman and recovery, especially achieving restful sleep. This Gilman's The Pharmacological Basis of Therapeuis the only report in the literature that describes tics, 10th edn. (JG Hardman, LE Limbird and the use of anticholinergics to treat the behavioral AG Gilman. Eds). McGrow-Hill, New York, NY, USA, pp 155–173

D'Arcy WG (1991) The Solanaceae since 1976, with a review of its biogeography. In: Hawkcs JG, Lester RN, Nee M, Estrada N (eds), Solanaceae III:taxonomy, chemistry, evolution. Royal Botanic Gardens, Kew

Daunay MC, Janick J (2007) History and iconography of eggplant. Chron Horticult 47(3):16–22

Davenport WA (1970) Progress report on the domestication of Capsicum (chili peppers). Proc Assoc Am Geogr 2:46–47

Davey M (1999) The European Tobacco Trade from the 15th to the 17th Centuries. University of Minnesota, James Ford Bell Library

Dupin J, Matzke NJ, Särkinen T, Knapp S, Olmstead RG, Bohs L, Smith SD (2017) Bayesian estimation of the global biogeographical history of the Solanaceae. J Biogeogr 44(4):887–899

FAO. (2018). http://www.fao.org/faostat/en/#data/QC. Accessed 26 Sept 2020

Goodman J (2005) Tobacco in history: the cultures of dependence. Routledge, London and New York

Hawkes JG, Francisco-Ortega J (1993) The early history of the potato in Europe. Euphytica 70(1–2):1–7

Heiser CB Jr, Pickersgill B (1969) Names for the cultivated Capsicum species (Solanaceae). Taxon 18 (3):277–283

Hu JY, Saedler H (2007) Evolution of the inflated calyx syndrome in Solanaceae. Mol Biol Evol 24(11):2443–2453

Hunziker AT (1979) South American Solanaceae: a synoptic survey. In: Hawkes JG, Lester RN, Skelding AD (Eds) The biology and taxonomy of the Solanaceae. Academic Press, London, UK, pp 49–85

Kermath BM, Bennett BC, Pulsipher LM (2014) Food plants in the Americas: a survey of the domesticated, cultivated, and wild plants used for human food in North, Central and South America and the Caribbean, Unpubl. Manuscript, University of Wisconsin Oshkosh, Oshkosh

Knapp S (2020) Biodiversity (Solanaceae) of Nicotiana. In: The Tobacco plant genome. Springer, Switzerland, p 21

Kumar R, Tewari AK (2018) Isolation of medicinally important constituents from rare and exotic medicinal plants. In Synthesis of medicinal agents from plants. Elsevier, pp 229–256

Lippert LF, Smith PG, Bergh BO (1966) Cytogenetics of the vegetable crops: garden pepper, Capsicum sp. Bot Rev 32:25–55

Maga JA (1994) Glycoalkaloids in Solanaceae. Food Rev Int 10(4):385–418

Musk AW, De Klerk NH (2003) History of tobacco and health. Respirology 8(3):286–290

Olmstead RG, Bohs L, Migid HA, Santiago-Valentin E, Garcia VF, Collier SM (2008) A molecular phylogeny of the Solanaceae. Taxon 57(4):1159–1181

Olmstead RG, Palmer JD (1992) A chloroplast DNA phylogeny of the Solanaceae: subfamilial relationships and character evolution. Ann Missouri Bot 346–360

Palchetti MV, Barboza GE, Cantero JJ (2020) Solanaceae diversity in South America and its distribution in Argentina. An Acad Bras Ciênc 92(2)

Peralta IE, Spooner DM, Knapp S (2008) Taxonomy of wild tomatoes and their relatives (Solanum sect.

Lycopersicoides, sect. Juglandifolia, sect. Lycopersicon; Solanaceae). Syst Bot Monogr 84:1–186

Petruzzello M. 2020 https://www.britannica.com/plant/tomato (Accessed 16 September 2020)

Petruzzello M (2020) https://www.britannica.com/plant/tomatillo. Accessed 16 Sept 2020

Prance G, Nesbitt M (eds) (2012) The cultural history of plants. Routledge, New York and London

Samuels J (2015) Biodiversity of food species of the Solanaceae family: a preliminary taxonomic inventory of subfamily Solanoideae. Resources 4(2):277–322

Sazima M, Buzato S, Sazima I (2003) Dyssochroma viridiflorum (Solanaceae): a reproductively bat-dependent epiphyte from the Atlantic Rainforest in Brazil. Ann Bot 92(5):725–730

Schultes RE (1979) Solanaceous hallucinogens and their role in the development of New World cultures. In: Linnean society symposium series

Sękara A, Cebula S, Kunicki E (2007) Cultivated eggplants–origin, breeding objectives and genetic resources, a review. Folia Hortic 19(1):97–114

Shah VV, Shah ND, Patrekar PV (2013) Medicinal plants from Solanaceae family. Res J Pharm Technol 6 (2):143–151

Small E (2011) Top 100 exotic food plants. CRC Press, pp 118–119

Spooner DM, Alvarez N, Peralta IE, Clausen AM (2016) Taxonomy of wild potatoes and their relatives in Southern South America (Solanum sect. Petota and Etuberosum). Syst Bot Monogr 100:1–240

Spooner DM, McLean K, Ramsay G, Waugh R, Bryan GJ (2005) A single domestication for potato based on multilocus amplified fragment length polymorphism genotyping. PNAS USA 102(41):14694–14699

Stevens PF (2015) Angiosperm phylogeny website, version 14. http://www.mobot.org/MOBOT/research/APweb/. Accessed 6 July 2020

Turner NJ, Łuczaj ŁJ, Migliorini P, Pieroni A, Dreon AL, Sacchetti LE, Paoletti MG (2011) Edible and tended wild plants, traditional ecological knowledge and agroecology. Crit Rev Plant Sci 30(1–2):198–225

Walsh BM, Hoot SB (2001) Phylogenetic relationships of Capsicum (Solanaceae) using DNA sequences from two noncoding regions: the chloroplast atpB-rbcL spacer region and nuclear waxy introns. Int J Plant Sci 162(6):1409–1418

Wang Y (2007) Chinese medicinal plant-Lycium barbarum. Sol Newsl 14:14

Wilf P, Carvalho MR, Gandolfo MA, Cúneo NR (2017) Eocene lantern fruits from Gondwanan Patagonia and the early origins of Solanaceae. Science 355 (6320):71–75

Yılmaz H, Akkemik Ü, Karagöz Ş (2013) Identification of plant figures on stone statues and sarcophaguses and their symbols: the hellenistic and roman periods of the eastern mediterranean basin in the Istanbul Archaeology Museum. Mediterr Archaeol Archaeom 13(2)

Cytogenetics of Potato and Tomato Wild Relatives

2

Paola Gaiero, Giovana A. Torres, and Marina Iovene

Abstract

Cytogenetics has historically contributed to the taxonomy, genetics, and breeding of cultivated and wild *Solanum* species. This chapter summarizes the contributions of cytogenetic research to our understanding of genome structure and evolution of potato and tomato wild relatives. We focus on the advances in cytogenetics, going from the classical chromosome morphological analysis of species and their hybrids to the recent oligonucleotide-based chromosome paints, which are helping to identify and compare chromosomes and genomes of the wild *Solanum* relatives, detect large-scale changes among these species, and clarify the parental origin of polyploid potatoes. Given the large number of species, comparative fluorescence in situ hybridization (FISH) mapping and genome size data are still sparse. However, these studies are helping uncover the karyotypic differences among cultivated and wild *Solanum* species, a diversity with a significant impact on introgression and pre-breeding programs, and characterize their rich repertoire of tandem satellite sequences. In addition, this chapter summarizes how the analysis of the centromeres of several *Solanum* species has provided a new model system to study the centromere evolution and the accumulation of satellite repeats in these specialized chromosomal regions.

2.1 Introduction

Solanum is one of the most abundant genera of the Angiosperms and comprises nearly half of the species of the Solanaceae family. The genus includes many landraces and wild crop relatives, which are an invaluable resource of genes and allelic variants critical for the genetic improvement of *Solanum* crops such as potato and tomato (see Chaps. 3, 4, 5, 6). The introgression of useful traits into the cultivated *Solanum* crops from their wild relatives is often challenging due to various pre- and postzygotic reproductive barriers among these species (reviewed in Camadro et al. 2004; Bedinger et al. 2011; Spooner et al. 2014; Bethke et al. 2017; Chetelat et al. 2019). Chromosomal rearrangements and other large-scale changes between parental genomes may represent an additional barrier to introgression, by compromising meiotic

P. Gaiero
Faculty of Agronomy, University of the Republic, Montevideo, Uruguay

G. A. Torres
Department of Biology, Institute of Natural Sciences, Federal University of Lavras, Lavras, MG, Brazil

M. Iovene (✉)
Institute of Biosciences and BioResources (CNR-IBBR), CNR, 80055 Portici, NA, Italy
e-mail: marina.iovene@ibbr.cnr.it

© Springer Nature Switzerland AG 2021

D. Carputo et al. (eds.), *The Wild Solanums Genomes*, Compendium of Plant Genomes,
https://doi.org/10.1007/978-3-030-30343-3_2

chromosome pairing and disjunction, and/or by decreasing the fitness of the interspecific hybrids (Seah et al. 2004; van der Knaap et al. 2004; Anderson et al. 2010; Verlaan et al. 2011; Gaiero et al. 2018). Cytogenetic research can contribute to point out the karyotypic differences among related species.

Historically, cytogenetics has had a profound impact on *Solanum* taxonomy, genetics, and breeding, especially for potatoes and tomatoes. Early cytological studies in potatoes contributed to the discovery of the widespread occurrence of unreduced (2n) gametes in the tuber-bearing Solanums and its underlying cytogenetic mechanisms (den Nijs and Peloquin 1977; Peloquin et al. 1999; Carputo et al. 2000), and to the endosperm balance number (EBN) hypothesis related to the fundamental role of endosperm in the outcome of crosses among *Solanum* species (Johnston et al. 1980; Hawkes and Jackson 1992; reviewed in Carputo et al. 1999; Bethke et al. 2017). These seminal studies, along with an efficient method to generate potato maternal haploids (Hougas et al. 1964), led to new potato breeding strategies based on ploidy level manipulation, thus facilitating crosses with wild species and diploid landraces (reviewed in Carputo et al. 2000; Bethke et al. 2017). Tomato cytogenetic stocks, including primary trisomics and radiation-induced deletion lines, have had a key role in the development of the first genetic linkage maps of tomato and the assignment of linkage groups to individual pachytene chromosomes (reviewed in Harper and Cande 2000; Chetelat and Ji 2007). These studies indicated an overall nonlinear relationship between genetic and cytological distances and that cytological chiasmata do not form in the heterochromatic regions of tomato (Barton 1951; Khush and Rick 1968).

Despite these advances, *Solanum* is not an amenable genus for cytogenetic studies. The wild relatives of potato (sect. *Petota*) and tomato (sect. *Lycopersicon*) and their close outgroup species (sects. *Etuberosum*, *Juglandifolium* and *Lycopersicoides*) all share the same base chromosome number x = 12 (Rodríguez and Spooner 2009; reviewed in Gavrilenko 2011; Grandillo et al. 2011). Most species are diploid (2n = 2x = 24), with polyploids restricted to the potato clade. Similar to cultivated potato (2n = 4x = 48) and tomato (2n = 2x = 24), most wild Solanums have small and condensed mitotic chromosomes that are poorly differentiated in morphology and size, and not suitable for a detailed cytogenetic analysis. Therefore, the identification of the chromosomes of cultivated potato and tomato and the initial comparisons with their wild relatives have been based on the morphology of their pachytene chromosomes (Barton 1950; Marks 1955, 1969; Sawant 1958; Rick 1960; Khush and Rick 1963; Yeh and Peloquin 1965; Ramanna and Prakken 1967; Ramanna and Wagenvoort 1976; Wagenvoort 1988; Matsubayashi 1991), an analysis feasible only for diploid genotypes.

A major breakthrough came from the development of chromosome-specific markers based on libraries of bacterial artificial clones (BAC) of potato and tomato species (reviewed in Szinay et al. 2010; Gavrilenko 2011) and, more recently, from chromosome paint probes based on collections of synthesized oligonucleotides that cover entire chromosomes or regions (Braz et al. 2018; He et al. 2018; reviewed in Jiang 2019; Pham et al. 2019). The chromosome-specific BACs were initially used to identify the chromosomes of cultivated *Solanum* crops, integrate genetic linkage maps of these species with their chromosomes, and contribute to the sequencing efforts of these crop species. In addition, chromosome markers and paints are currently helping to elucidate the genome organization of both wild and cultivated Solanums and the extent of synteny and large-scale chromosome rearrangements among these species. Given the large number of wild potatoes and tomatoes, these comparative studies are sketchy, that is, focused on a few species and/or a few chromosomes. Besides, we lack a comprehensive dataset on the genome size values and repeat contents, especially across the numerous wild potato relatives.

The present chapter summarizes the contributions of cytogenetic research to our understanding of the genome structure and evolution of potato and tomato wild relatives, with a focus on the extent of chromosomal rearrangements and differences in composition and chromosomal location of repetitive DNA among *Solanum* species. Emphasis is also given to centromeric DNA because the potato and its close relatives have recently emerged as a model system to study the evolution of centromere-associated sequences. Prospective applications of cytogenetics to comparative studies in *Solanum* are briefly discussed.

2.2 Chromosome Identification in *Solanum*

Chromosome identification is at the basis of any cytogenetic investigation and is a useful tool for plant breeding and comparative studies. This section summarizes resources and tools that were initially established to identify potato and tomato chromosomes, going from chromosome morphological analysis to the recent oligonucleotide-based chromosome paints, and that were subsequently applied to study wild *Solanum* relatives.

Meiotic pachytene chromosomes provide sufficient morphological variations in length, arm ratio, and amount and distribution of heterochromatin to distinguish all 12 chromosome pairs of tomato and diploid potato clones (Barton 1950; Yeh and Peloquin 1965; Ramanna and Prakken 1967; Marks 1969; Ramanna and Wagenvoort 1976; Wagenvoort 1988). The pachytene analysis was extended to several wild tomato relatives and interspecific hybrids as well as a few wild potatoes (Sawant 1958; Menzel 1962; Khush and Rick 1963; Marks 1969; Hermsen and Ramanna 1973). However, this classical analysis had limited power to detect structural rearrangements, and, in addition, it was not easy to perform on polyploid potatoes.

Fluorescence in situ hybridization (FISH) and the use of large-insert genomic libraries have provided a robust tool for chromosome identification, karyotyping, and integration of the chromosomal features in the genetic linkage maps of both potato and tomato (Fuchs et al. 1996; Dong et al. 2000; Song et al. 2000; Tang et al. 2009; Choudhary et al. 2020). Various BAC libraries are available for potato (de Boer et al. 2011; Yang et al. 2015; Chen et al. 2019), tomato (Fulton et al. 2002; Budiman et al. 2004), and several wild relatives, e.g. *S. bulbocastanum* (Song et al. 2000), *S. habrochaites* (Wolters et al. 2015), and *S. pinnatisectum* (Chen et al. 2004). For each potato and tomato chromosome, a considerable number of BACs were selected by screening these libraries with genetically mapped molecular markers. In turn, many of these potato and tomato map-anchored BACs were FISH-mapped on the chromosomes of the corresponding species. Because the initial sequencing efforts for these crops used a BAC-by-BAC approach, BAC FISH provided valuable support to validate the assemblies of both species genomes (The Potato Genome Sequencing Consortium 2011; The Tomato Genome Consortium 2012). Chromosome-specific BACs have also provided a powerful tool to reveal collinearity and chromosome rearrangements between *Solanum* crops and their wild relatives in comparative FISH mapping studies. This is because most BACs located in either potato or tomato euchromatic regions generate distinct FISH signals not only within *Solanum* genus but also in related genera such as *Capsicum*. In addition, the availability of many stable fluorochromes has enabled the mapping of multiple probes at once, avoiding the need for time-consuming re-probing experiments of the same slides (Peters et al. 2009; Szinay et al. 2010). However, potato and tomato BACs located in heterochromatic regions generally work exclusively in close relatives (Iovene et al. 2008; Tang et al. 2008; Lou et al. 2010; Peters et al. 2012; Szinay et al. 2012; Gaiero et al. 2016), thus limiting the detection of rearrangements with breaks in these regions.

Similarly, several satellite repeats generate different hybridization patterns in terms of abundance, distribution, and number of FISH signals among closely related species and even at intraspecies level among accessions (Tek et al. 2005; Torres et al. 2011; Gong et al. 2012; Zhang et al. 2014). Comparative FISH mapping using these satellite repeats has often provided insights into the evolutionary dynamics of these repetitive elements rather than highlighting rearrangements per se. This topic is reviewed in the last part of the chapter.

The availability of reference genomes of various *Solanum* crops and the technical advances in DNA synthesis have opened to a new strategy to paint individual chromosomes (Beliveau et al. 2012; Han et al. 2015; Braz et al. 2018; reviewed in Jiang 2019). This new strategy relies on FISH probes made of pools of thousands of custom-synthesized oligonucleotides, which are designed based on single-copy sequences associated with a specific chromosome or chromosome region (Beliveau et al. 2012; Han et al. 2015; Braz et al. 2018; Pham et al. 2019; do Vale Martins et al. 2019). Braz et al. (2018) selected as FISH probes a large set of oligos from the single-copy sequences associated with 26 specific chromosome regions in the potato genome. These oligo-probes produced 26 distinct FISH signals that uniquely labeled each of the twelve potato chromosomes with a sort of barcode/banding pattern, which, in turn, allowed the karyotyping of all mitotic metaphase chromosomes at once in diploid, tetraploid, and hexaploid potatoes. Along with oligo-based whole chromosome paints, these probes were successfully used for comparative FISH mapping among distantly related *Solanum* species (Braz et al. 2018). The authors showed that the oligo-FISH barcode approach enables to pinpoint rearranged chromosomes among related species even by using low-resolution mitotic karyotypes (Braz et al. 2018; see below). In addition, He et al. (2018) demonstrated that oligo-based FISH is a robust tool to visualize specific chromosomes of *Solanum* polyploids during various meiotic stages, which opens new opportunities for cytogenetic investigations in polyploids and hybrids (see next section).

Finally, genomic in situ hybridization (GISH), alone or in combination with FISH, has been extensively applied in *Solanum* to identify specific genomes or alien chromosomes in interspecific hybrids and natural polyploids, as well as to study their meiotic behavior (Parokonny et al. 1997; Garriga-Calderé et al. 1998; Dong et al. 2001; Ji and Chetelat 2003; Ji et al. 2004; Pendinen et al. 2012; Rakosy-Tican et al. 2020). The next sections of this chapter will summarize the main findings of these cytogenetic investigations.

2.3 Chromosome Number, Ploidy, and Genome Differentiation

All species of sect. *Petota*, *Lycopersicon,* and their closely related outgroups (sects. *Etuberosum*, *Juglandifolium,* and *Lycopersicoides*) share the same basic chromosome number x = 12. Polyploidy is confined to potatoes, whereas all 13 wild tomato taxa are diploid. Ploidy level has been one of the most important taxonomic characters for the identification of cultivated potatoes; these show various ploidies, from the diploid (2n = 2x = 24) to the pentaploid (2n = 5x = 60) level (reviewed in Spooner et al. 2014). Chromosome number has been determined for most of the 107 wild potato species recognized by Spooner et al. (2014). Hijmans et al. (2007), in a comprehensive survey of ploidy reports in sect. *Petota*, concluded that over 60% of the wild potatoes exist exclusively at diploid level. Tetraploids are the most common polyploids among wild potatoes, followed by hexaploids and triploids, whereas pentaploids are rare. In addition, many species [19–21, depending on the taxonomic treatment; see Hijmans et al. (2007); Spooner et al. (2014)] have multiple cytotypes.

The determination of the type of polyploidy (that is, auto- or allopolyploids) has traditionally relied on the analysis of meiotic chromosome configurations in natural polyploids and interspecific hybrids, as well as on interspecific crossability and on hybrid fitness (Matsubayashi 1991; reviewed in Gavrilenko 2007; Gavrilenko 2011). Clearly, this analysis had to face several challenges due to the large number of small chromosomes, the difficulty to trace various types of meiotic configurations, and the diverse origin and cytological behavior of the polyploids of sect. *Petota* (see He et al. 2018). Matsubayashi (1991) proposed a five-genome hypothesis to explain the cytological and crossability data and to distinguish the different ploidy types across sect. *Petota* (reviewed in Gavrilenko 2007, 2011; Spooner et al. 2014). According to this hypothesis, all diploid potato species shared a similar genome denoted with an A, which represented the main genomic group and included slightly different variants to account for minor structural differences (denoted with superscript letters). A different genome E was hypothesized for the distantly related diploid species of sect. *Etuberosum* (Matsubayashi 1991). Allopolyploid potatoes always contained one genome A component and differed from each other for the second genome denoted as B, C, D, or P (Matsubayashi 1991; reviewed in Gavrilenko 2007, 2011; Spooner et al. 2014). However, the origin of the allopolyploid species has been much debated because, while the diploid *S. verrucosum* (or its progenitor) was indicated as a donor for the genome A, the other genomes had no known extant diploid species representatives (Spooner et al. 2008; reviewed in Gavrilenko 2007, 2011; Spooner et al. 2014). GISH helped clarify the controversial origin of several allopolyploid potatoes (Pendinen et al. 2008, 2012), by testing candidate donor species, which were identified through extensive molecular phylogenetic studies (reviewed in Spooner et al. 2014). GISH confirmed that the North and Central American *S. hjertingii* and *S. stoloniferum* ($2n = 4x = 48$) derived from two different genomes (genome A of *S. verrucosum* and genome B of diploid Mexican species; Pendinen et al. 2008). Based on the observation of exclusively intragenomic pairing at meiosis, *S. hjertingii* and *S. stoloniferum* should be considered strict allotetraploids (Pendinen et al. 2008). In addition, GISH revealed a complex genomic constitution for the Mexican allohexaploids *S. hougasii, S. iopetalum,*, and *S. schenckii* ($2n = 6x = 72$), which involved the contributions of at least three genomic components (A, B, and P genomes; Pendinen et al. 2012).

Conversely, GISH indicated a different origin for the Mexican hexaploid *S. demissum* ($2n = 6x = 72$), a species often used in potato breeding as a source of disease resistance. Various authors postulated different genome formulae for *S. demissum*, although there was a general agreement that *S. demissum* had two similar genomes differing from the third one (reviewed in Matsubayashi 1991). However, the GISH results of Pendinen et al. (2012) supported an autopolyploid origin of *S. demissum*, containing three chromosome sets similar to the basic A genome. Previous sequencing data indicated that *S. demissum* likely comprises two types of slightly differentiated genome A (Spooner et al. 2008; Rodríguez and Spooner 2009). Additional support came from the karyotype analysis of this species using oligo-FISH barcode approach (described above) which indicated that *S. demissum* contains six copies of each of the 12 potato chromosomes because the FISH signal pattern of the six homeologous chromosomes was identical to those of the reference potato species (Braz et al. 2018). Moreover, He et al. (2018) used oligo-based chromosome painting probes to monitor the chromosome pairing of four different *S. demissum* chromosomes (namely, chromosome 2, 4, 7, and 11) at meiotic prophase I. The authors demonstrated that during male meiosis, these *S. demissum* chromosomes have a diploid-like pairing behavior (Fig. 2.1). No hexavalent pairing was detected (He et al. 2018). Indeed, the analysis of chromosome pairing at pachytene using chromosome 7 and 11 probes demonstrated three independent bivalents in 80% and 98% of the cells observed, respectively (Fig. 2.1a1–c1; He et al. 2018). In addition, the prevalent configuration at diakinesis/metaphase I was of three independent bivalents for each of the four *S. demissum* chromosomes analyzed (He et al. 2018). Therefore, other mechanisms, independent from genome differentiation, are at the base of the bivalent pairing of the putative autohexaploid *S. demissum* (He et al. 2018).

The work of Pendinen et al. (2008; 2012) provided implicit evidence for a significant genome differentiation among the diploid representatives of various genome groups, that is, among *S. verrucosum* (genome A); the Mexican *S. cardiophyllum*, *S. ehrenbergii*, *S. jamesii*, and *S. bulbocastanum* (all genome B); and *S. andreanum* and *S. piurae* (both genome P). Similar indirect evidence comes from the GISH analysis of various hybrids involving both potatoes and

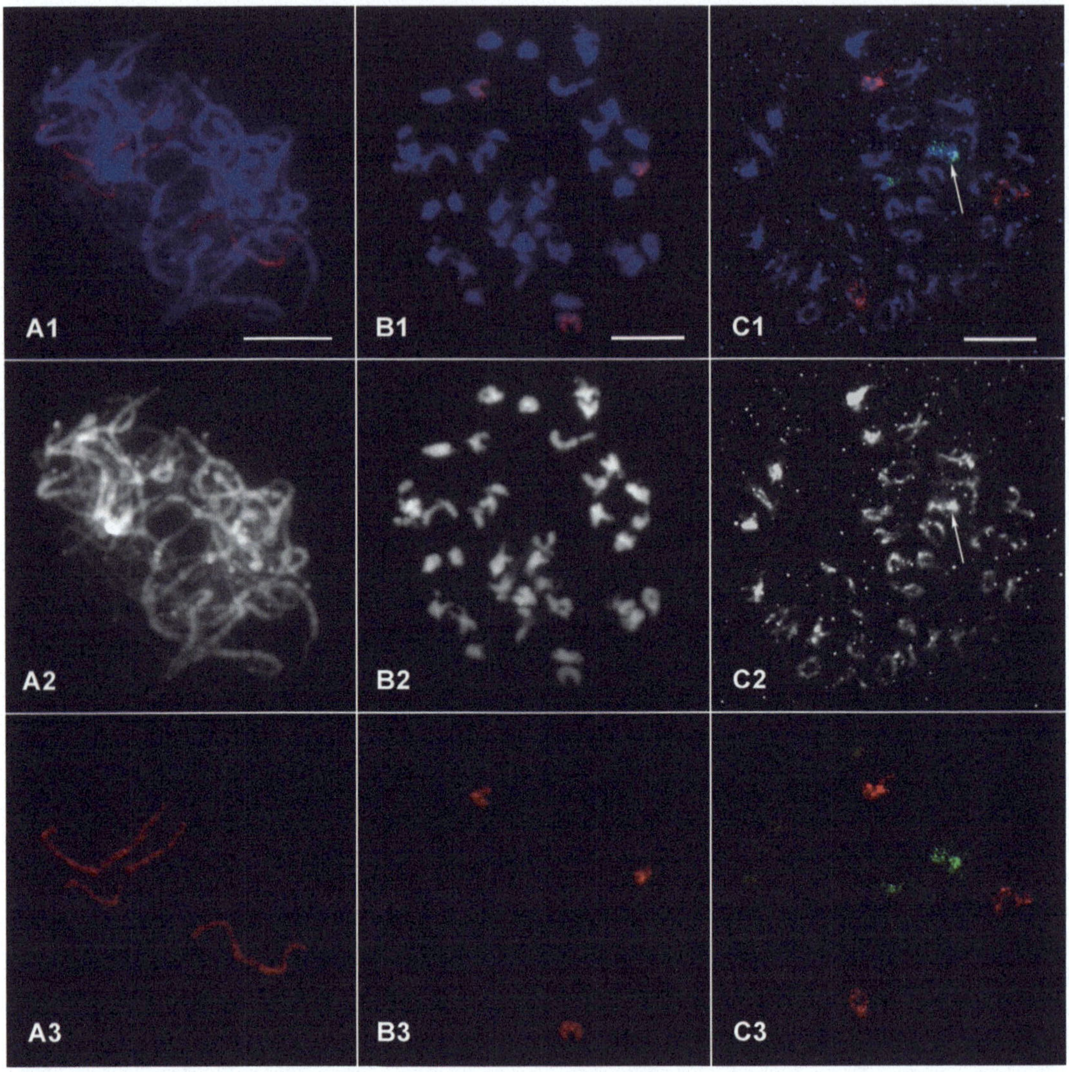

Fig. 2.1 Chromosome painting in meiotic cells from *Solanum demissum* (2n = 6x = 72; He et al. 2018). Photographs by He L. and Jiang J. **a1** Three chromosome 11 bivalents are observed after hybridization of a pachytene cell with chromosome 11 specific probe. **b1** Three chromosome 11 bivalents are observed after hybridizaton of a diakinesis cell with chromosome 11 probe. **c1** Six copies of chromosome 7 paired as three bivalents and six copies of chromosome 2 paired as one quadrivalent (white arrow) and one bivalent after hybridization of a diakinesis cell with chromosome 2 (green) and chromosome 7 (red) probes, respectively. **a2–c2** Chromosome images that were digitally separated from **a1–c1**, respectively. **a3–c3** FISH signals that were digitally separated from **a1–c1**, respectively. Bars = 10 μm

tomatoes. Apart from wide hybrids between cultivated potato and tomato (A and L genomes, respectively; Garriga-Calderé et al. 1997), cultivated potato (or tomato) and E genome species from sect. *Etuberosum* (Dong et al. 1999, 2001; Gavrilenko et al. 2001, 2002), and between tomato and species from sect. *Lycopersicoides* (Pertuzé et al. 2003; Ji et al. 2004), GISH differentiated the parental chromosomes of hybrids between cultivated tomato and *S. pennellii* (Haider Ali et al. 2001) and *S. peruvianum* (Parokonny et al. 1997), as well as between cultivated potato and *S. bulbocastanum* (A and B genomes, respectively; Iovene et al. 2007). Conversely, GISH could not distinguish the parental genomes of hybrids between cultivated potato and *S. commersonii*. This result suggested poor divergence between the bulks of the repetitive sequences of these species (Gaiero et al. 2017).

2.4 Nuclear Genome Size of Wild and Cultivated *Solanum* Species

The nuclear genome size of a species is an important taxonomic character with several practical and predictive applications (Bennett and Leitch 2011). Its knowledge contributes to identify species and uncover polyploidization/aneuploidization/diploidization events as well as large-scale differential repeat amplification among close relatives. Nuclear genome size estimates are available for a limited number of potato and tomato wild relatives (http://data.kew.org/cvalues/). However, because *Lycopersicon* is a relatively small section, the available data provide an idea of the extent of variation in nuclear genome size among wild tomatoes. The sizes of their genomes are about 1 pg/1C, with a variation among species of up to about 30%. DNA content of cultivated tomato is estimated at 0.94–1.03 pg/1C based on different studies and cultivars (Arumuganathan and Earle 1991; Michaelson et al. 1991; Valkonen et al. 1994), equal to approximately 907–1000 Mb/1C. The closely related *S. cheesmaniae* and the more distant *S. habrochaites* (=*Lycopersicon hirsutum*) have estimates comparable to that of tomato (Bennett and Smith 1976; Arumuganathan and Earle 1991). Two different values (0.85 pg/1C and 1.15 pg/1C) are reported for *S. pimpinellifolium*, which is another close relative of cultivated tomato (Bennett and Smith 1976; Barow and Meister 2002), whereas *S. peruvianum* and *S. pennellii* have larger genomes of 1.15 pg/1C (1095 Mb) and 1.23–1.38 pg/1C (1192–1337 Mb), respectively (Arumuganathan and Earle 1991). The genome sizes of the outgroup species from sect. *Juglandifolia* are comparable to that of tomato, whereas the genomes of *S. lycopersoides* and *S. sitiens* (sect. *Lycopersicoides*) are about 30% larger than tomato (Chetelat 2009). The larger genome size of *S. pennellii* and *S. lycopersicoides* are consistent with earlier cytological observations of the pachytene complements of both species, which indicated that several *S. pennellii* chromosomes have longer heterochromatic regions than those of tomato (Khush and Rick 1963) and that the pachytene karyotype length of *S. lycopersicoides* is 1.5 fold longer than in tomato (Menzel 1962).

Repetitive sequences may underlie the genome size variation observed among some *Solanum* species. Repeat abundance and genome size are correlated in representative species of the potato and tomato clade. The tomato relatives with larger genomes contain different amounts of some satellite repeats and a significantly higher proportion of unclassified elements consisting of degraded or truncated elements (Gaiero et al. 2019). On the other hand, although the intraspecific variation in nuclear DNA content may reflect real variability (Doležel and Greilhuber 2010), the different values reported for *S. pimpinellifolium* (as well as for other *Solanum* species) could be due to different methodologies and reference standards (Bennett and Smith 1976; Barow and Meister 2002).

In contrast, the genome size estimates for sect. *Petota* are sparse in proportion to the large number of species of this section. Several estimates rely on a single study and accession. In addition, a reassessment of the genome size data is highly desirable, considering the revised taxonomy of the section. Haploid potato clones have a genome size estimated at 0.8–0.88 pg/1C (that is, 831–856 Mb). The genomes of several wild diploid species, including *S. berthaultii*, *S. pinnatisectum*, *S. sparsipilum*, *S. vernei*, and the recently sequenced *S. commersonii* and *S. chacoense* are in the same size range of the haploid potato clones (Anderson et al. 1985; Arumuganathan and Earle 1991; Bennett and Smith 1991; Valkonen et al. 1994; Aversano et al. 2015; Leisner et al. 2018). However, there are also reports of diploid species with larger genomes, such as 1.1 pg/1C for the Mexican *S. polyadenium* and 1.2 pg/1C for the landrace

S. stenotomum (Anderson et al. 1985; Bennett and Smith 1991). For comparison, species of sect. *Etuberosum* have genomes of similar or slightly smaller size than that of haploid potato clones (Valkonen et al. 1994).

2.5 Collinearity and Rearrangements Revealed by Meiotic Analyses

Disturbances in pairing between homeologous in interspecific hybrids may provide evidence for chromosomal rearrangements. Early studies on the meiosis of several interspecific tomato hybrids suggested no large-scale structural rearrangements between parental chromosomes and minor differences in the lengths of the heterochromatic regions of some chromosomes of certain species (e.g., tomato and *S. pennellii*; Sawant 1958; Khush and Rick 1963; reviewed in Chetelat and Ji 2007). Thanks to the higher resolution, electron microscope analyses of pachytene spread synaptonemal (SC) complexes from five F_1 tomato interspecific hybrids revealed numerous synaptic irregularities occurring mostly in heterochromatin but also in euchromatin (Anderson et al. 2010). The irregularities consisted primarily of mismatched kinetochores, paracentric inversion loops, and a large reciprocal translocation (exclusively in the tomato x *S. chmielewskii* hybrid). Mismatched kinetochores were the most common irregularity observed in all hybrids and were interpreted as the result of pericentric inversions that occurred in some of the lineages and/or of differences in genome size (Anderson et al. 2010). The F_1 hybrid tomato x *S. chmielewskii* (one of the closest tomato relatives) had the highest number of irregularities. However, in general, the synaptic irregularities increased with the phylogenetic distance of the wild parent from cultivated tomato (Anderson et al. 2010). Preferential pairing was observed in the allohexaploids produced by doubling 3 × fusion hybrids between tomato, potato, and *S. pennellii,* whereas tomato chromosomes readily paired with their homeologues from *S. pennellii* before doubling (Haider Ali et al. 2001). Preferential pairing was also observed in hybrids and substitution lines with more distantly related species *S. lycopersicoides* and *S. sitiens* (Ji and Chetelat 2003; Ji et al. 2004). However, in most cases, tomato chromosomes show regular homoeologous pairing with chromosomes from its wild relatives.

Analysis of pairing and segregation has also been performed in interspecific hybrids within the potato clade. The analysis of several interspecific potato hybrids provided evidence for cryptic structural differences as well as "definite structural differences" (such as translocations) among diploid wild potatoes genomes (reviewed in Matsubayashi 1991). For example, the genomes of *S. jamesii* and *S. bulbocastanum* were denoted by different superscript letters because their diploid F1 hybrids formed ten bivalents and one tetravalent in metaphase I, indicative of a reciprocal translocation (Matsubayashi 1991). Gaiero et al. (2017) reported that the pachytene chromosomes of 3 × hybrids between *S. commersonii* and *S. tuberosum* Group Phureja were paired both as bivalents and trivalents (Fig. 2.2a1). Some pairing breakpoints were observed, which could be evidence of small-scale rearrangements (Fig. 2.2a1 and inset). At diakinesis/metaphase I of these triploids, configurations of 7III + 5II + 5I suggested a near autotriploid behavior (Fig. 2.2a2–3). The univalents occurred at random, as indicated by BAC FISH chromosome identification (Fig. 2.2b). In backcross (BC) progenies, homoeologous pairing was maintained, as evidenced by the formation of multivalents (Fig. 2.2c). Analysis of fertile BC1 pentaploid/near-pentaploid *S. commersonii–S. tuberosum* hybrids obtained from a different breeding scheme also indicated intergenomic pairing with multivalent associations of up to five chromosomes, even though most chromosomes paired as bivalents (Barone et al. 1999). Similarly, analysis of tetraploid somatic hybrids between a haploid potato clone and *S. bulbocastanum* detected multivalent pairing at pachytene as well as at diakinesis. However, most chromosomes formed bivalents, likely an indication of preferential intragenomic pairing (Iovene et al. 2012). Comparative FISH mapping studies can shed light on the underlying reasons for these pairing behaviors.

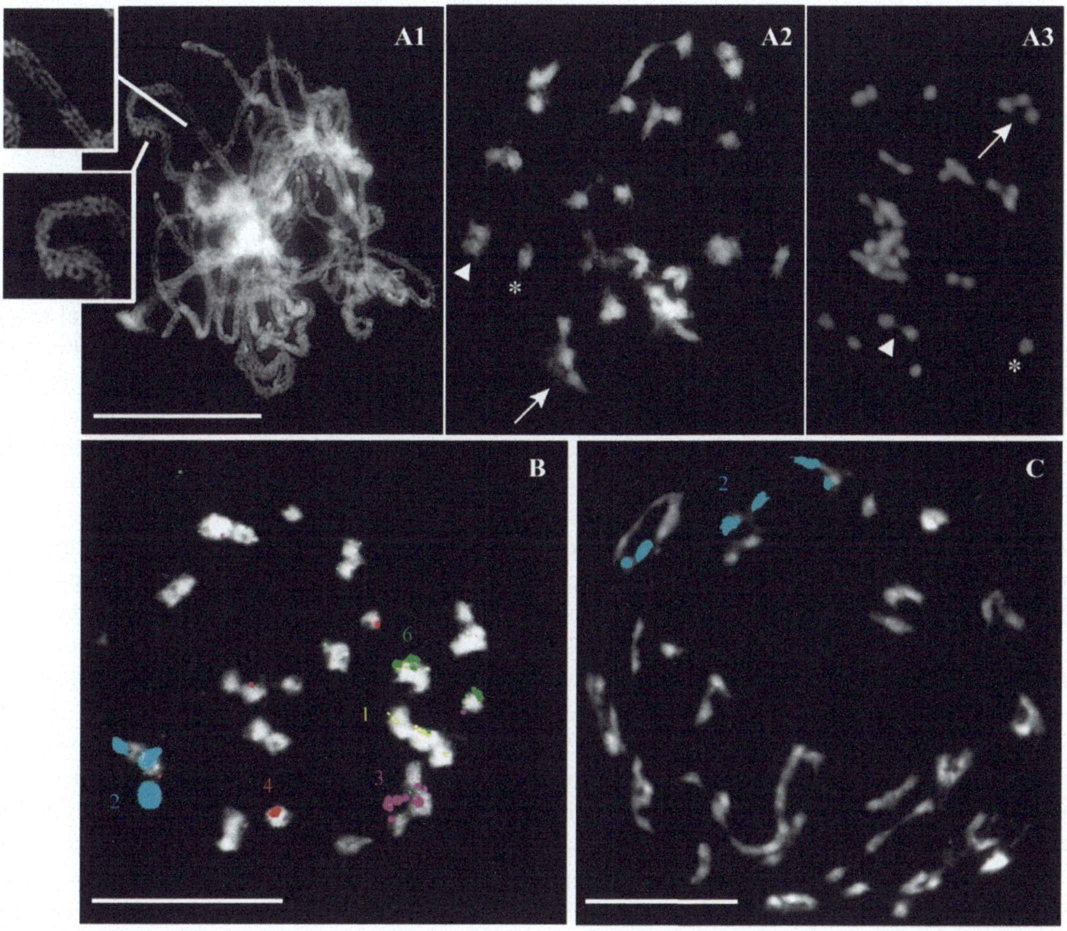

Fig. 2.2 **a** Homoeologous pairing in pollen mother cells (PMC) from *Solanum commersonii* and *S. tuberosum* Group Phureja 3 × hybrids (2n = 3*x* = 36). **a1** Pachytene complement with bivalents and trivalents and a loop (see insets). **a2** Diakinesis with five trivalents (arrow), seven bivalents (arrowhead), and seven univalents (asterisk). **a3** Metaphase/early anaphase I complement showing migration of trivalents (arrow), bivalents (arrowhead), and univalents (asterisk). **b** Homoeologous pairing in pollen mother cells (PMC) from the same 3 × hybrids hybridized with rDNA and potato chromosome-specific BAC probes specific to chromosomes: 1 (yellow, 5S rDNA), 2 (blue, 18S-25S rDNA), 3 (purple), 4 (red), and 6 (green). Chromosome 1, 2, and 3 paired as trivalents; chromosomes 4 and 6 paired as bivalents/univalents. **c** Homoeologous pairing in pollen mother cells (PMC) at diplotene/diakinesis of a genotype (2n = 5*x* + 5 = 65) derived from the backcross progeny of the same 3 × hybrid using 18S-25S rDNA (blue) as probe. The six identified chromosomes are forming multivalents (a quadrivalent + a bivalent). Scale bars represent 10 μm. Adapded from Gaiero et al. (2017)

2.6 Collinearity and Rearrangements Revealed by Comparative FISH

Synteny and collinearity among Solanaceae crops have been mainly studied through comparative genetic linkage mapping. In addition, considerable efforts have been devoted to integrating genetic, cytogenetic and high-throughput sequencing approaches in the assessment of collinearity. These studies showed that tomato and potato are differentiated by nine major inversions involving five whole arm paracentric inversions on chromosomes 5, 9, 10, 11 and 12, one inversion encompassing the euchromatic portion of 6S, two additional inverted chromosome segments on the long arm of chromosome 2, and another one on the long arm of chromosome 12 (Bonierbale et al. 1988; Tanksley et al. 1992; Iovene et al. 2008; Tang et al. 2008; The Potato Genome Sequencing Consortium 2011; The Tomato Genome Consortium 2012; Peters et al. 2012; Szinay et al. 2012).

Comparative FISH mapping has been extended to wild potato and tomato relatives, and it is bringing to light several previously undescribed rearrangements (see Table 2.1, Achenbach et al. 2010; Lou et al. 2010; Peters et al. 2012; Szinay et al. 2012). However, relatively few species and/or few chromosomes have been analyzed, especially among wild potatoes. One of such comparative FISH studies analyzed the order of several potato and tomato BACs along seven chromosome arms (5S, 6S, 7S, 9S, 10L, 11S, and 12S) of potato, tomato as well as selected wild relatives and outgroups (Szinay et al. 2012). The authors noted that potato and its wild relatives (*S. bulbocastanum*, *S. pinnatisectum*, *S. tarijense*, *S. megistacrolobum*) had an identical hybridization pattern on those chromosome arms, and therefore these species were regarded as syntenic group A (Szinay et al. 2012). Similarly, tomato and its wild relatives *S. pimpinellifolium*, *S. peruvianum* and the distantly related *S. habrochaites* had identical BAC order and were thus regarded as syntenic group B. Based on the meiotic pairing analysis of hybrids *S. habrochaites*-tomato (Anderson et al. 2010; see the previous section), it is likely that *S. habrochaites* differs from tomato for rearrangements involving chromosome arms not studied by Szinay et al. (2012). Other distantly related wild tomatoes showed synteny with group B in some chromosome arms but not in others (Table 2.1). For example, *S. pennellii*, *S. chilense* and the outgroups *S. ochrantum* (sect. *Juglandifolia*) and *S. lycopersicoides* (sect. *Lycopersicoides*) were collinear with the rest of the tomato species in the short arm of chromosome 5 (5S), as well as in 9S, 10L, and 11S. However, there was a small inversion close to the pericentromeric heterochromatin on 12S that differentiated *S. chilense*, while a small terminal inversion on 6S separates *S. pennellii* from syntenic species B (Table 2.1, Szinay et al. 2012). Non-tuber-bearing *S. etuberosum*, in many cases, shared chromosome collinearity with potato and its relatives. However, *S. etuberosum* had large inversions in 7S and 9S compared to both potatoes and tomatoes, whereas it was collinear with syntenic species B for 10L (Table 2.1, Szinay et al. 2012). High resolution cytogenetic mapping was also employed to detect potential rearrangements among selected wild and cultivated *Solanum* species along the entire length of chromosome 6 (Lou et al. 2010). The authors were able to elucidate the ancestral structure of this chromosome and the different steps in chromosomal evolution through cross-species BAC FISH. The ancestral chromosome 6 should resemble that of *S. melongena* (eggplant), and it is conserved across potato and wild relatives *S. bulbocastanum* and *S. chromatophylum*. The non-tuber-bearing *S. etuberosum* displays a large pericentric inversion, while tomato differs in the previously-reported paracentric inversion in the short arm (Table 2.1, Lou et al. 2010).

Additional smaller inversions between tomato and its wild relatives were uncovered because these structural rearrangements had a negative impact on breeding (see Table 2.1; van der Knaap et al. 2004; Verlaan et al. 2011). BAC FISH revealed two rearrangements between tomato and *S. chilense* in the pericentromere of the long arm of chromosome 6, that is the region where the resistance gene *Ty-1* is introgressed, causing suppression of recombination and linkage drag (Verlaan et al. 2011). The locus *sun*, which controls the tomato fruit shape, was accurately located on the short arm of tomato chromosome 7 (van der Knaap et al. 2004) through FISH on extended DNA fibers. The authors suggested that because *sun* is located in a highly dynamic region of the tomato genome, the allelic variation found at this locus may be due to an insertion/deletion event. Therefore, comparative FISH helped clarify the causes of suppressed recombination around the gene of interest. On the other hand, Gaiero et al. (2016) found high collinearity at the chromosomal scale between potato and its wild relatives *S. commersonii* and

Table 2.1 Overview of chromosome rearrangements between tomato, potato, and their wild relatives discovered through fluorescent *in* situ hybridization (FISH). Tomato is compared to its wild relatives from *Solanum* Sect. *Lycopersicon* and its more distant relatives from Sect. *Lycopersicoides* (*S. lycopersicoides)* and Sect. *Junglandifolia* (*S. ochrantum*), as well as to *S. etuberosum*. Potato and its wild relatives from Sect. *Petota* for which there are reports (*S. bulbocastanum, S. chromatophylum* only for chr 6, *S. chacoense, S. commersonii, S. megistacrolobum, S. pinnatisectum, S. tarijense*) are taken as a group and compared to both tomato and the outgroup species *S. etuberosum.* Chromosome (Chr) arms involved in the rearrangements are indicated by S (short) or L (long). Superscripts indicate source

Chr	Tomato versus					Potato and wild relatives versus	
	S. pennellii	*S. chilense*	*S. lycopersicoides*	*S. ochrantum*	*S. etuberosum*	Tomato	*S. etuberosum*
2	nd				Reciprocal translocation between 2L and 7S [d]	Large distal 2L inversion [f]	Reciprocal translocation between 2L and 7S [d]
5	–				Large 5S inversion [a]	Large 5S inversion [a] and small 5L inversion [g]	–
6	Small distal 6S inversion.[a]	Small proximal 6L inversion [c]	Small proximal 6S inversion [a]	Large 6S inversion [a]	Large pericentric inversion [e]	Large 6S inversion [a, e, f, h, i]	Large pericentric inversion [e]
7	Small 7S inversion [a, b]	–	Small 7S inversion [a]	–	Large 7S inversion [a]	–	Large 7S inversion [a]
9	–				Small proximal 9S inversion [a]	Large 9S inversion [a]	Small distal 9S inversion [a]
10	–				–	Large 10L inversion [a, f]	Large 10L inversion [a]
11	–				Large distal 11S inversion [a]	Large distal 11S inversion [a]	–
12	–	Small proximal 12S inversion [a]	–	–	Large distal 12S inversion [a]	Large distal 12S inversion [a]	–

[a]Szinay et al. 2012; [b]van der Knapp et al. 2004; [c]Verlaan et al 2011; [d]Braz et al. 2018; [e]Lou et al. 2010; [f]Peters et al. 2012; [g]Achenbach et al. 2010; [h]Tang et al. 2008; [i]Iovene et al. 2008; nd = not determined; – = no rearrangement detected by FISH

S. chacoense when they compared the cytogenetic positions of potato BACs previously located on RH potato (Tang et al. 2009). Altogether, the few comparative BAC FISH studies available for tuber-bearing Solanums reported no evidence for large-scale chromosome rearrangements between potato and its wild relatives (Lou et al. 2010; Szinay et al. 2012; Gaiero et al. 2016; Braz et al. 2018).

All the genetic and cytogenetic studies have indicated that inversions are the main mode of chromosome differentiation among potatoes and tomatoes. However, a comparative oligo-based chromosome painting study showed that *S. etuberosum* differs from potato and its wild relative *S. bulbocastanum*, tomato, and eggplant for a reciprocal translocation between 2L and 7S (Fig. 2.3; Braz et al. 2018). Similarly, *S.*

caripense, a more distant relative of tomato and potato, differs from the same set of species for another reciprocal translocation between 4L and 11S (Braz et al. 2018). On the other hand, the oligo-FISH pattern on *S. bulbocastanum* chromosomes was identical to that on potato (Braz et al. 2018). Therefore, the chromosomes of the potato species analyzed seem not to be affected by gross rearrangements, differently to tomatoes (Lou et al. 2010; Peters et al. 2012; Szinay et al. 2012; Braz et al. 2018). However, additional comparative mapping studies comprising more tuber-bearing species are needed to confirm collinearity between potato and its wild relatives.

2.7 Cytogenetics of Satellite Repeats in *Solanum*

A significant portion of the potato and tomato genomes is occupied by satellite DNA, which consists of long arrays of nearly identical tandem repeat units (called monomers) spanning up to several megabases. Cytogenetics has contributed significantly to the characterization of the satellite repeats repertoire in *Solanum*, which is the focus of this last section. Satellite families isolated from various potatoes and tomatoes are typically located in the heterochromatin, at subtelomeric and (peri)centromeric regions but also at interstitial chromosomal sites. Many of these repeats are widespread throughout potatoes and tomatoes (Stupar et al. 2002; Tek and Jiang 2004; Jo et al. 2009; Torres et al. 2011; Tang et al. 2014). However, many satellite families display a remarkable inter- and even intraspecific variation, especially in their abundance and chromosomal distribution, with extreme patterns of presence/absence, which have suggested that these sequences evolve rapidly (Tek et al. 2005; Gong et al. 2012; Wang et al. 2014; Zhang et al. 2014). For this reason, several repeats that were initially identified provided useful species-specific RFLP-based markers to identify *Solanum* species and their interspecific somatic hybrids (Pehu et al. 1990; Schweizer et al. 1993; Stadler et al. 1995). The following paragraphs provide an overview of the main satellite repeats among potato and tomato wild relatives, with the satellite repeats grouped for their similarity with "universal" repeats (rDNA and telomeric sequences) and/or for their chromosomal distribution.

2.7.1 rDNA Gene Clusters and Related Satellite Repeats

The ribosomal DNA gene clusters are among the best-characterized satellite arrays in eukaryotes. In potato, tomato, and their wild relatives, the 18S-25S rDNA cluster was mapped at the end of the short arm of chromosome 2, whereas the 5S rDNA was located interstitially on the short arm and next to the centromere of chromosome 2 (Ganal et al. 1988; Visser and Hoekstra 1988; Lapitan et al. 1989; Xu and Earle 1996; Dong et al. 2000; Stupar et al. 2002; Chang et al. 2008; Jo et al. 2009; Gaiero et al. 2016; Choudhary et al. 2020). Minor rDNA sites have been reported on other chromosomes in various species or accessions (Xu and Earle 1996; Brasileiro-Vidal et al. 2009). Conversely, two of the six copies of chromosome 2 in the autohexaploid *S. demissum* ($2n = 6x = 72$) lacked the 18S-25S rDNA sites (Braz et al. 2018), whereas in the allotetraploid *S. stoloniferum* the two pairs of 18S-25S rDNA sites derived from two different parental genomes had a very different size, possibly a result of the allopolyploidization process (Pendinen et al. 2008).

Ribosomal DNA gene clusters may be a source of novel satellite families. Indeed, satellite repeats made of tandem monomers with high sequence similarity to the intergenic spacer (IGS) of the 18S-25S rDNA are widespread across potatoes and tomatoes (Table 2.2; Stupar et al. 2002; Jo et al. 2009). These sequences were initially isolated from *S. bulbocastanum* and tomato BAC libraries. Some of these sequences had similarity to both the IGS and portions of the rDNA coding sequences (Stupar et al. 2002; Jo et al. 2009), which could explain the minor interstitial 18S-25S rDNA signals reported for tomatoes (Xu and Earle 1994; Brasileiro-Vidal

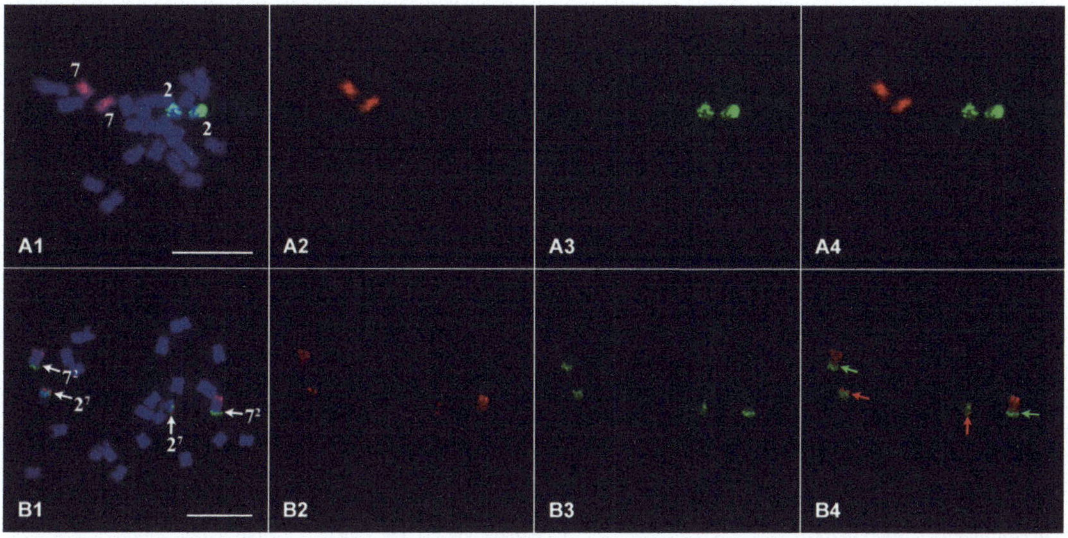

Fig. 2.3 A chromosomal translocation between potato and *Solanum etuberosum* detected by chromosome painting (Braz et al. 2018). Photographs by Braz G. T., He L., and Jiang J. **a1–a4** Painting of chromosome 2 (green) and 7 (red) of the diploid potato clone DM. Red (**a2**), green (**a3**), and both red and green (**a4**) fluorescence signals were digitally separated from **a1**. **b1–b4** Painting of chromosomes 2 (green) and 7 (red) in *S. etuberosum*. Red (**b2**), green (**b3**), and both red and green (**b4**) fluorescence signals were digitally separated from **b1**. Red arrows in **b4** point to the breakpoint where a small chromosome 7 fragment attached to chromosome 2 (2^7). Green arrows in **b4** point to the breakpoint where a large chromosome 2 fragment attached to chromosome 7 (7^2). Bars = 10 μm

et al. 2009). FISH using these IGS-related repeats in potatoes revealed a variable number of pericentromeric sites in hemizygous condition (Stupar et al. 2002). In addition, FISH indicated that different IGS-related repeats occupied adjacent and distinct pericentromeric heterochromatic domains. Southern blot analysis indicated that IGS-related DNA sequences are present in a wide range of *Solanum* species and that this repeat family is evolutionary dynamic and capable of rapid structural and copy number changes (Stupar et al. 2002). Similar to what found in potatoes, IGS-related repeats are associated with the heterochromatic pericentromeric regions of several pachytene chromosomes of tomato and close wild relatives (Table 2.2; Jo et al. 2009).

2.7.2 Telomeric, Subtelomeric, and Related Repeats

The chromosomal ends of *Solanum* spp. contain typical Arabidopsis-type telomeric tandem repeats (TT[T/A]AGGG), which are organized in long arrays at the end of the chromosomes in association with subtelomeric satellite repeats (Ganal et al. 1991; Zhong et al. 1998; Torres et al. 2011).

A number of subtelomeric satellite sequences have been identified in *Solanum* (Table 2.2), including TGRI in cultivated tomato, CL14/PRG1 and CL34 in potato, and Sb4/2 in *S. brevidens* (Ganal et al. 1988; Preiszner et al. 1994; Torres et al. 2011; Tang et al. 2014). Four repeats (TGR1, Sb4/2, Cl14, and PRGI) show sequence similarity to each other and are widespread across tomatoes and potatoes (Table 2.2). TGRI, one of the most abundant satellites of the tomato genome, is located at almost all chromosome ends in association with canonical telomeric repeats to form arrays up to 1.3 Mb long, as well as at interstitial sites on some chromosomes (Ganal et al. 1988; Zhong et al. 1998). A Southern analysis indicated that TGRI is widespread across the tomato clade (Ganal et al. 1988). On the other hand, CL14/PRG1,

Table 2.2 Partial list of *Solanum* satellite repeats characterized by cytogenetic tools. Repeats are grouped according to their chromosomal distribution and for their similarity with "universal" repeats (rDNA and telomeric sequences). References are indicated with superscripts

Repeat type	Repeat name (monomer length, bp)	Species	Chromosome distribution	Similarity with other repeats	Hybridization in another species
Subtelomeric	TGRI (162)[a,b,c]	*S. lycopersicum*	Most chrs; few interstitial sites	Partial with Sb4AX	Widespread among wild tomatoes
	Sb4AX (1728)[d]	*S. brevidens*	Most chrs	Partial with TGRI	nd
	CL14 (182)[e]; PGR1 (182)[f]	*S. tuberosum*	Most chrs	Partial with TGRI and Sb4AX	*S. verrucosum, S. cardiophyllum, S. chomatophilum, S. lycopersicum* (only CL14)
	CL34 (339)[e]	*S. tuberosum*	Most chrs	nd	*S. verrucosum*
Pericentromeric	Sobo (4700)[g]	*S. bulbocastanum*	1 hemizygous site on chr 7	LTR *Sore1*	No hybridization
Pericentromeric, IGS-related	2D8 (5900)[h]	*S. bulbocastanum*	4 hemizygous sites on 4 chrs	IGS of 18S-25S rDNA	Widespread among wild potatoes and tomatoes
	26J19 (5900)[h]	*S. bulbocastanum*	2 hemizygous sites on 2 chrs	IGS	nd
	4A4 (nd)[h]	*S. bulbocastanum*	1 hemizygous site	IGS and 18S-25S rDNA coding sequences	nd
	pIGS (nd)[i]	*S. lycopersicum*	Sites on 3 pachytene chrs	IGS and 18S-25S rDNA coding sequences	*S. lycopersicum* var. *cerasiforme*, *S. pimpinellifolium*
Pericentromeric, ITRs	pSbTC1 (2800)[j]; oligonucleotide telomeric probe[k]	*S. bulbocastanum* –	Most chrs	Perfect and degenerated telomeric motifs	Cultivated potato and many wild potato relatives
Centromeric	St24 (979)[l]; Sv161.5 (nd)[m]	*S. tuberosum; S. verrucosum*	Cen1	nd	*S. verrucosum* Cen7; weak or no signals in other species
	St3-58 (2957)[l]	*S. tuberosum*	Cen2	Ty3/gypsy, Chromovirus	No signals
	St3-294 (5390)[l]	*S. tuberosum*	Cen3 and Cen9	Ty3/gypsy, Chromovirus	Weak or no signals
	St3-238 (3814)[l]	*S. tuberosum*	Cen8	Ty3/gypsy, Chromovirus	Weak or no signals
	St49 (2754)[l]; Sv209 (nd)[m]	*S. tuberosum*; *S. verrucosum*	Cen5	ITR	Multiple centromeres of various wild potatoes: telomeric signals in tomato

(continued)

Table 2.2 (continued)

Repeat type	Repeat name (monomer length, bp)	Species	Chromosome distribution	Similarity with other repeats	Hybridization in another species
	St57 (1924) [l]; Sv161.6 (nd) [m]	*S. tuberosum*; *S. verrucosum*	Cen7	nd	*S. verrucosum* Cen7; no signals in other species
	St18 (1180) [l]	*S. tuberosum*	Cen9	Ty3/gypsy, Chromovirus	Unidentified Cen of *S. verrucosum*; no signals in other species
	Sv14 (nd)[m]	*S. verrucosum*	Cen4	LTR retrotransposon chromodomain	No signals
	Sv44 (nd)[m]	*S. verrucosum*	Cen4	nd	No signals
	Sv54 (nd)[m]	*S. verrucosum*	Cen2 and Cen10	nd	*S. tuberosum* Cen9; *S. chomaophilum* Cen2; no signals in other species
	Sv123 (nd)[m]	*S. verrucosum*	Cen2 and Cen10	nd	Cen1, Cen2 and unknown Cen of *S. chomatophilum*; no signals in other species
	Sv98 (nd)[m]	*S. verrucosum*	Most centromeres	Ty3/gypsy, Chromovirus	Most centromeres of *S. tuberosum*; no signals in other species
	Sv43 (nd)[m]	*S. verrucosum*	Weak signals in most centromeres	nd	Weak signals in most centromeres of *S. tuberosum* and *S. chomaophilum*; no signals in other species
	Sv132 (nd)[m]	*S. verrucosum*	Weak signals in most centromeres	Ty3/gypsy retrotransposon	Weak signals in most centromeres of *S. tuberosum* and *S. jamesii*; no signals in other species
	TGRIV (7000)[c]	*S. lycopresicum*	All centromeres	GYPSODE1 retrotransposon	nd

[a] Ganal et al. 1988; [b] Zhong et al. 1998; [c] Chang et al. 2008; [d] Preiszner et al. 1994; [e] Torres et al. 2011; [f] Tang et al. 2014; [g] Tek et al. 2005; [h] Stupar et al. 2002; [i] Jo et al. 2009; [j] Tek and Jiang 2004; [k] He et al. 2013; [l] Gong et al. 2012; [m] Zhang et al. 2014; nd = not determined; Chr = chromosome; IGS = intergenic spacer of the 18S-25S rDNA; ITR = interstitial telomeric repeats

identified in potato, is located exclusively at the subtelomeric regions of about half of the potato chromosomes (Fig. 2.4; Torres et al. 2011; Tang et al. 2014). CL14 repeat generated similar hybridization patterns even among *Solanum* species distantly related to potato, although few non-subtelomeric FISH signals were observed in several species (Torres et al. 2011). These findings suggested that CL14/PRG1, along with the related TGRI and Sb4/2, belong to an ancient repeat family that has maintained its (predominant) subtelomeric positions in all *Solanum* species (Torres et al. 2011; Tang et al. 2014). By contrast, a FISH survey of CL34, also identified in potato, indicated that this repeat had emerged recently, since it hybridized to about half of the chromosome ends of cultivated potato and its close relative *S. verrucosum*, whereas it generated very weak or no signals in more distantly related wild potato species (Fig. 2.4; Torres et al. 2011).

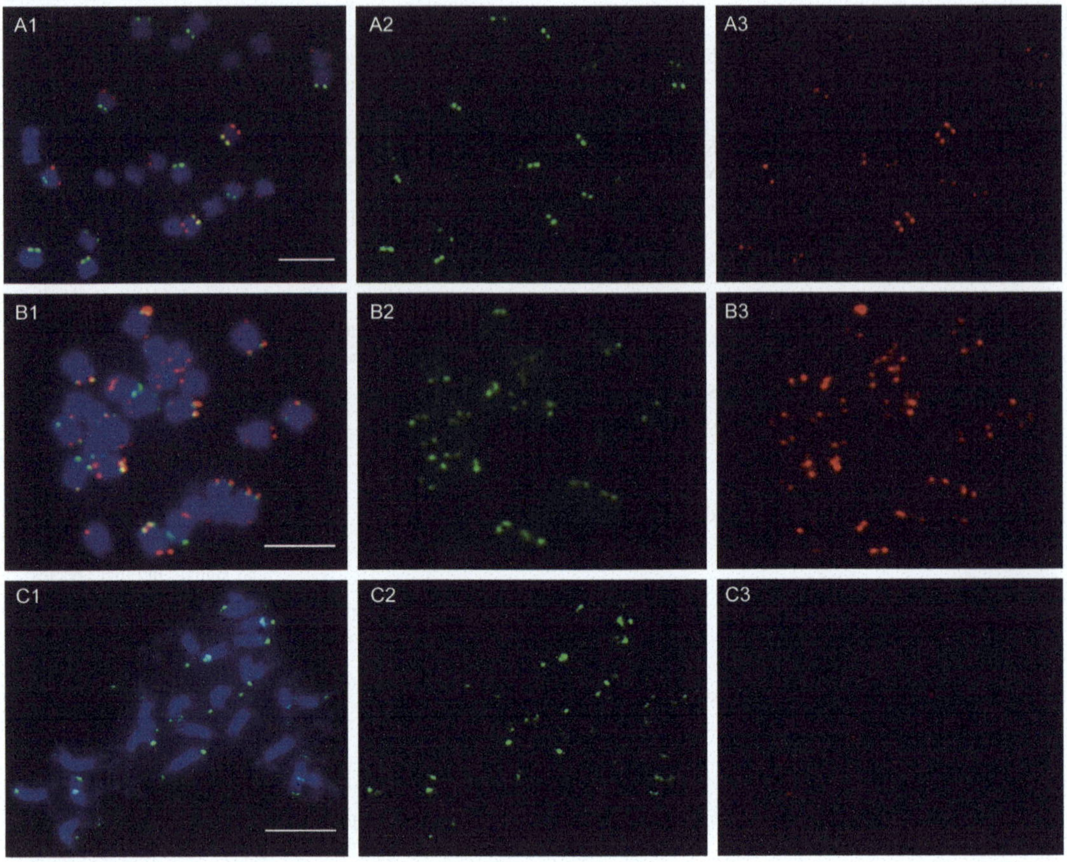

Fig. 2.4 Comparative FISH mapping of the subtelomeric repeats CL14 (green) and CL34 (red), among three different *Solanum* species (Torres et al. 2011). Photographs by Torres G. A. and Jiang J. **a1–a3** DM1-3 potato (A genome); **b1–b3** *S. verrucosum* (A genome); **c1–c3** *S. palustre* (E genome). Bars = 5 μm

2.7.3 Pericentromeric Repeats in *Solanum*

The pericentromeric heterochromatin of most eukaryotic organisms contains large amounts of satellite repeats characterized by different degrees of lineage-specificity. In addition to the IGS-related repeats, telomeric-like sequences are located in the (peri)centromeric regions of several chromosomes of many *Solanum* species. An early genetic linkage mapping study detected short arrays of interstitial telomeric repeats (ITRs) in the centromeric regions of several tomato chromosomes (Presting et al. 1996). In potatoes, pSbTC1, an ITR sequence isolated from *S. bulbocastanum*, generated strong FISH signals in the pericentromeric heterochromatin as well as weak signals in the telomeric regions of several chromosomes of potato and various wild relatives (Table 2.2; Tek and Jiang 2004). The 2.8 kb monomers of pSbTC satellite consisted of exclusively degenerated telomeric DNA sequences, and their long tandem clusters spanned several megabases (Tek and Jiang 2004). This suggested that, differently from tomato (Presting et al. 1996), ITR-like repeats in potatoes have undergone massive local amplification, and therefore, they are not simple footprints of ancient events of chromosome rearrangements (Tek and Jiang 2004). In addition, a FISH survey using a telomeric DNA probe indicated that species with B (*S. bulbocastanum* and *S. pinnatisectum*) and P (*S. paucissectum*) genomes are particularly rich of ITR-like satellites, mainly in the centromeric/pericentromeric regions of several chromosomes (He et al. 2013). By contrast, tomato did not show any distinct interstitial telomeric signals which corroborated the finding that centromeric and pericentromeric regions of the wild potato relatives, but not tomato, contain megabase-sized arrays of telomeric-like sequences (He et al. 2013). Additional repetitive sequences have been mapped to the pericentromeric heterochromatin of various *Solanum* species (Table 2.2), including a species-specific satellite repeat (Sobo) identified in the *S. bulbocastanum* genome (Tek et al. 2005). Sobo mapped on chromosome 7 of some *S. bulbocastanum* accessions in hemizygous condition, spanning > 350 kb of pericentromeric heterochromatin (Tek et al. 2005). Interestingly, the Sobo repeat was not detected in any other *Solanum* species, which suggested that Sobo repeat is a recently amplified satellite repeat, pointing to the dynamic nature of the satellite DNA (Tek et al. 2005).

2.7.4 Centromeric Repeats in *Solanum*

The centromeres are essential for the faithful segregation of sister chromatids during cell divisions. Centromeric DNA in most eukaryotes consists of long arrays of satellite repeats and/or retrotransposons, and it is among the most rapidly evolving sequences in the genome. Repeat–based centromeres are thought to have evolved from "neocentromeres" arose in novel sites, usually in gene-poor environment, by an accumulation of the specific histone variant CenH3. Given time, these "neocentromers" are believed to evolve into mature centromeres through the "invasion" of satellite DNA (reviewed in Kalitsis and Choo 2012; Jiang 2013; Plohl et al. 2014; Oliveira and Torres 2018).

Potato and its relatives have recently provided a new model system to support the hypothesis of centromere evolution from neocentromere (Gong et al. 2012; Zhang et al. 2014). A genome-wide characterization of the DNA sequences associated to CENH3 nucleosomes has shown that each potato centromeres contains distinct DNA sequences (Gong et al. 2012). Five potato chromosomes did not include any satellite DNA, but consisted primarily of single- or low-copy DNA sequences, including active genes (Gong et al. 2012). Thus, the DNA structure of these five potato centromeres is thought to resemble "immature" neocentromeres (reviewed in Jiang 2013; Gong et al. 2012). In contrast, the centromere of the other six potato chromosomes (1, 2, 3, 5, 7, and 8) are composed of megabase-sized satellite repeat arrays that are specific to individual chromosomes (Table 2.2). The centromere of potato chromosome 9 contains two different satellites, as well as single-copy sequences (Table 2.2). The monomer sizes of these satellite repeats range from ~980 bp to >5.3 kb, and the satellites form long arrays from ~900 kb to > 4 Mb, likely occupying the entire functional cores of the centromeres (Gong et al. 2012). Comparative FISH mapping of the potato centromeric satellites in wild *Solanum* representatives of the genomes A (*S. verrucosum*), B (*S. jamesii*), P (*S. chromatophilum*), and E (*S. etuberosum* and *S. palustre*) indicated that St49 is likely an ancient repeat belonging to an ITR family, and it is present in all species analyzed (Gong et al. 2012). By contrast, the other potato satellite repeats appeared to be amplified recently from retrotransposon-related sequences. These repeats either hybridized only to the closely related *S. verrucosum*, or were absent in all species, suggesting a rapid evolution from repeatless neocentromeres to repeat-based centromeres (Gong et al. 2012). The sequence specificity of the potato centromeres has opened opportunities to comparative analysis of homoeologous centromeres among related species. Isolation of satellite repeats associated with CENH3 in *S. verrucosum* revealed that homoeologous centromeric sequences between *S. verrucosum* and potato were restricted to a single centromere (Cen9). Four *S. verrucosum* centromeres (Cen2, Cen4, Cen7, and Cen10) contained distinct satellite repeats (Table 2.2; Zhang et al. 2014). Strikingly, the same four centromeres in potato contained either different satellite repeats (Cen2 and Cen7) or exclusively single/low-copy sequences (Cen4 and Cen10). Comparative FISH mapping among *Solanum* species representatives of the genomes A, B, P, and E revealed the absence of the *S. verrucosum* centromeric repeats in most species analyzed, confirming the rapid divergence of the centromeric sequences and suggesting a recent emergence of these centromeric satellites in the *S. verrucosum* genome.

There are no such genome-wide studies in tomato and its relatives. Chang et al. (2008) cytologically mapped a repeat element named TGRIV to the primary constrictions of all tomato pachytene chromosomes. Similar to the sequences of other satellite repeats identified in *Solanum*, TGRIV likely derived from a retrotransposon, the Ty3-Gypsy GYPSODE1 (Chang et al. 2008). However, the association of TGRIV with the tomato centromeric chromatin and its distribution in the tomato relatives remain to be explored.

2.8 Conclusions and Perspectives

Comparative linkage maps and classical cytogenetic studies indicated a conserved genome structure and high collinearity within potato and tomato wild relatives. However, along with high-throughput comparative sequencing, comparative BAC FISH mapping and an increase of resolution in cytogenetic technologies (such as that achieved with comparative genome mapping through nanochannels) are providing evidence for substantial structural rearrangements as well as striking differences in the repeat composition of heterochromatic domains among these species. Many rearrangements found among tomato species involve inversions located in heterochromatic pericentromeric regions, which would be difficult to detect by genetic mapping. Such comparative FISH studies are still sparse among potatoes. However, oligo-based chromosome painting is expected to facilitate these studies by avoiding technical difficulties due to repeat-rich BACs and allowing reciprocal comparative analysis using oligo-probes designed on any sequenced *Solanum* genome. In addition, oligo-based chromosome painting enables the monitoring of pairing of homologous/homeologous chromosomes during meiosis, which provides useful insights into differentiation and recombination between the parental genomes of species and experimental hybrids. Cytogenomic studies have demonstrated the complex structure of the centromeric and pericentromeric regions of several *Solanum* species, associated with impressive variation in the sequence composition of homologous centromeres/pericentromeres within accessions and cultivars as well as homeologous centromeres/pericentromeres among closely related species. In the future, similar studies in tomato and its relatives may reveal whether such sequence diversity at the (peri)centromeres is common across the genus *Solanum,* or it is a phenomenon restricted within potatoes. Cytogenetics will continue to support high-throughput comparative sequencing studies to identify the landscape of structural variants and reveal genome diversity across potato and tomato wild relatives.

Acknowledgements The authors wish to thank Dr. Li He, Dr. Guilherme T. Braz, and Prof. Jiming Jiang for providing Figs. 2.1 and 2.3.

References

Anderson L, Covey P, Larsen L et al (2010) Structural differences in chromosomes distinguish species in the tomato clade. Cytogenet Genome Res 129:24–34. https://doi.org/10.1159/000313850

Anderson L, Stack S, Fox M, Chuanshan Z (1985) The relationship between genome size and synaptonemal complex length in higher plants. Exp Cell Res 156:367–378. https://doi.org/10.1016/0014-4827(85)90544-0

Arumuganathan K, Earle ED (1991) Nuclear DNA content of some important plant species. Plant Mol Biol Report 9:208–218. https://doi.org/10.1007/BF02672069

Aversano R, Contaldi F, Ercolano MR et al (2015) The *Solanum commersonii* genome sequence provides insights into adaptation to stress conditions and genome evolution of wild potato relatives. Plant Cell 27:954–968. https://doi.org/10.1105/tpc.114.135954

Barone A, Sebastiano A, Carputo D (1999) Chromosome pairing in *Solanum commersonii*- *S. tuberosum* sexual hybrids detected by *commersonii*-specific RAPDs and cytological analysis. Genome 42:218–224. https://doi.org/10.1139/gen-42-2-218

Barow M, Meister A (2002) Lack of correlation between AT frequency and genome size in higher plants and the effect of nonrandomness of base sequences on dye binding. Cytometry 47:1–7. https://doi.org/10.1002/cyto.10030

Barton D (1951) Localized chiasmata in the differentiated chromosomes of the tomato. Genetics 36:374–381

Barton D (1950) Pachytene morphology of the tomato chromosome complement. Am J Bot 37:639–643. https://doi.org/10.1002/j.1537-2197.1950.tb11053.x

Bedinger PA, Chetelat RT, McClure B et al (2011) Interspecific reproductive barriers in the tomato clade: Opportunities to decipher mechanisms of reproductive isolation. Sex Plant Reprod 24:171–187. https://doi.org/10.1007/s00497-010-0155-7

Beliveau BJ, Joyce EF, Apostolopoulos N et al (2012) Versatile design and synthesis platform for visualizing genomes with Oligopaint FISH probes. Proc Natl Acad Sci U S A 109:21301–21306. https://doi.org/10.1073/pnas.1213818110

Bennett M, Smith J (1991) Nuclear DNA amounts in angiosperms. Philos Trans R Soc London Ser B Biol Sci 334:309–345. https://doi.org/10.1098/rstb.1991.0120

Bennett MD, Leitch IJ (2011) Nuclear DNA amounts in angiosperms: Targets, trends and tomorrow. Ann Bot 107:467–590. https://doi.org/10.1093/aob/mcq258

Bennett MD, Smith JB (1976) Nuclear DNA Amounts in Angiosperms. Philos Trans R Soc B Biol Sci 274:227–274. https://doi.org/10.1098/rstb.1976.0044

Bethke P, Halterman DA, Jansky S (2017) Are we getting better at using wild potato species in light of new tools? Crop Sci 57:1241–1258. https://doi.org/10.2135/cropsci2016.10.0889

Bonierbale MW, Plaisted RL, Tanksley SD (1988) RFLP maps based on a common set of clones reveal modes of chromosomal evolution in potato and tomato. Genetics 120

Brasileiro-Vidal A, Melo-Oliveira M, Carvalheira G, Guerra M (2009) Different chromatin fractions of tomato (*Solanum lycopersicum* L.) and related species. Micron 40:851–859. https://doi.org/10.1016/j.micron.2009.06.004

Braz G, He L, Zhao H et al (2018) Comparative Oligo-FISH mapping: an efficient and powerful methodology to reveal karyotypic and chromosomal evolution. Genetics 208:513–523. https://doi.org/10.1534/genetics.117.300344

Budiman M, Chang S-B, Lee S et al (2004) Localization of *jointless-2* gene in the centromeric region of tomato chromosome 12 based on high resolution genetic and physical mapping. Theor Appl Genet 108:190–196. https://doi.org/10.1007/s00122-003-1429-3

Camadro EL, Carputo D, Peloquin SJ (2004) Substitutes for genome differentiation in tuber-bearing *Solanum*: interspecific pollen-pistil incompatibility, nuclear-cytoplasmic male sterility, and endosperm. Theor Appl Genet 109:1369–1376. https://doi.org/10.1007/s00122-004-1753-2

Carputo D, Barone A, Frusciante L (2000) 2N gametes in the potato: essential ingredients for breeding and germplasm transfer. Theor Appl Genet 101:805–813. https://doi.org/10.1007/s001220051547

Carputo D, Monti L, Werner J et al (1999) Uses and usefulness of endosperm balance number. Theor Appl Genet 98:478–484. https://doi.org/10.1007/s001220051095

Chang S-B, Yang T-J, Datema E et al (2008) FISH mapping and molecular organization of the major repetitive sequences of tomato. Chromosome Res 16:919–933. https://doi.org/10.1007/s10577-008-1249-z

Chen N, Zhu W, Xu J, et al. (2019) Molecular marker development and primary physical map construction for the tuber shape *Ro* gene locus in diploid potato (*Solanum tuberosum* L.). Mol Breeding 39:6. https://doi.org/10.1007/s11032-018-0913-z

Chen Q, Sun S, Ye Q et al (2004) (2004) Construction of two BAC libraries from the wild Mexican diploid potato, *Solanum pinnatisectum,* and the identification of clones near the late blight and Colorado potato beetle resistance loci. Theor Appl Genet 108:1002–1009. https://doi.org/10.1007/s00122-003-1513-8

Chetelat R (2009) Nuclear DNA content in *Solanum* sect. Juglandifolium and *Solanum* sect. Lycopersicoides. Tomato Genet Coop Rep 59:11–13

Chetelat R, Ji Y (2007) Cytogenetics and Evolution. In: Razdan MK, Mattoo A (eds) Genet. Improv. solanaceous Crop. Vol. 2, Tomato. CRC "Press, Taylor & Francis Group, Boca Raton, FL, pp 77–112

Chetelat RT, Qin X, Tan M et al (2019) Introgression lines of *Solanum sitiens*, a wild nightshade of the Atacama Desert, in the genome of cultivated tomato. Plant J 100:836–850. https://doi.org/10.1111/tpj.14460

Choudhary A, Wright L, Ponce O et al (2020) Varietal variation and chromosome behaviour during meiosis in *Solanum tuberosum*. Heredity 125:212–226. https://doi.org/10.1038/s41437-020-0328-6

de Boer JM, Borm TJA, Jesse T et al (2011) A hybrid BAC physical map of potato: a framework for sequencing a heterozygous genome. BMC Genomics 12:594. https://doi.org/10.1186/1471-2164-12-594

den Nijs TPM, Peloquin SJ (1977) 2n gametes in potato species and their function in sexual polyploidization. Euphytica 26:585–600. https://doi.org/10.1007/BF00021684

do Vale Martins L, Yu F, Zhao H et al (2019) Meiotic crossovers characterized by haplotype-specific chromosome painting in maize. Nat Commun 10:4604. https://doi.org/10.1038/s41467-019-12646-z

Doležel J, Greilhuber J (2010) Nuclear genome size: Are we getting closer? Cytom Part A 77A:635–642. https://doi.org/10.1002/cyto.a.20915

Dong F, McGrath J, Helgeson J, Jiang J (2001) The genetic identity of alien chromosomes in potato breeding lines revealed by sequential GISH and FISH analyses using chromosome-specific cytogenetic DNA markers. Genome 44:729–734. https://doi.org/10.1139/g01-043

Dong F, Novy RG, Helgeson JP, Jiang J (1999) Cytological characterization of potato - *Solanum etuberosum* somatic hybrids and their backcross progenies by genomic in situ hybridization. Genome 42:987–992. https://doi.org/10.1139/gen-42-5-987

Dong F, Song J, Naess SK et al (2000) Development and applications of a set of chromosome-specific cytogenetic DNA markers in potato. Theor Appl Genet 101:1001–1007. https://doi.org/10.1007/s001220051573

Fuchs J, Kloos D-U, Ganal MW, Schubert I (1996) *In situ* localization of yeast artificial chromosome sequences on tomato and potato metaphase chromosomes. Chromosom Res 4:277–281. https://doi.org/10.1007/BF02263677

Fulton T, van der Hoeven R, Eannetta N, Tanksley S (2002) Identification, analysis, and utilization of conserved ortholog set markers for comparative genomics in higher plants. Plant Cell 14:1457–1467. https://doi.org/10.1105/tpc.010479

Gaiero P, Mazzella C, Vilaró F, et al. (2017) Pairing analysis and *in situ* Hybridisation reveal autopolyploid-like behaviour in *Solanum commersonii* × *S. tuberosum* (potato) interspecific hybrids. Euphytica 213:137. https://doi.org/10.1007/s10681-017-1922-4

Gaiero P, Speranza P, de Jong H (2018) Introgressive hybridization in potato revealed by novel cytogenetic and genomic technologies. Am J Potato Res 95:607–621. https://doi.org/10.1007/s12230-018-9669-6

Gaiero P, Vaio M, Peters SA et al (2019) Comparative analysis of repetitive sequences among species from the potato and the tomato clades. Ann Bot 123:521–532. https://doi.org/10.1093/aob/mcy186

Gaiero P, van de Belt J, Vilaró F et al (2016) Collinearity between potato (*Solanum tuberosum* L.) and wild relatives assessed by comparative cytogenetic mapping. Genome 60:228–240. https://doi.org/10.1139/gen-2016-0150

Ganal M, Lapitan N, Tanksley S (1988) A molecular and cytogenetic survey of major repeated DNA sequences in tomato (*Lycopersicon esculentum*). Mol Gen Genet 213:262–268

Ganal MW, Lapitan NLV, Tanksley SD (1991) Macrostructure of the tomato telomeres. Plant Cel 3:87–94

Garriga-Calderé F, Huigen D, Angrisano A et al (1998) BC1 to BC2 progenies derived from backcrossing potato (+) tomato fusion hybrids to potato: the selection of single additions for seven different tomato chromosomes. Theor Appl Genet 96:155–163

Garriga-Calderé F, Huigen D, Filotico F et al (1997) Identification of alien chromosomes through GISH and RFLP analysis and the potential for establishing potato lines with monosomic additions of tomato chromosomes. Genome 40:666–673

Garriga-Calderé F, Huigen DJ, Jacobsen E, Ramanna MS (1999) Prospects for introgressing tomato chromosomes into the potato genome: An assessment through GISH analysis. Genome 42:282–288. https://doi.org/10.1139/gen-42-2-282

Gavrilenko T (2011) Application of molecular cytogenetics in fundamental and applied research of potato. In: Bradeen J, Kole C (eds) Genet. Genomics Breed. Potato. CRC Press, Taylor & Francis Group, pp 184–206

Gavrilenko T (2007) Potato cytogenetics. In: Vreugdenhil D, Bradshaw J, Gebhardt C et al (eds) Potato Biol. Elsevier, Biotechnol. Adv. Perspect, pp 203–216

Gavrilenko T, Larkka J, Pehu E, Rokka V-M (2002) Identification of mitotic chromosomes of tuberous and non-tuberous *Solanum* species (*Solanum tuberosum* and *Solanum brevidens*) by GISH in their interspecific hybrids. Genome 45:442–449. https://doi.org/10.1139/g01-136

Gavrilenko T, Thieme R, Rokka V (2001) Cytogenetic analysis of *Lycopersicon esculentum* (+) *Solanum etuberosum* somatic hybrids and their androgenetic regenerants. Theor Appl Genet 103:231–239. https://doi.org/10.1007/s001220100626

Gong Z, Wu Y, Koblizkova A et al (2012) Repeatless and repeat-based centromeres in potato: implications for centromere evolution. Plant Cell 24:3559–3574. https://doi.org/10.1105/tpc.112.100511

Grandillo S, Chetelat R, Knapp S, et al. (2011) *Solanum* sect. Lycopersicon. Wild crop relations genomic breeding resource temperature fruits. Springer, Berlin, Heidelberg, pp 1–247

Haider Ali S, Huigen D, Ramanna M et al (2001) *Genomic in situ* hybridization analysis of a trigenomic hybrid involving *Solanum* and *Lycopersicon* species. Genome 44:299–304. https://doi.org/10.1139/gen-44-2-299

Han Y, Zhang T, Thammapichai P et al (2015) Chromosome-specific painting in *Cucumis* species using bulked oligonucleotides. Genetics 200:771–779. https://doi.org/10.1534/genetics.115.177642

Harper L, Cande W (2000) Mapping a new frontier; development of integrated cytogenetic maps in plants. Funct Integr Genomics 1:89–98. https://doi.org/10.1007/s101420000013

Hawkes JG, Jackson MT (1992) Taxonomic and evolutionary implications of the Endosperm Balance Number hypothesis in potatoes. Theor Appl Genet 84:180–185. https://doi.org/10.1007/BF00223998

He L, Braz GT, Torres GA, Jiang J (2018) Chromosome painting in meiosis reveals pairing of specific chromosomes in polyploid *Solanum* species. Chromosoma 127:505–513. https://doi.org/10.1007/s00412-018-0682-9

He L, Liu J, Torres GA et al (2013) Interstitial telomeric repeats are enriched in the centromeres of chromosomes in *Solanum* species. Chromosom Res 21:5–13. https://doi.org/10.1007/s10577-012-9332-x

Hermsen J, Ramanna M (1973) Double-bridge hybrids of *Solanum bulbocastanum* and cultivars of *Solanum tuberosum*. Euphytica 22:457–466. https://doi.org/10.1007/BF00036641

Hijmans RJ, Gavrilenko T, Stephenson S et al (2007) Geographical and environmental range expansion through polyploidy in wild potatoes (*Solanum* section Petota). Glob Ecol Biogeogr 16:485–495. https://doi.org/10.1111/j.1466-8238.2007.00308.x

Hougas RW, Peloquin SJ, Gabert AC (1964) Effect of seed-parent and pollinator on frequency of haploids in *Solanum tuberosum1*. Crop Sci 4:593. https://doi.org/10.2135/cropsci1964.0011183X000400060013x

Iovene M, Aversano R, Savarese S et al (2012) Interspecific somatic hybrids between *Solanum bulbocastanum* and *S. tuberosum* and their haploidization for potato breeding. Biol Plant 56:1–8. https://doi.org/10.1007/s10535-012-0008-3

Iovene M, Savarese S, Frusciante L et al (2007) Nuclear and cytoplasmic genome composition of *Solanum bulbocastanum* (+) *S. tuberosum* somatic hybrids. Genome 50:443–450. https://doi.org/10.1139/g07-024

Iovene M, Wielgus S, Simon P et al (2008) Chromatin structure and physical mapping of chromosome 6 of potato and comparative analyses with tomato. Genetics 180:1307–1317. https://doi.org/10.1534/genetics.108.093179

Jacobsen E, de Jong J, Kamstra S (1995) Genomic in situ hybridization (GISH) and RFLP analysis for the identification of alien chromosomes in the backcross progeny of potato (+) tomato fusion hybrids. Heredity (edinb) 74:250–257

Jacobsen E, Schouten HJ (2008) Cisgenesis, a new tool for traditional plant breeding, should be exempted from the regulation on genetically modified organisms in a step by step approach. Potato Res 51:75–88. https://doi.org/10.1007/s11540-008-9097-y

Ji Y, Chetelat RT (2003) Homoeologous pairing and recombination in *Solanum lycopersicoides* monosomic addition and substitution lines of tomato. TAG Theor Appl Genet 106:979–989. https://doi.org/10.1007/s00122-002-1090-2

Ji Y, Pertuzé R, Chetelat RT (2004) Genome differentiation by GISH in interspecific and intergeneric hybrids of tomato and related nightshades. Chromosome Res 12:107–116

Jiang J (2019) Fluorescence in situ hybridization in plants: recent developments and future applications. Chromosome Res 27:153–165. https://doi.org/10.1007/s10577-019-09607-z

Jiang J (2013) Centromere evolution. Plant Centromere Biol 159–168. https://doi.org/10.1002/9781118525715.ch12

Jo SH, Koo DH, Kim JF et al (2009) Evolution of ribosomal DNA-derived satellite repeat in tomato genome. BMC Plant Biol 9:1–14. https://doi.org/10.1016/j.pbiomolbio.2017.06.018

Johnston S, den Nijs T, Peloquin S, Hanneman R (1980) The significance of genic balance to endosperm development in interspecific crosses. Theor Appl Genet 57:5–9

Kalitsis P, Choo KHA (2012) The evolutionary life cycle of the resilient centromere. Chromosoma 121:327–340. https://doi.org/10.1007/s00412-012-0369-6

Khush GS, Rick CM (1968) Cytogenetic analysis of the tomato genome by means of induced deficiencies. Chromosoma 23:452–484. https://doi.org/10.1007/BF00625288

Khush GS, Rick CM (1963) Meiosis in hybrids between *Lycopersicon esculentum* and *Solanum pennellii*. Genetica 33:167–183. https://doi.org/10.1007/BF01725760

Lapitan N, Ganal M, Tanksley S (1989) Somatic chromosome karyotype of tomato based on *in situ* hybridization of the TGRI satellite repeat. Genome 32:992–998

Leisner C, Hamilton J, Crisovan E, et al. (2018) Genome sequence of M6, a diploid inbred clone of the high glycoalkaloid-producing tuber-bearing potato species *Solanum chacoense*, reveals residual heterozygosity. Plant J 94:2. https://doi.org/10.1111/tpj.13857

Lou Q, Iovene M, Spooner DM et al (2010) Evolution of chromosome 6 of *Solanum* species revealed by comparative fluorescence in situ hybridization mapping. Chromosoma 119:435–442. https://doi.org/10.1007/s00412-010-0269-6

Marks G (1955) Cytogenetic studies in tuberous *Solanum* species I. Genomic differentiation in the Group Demissa. J Genet 53:262–269

Marks G (1969) The pachytene chromosomes of *Solanum clarum*. Caryologia 22:161–167. https://doi.org/10.1080/00087114.1969.10796334

Matsubayashi M (1991) Phylogenetic relationships in the potato and its related species. In: Tsuchiya T, Gupta P (eds) Chromosom. Engineering plants genetics breeding evoluation Part B. Elsevier, pp 93–118

Menzel M (1962) Pachytene chromosomes of the intergeneric hybrid *Lycopersicon esculentum* x *Solanum lycopersicoides*. Am J Bot 49:605–615. https://doi.org/10.1002/j.1537-2197.1962.tb14988.x

Michaelson M, Price H, Ellison J, Johnston J (1991) Comparision of plant DNA contents determined by Feulgen microspectrophotometry and laser flow cytometry. Am J Bot 78:183–188. https://doi.org/10.1002/j.1537-2197.1991.tb15745.x

Oliveira LC, Torres GA (2018) Plant centromeres: genetics, epigenetics and evolution. Mol Biol Rep 45:1491–1497. https://doi.org/10.1007/s11033-018-4284-7

Parokonny A, Marshall J, Bennett M et al (1997) Homoeologous pairing and recombination in backcross derivatives of tomato somatic hybrids (*Lycopersicon esculentum* (+) *L. peruvianum).* Theor Appl Genet 94:713–723

Pehu E, Thomas M, Poutala T et al (1990) Species-specific sequences in the genus *Solanum*: identification, characterization, and application to study somatic hybrids of *S. brevidens* and *S. tuberosum*. Theor Appl Genet 80:693–698

Peloquin SJ, Boiteux LS, Carputo D (1999) Meiotic mutants in potato: valuable variants. Genetics 153:1493–1499

Pendinen G, Gavrilenko T, Jiang J, Spooner DM (2008) Allopolyploid speciation of the Mexican tetraploid potato species *Solanum stoloniferum* and *S. hjertingii* revealed by genomic in situ hybridization. Genome 51:714–720. https://doi.org/10.1139/G08-052

Pendinen G, Spooner DM, Jiang J, Gavrilenko T (2012) Genomic *in situ* hybridization reveals both auto- and allopolyploid origins of different North and Central American hexaploid potato (*Solanum* sect. Petota) species. Genome 55:407–415. https://doi.org/10.1139/g2012-027

Pertuzé RA, Ji Y, Chetelat RT (2003) Transmission and recombination of homeologous *Solanum sitiens* chromosomes in tomato. Theor Appl Genet 107:1391–1401. https://doi.org/10.1007/s00122-003-1384-z

Peters SA, Bargsten JW, Szinay D et al (2012) Structural homology in the Solanaceae: analysis of genomic regions in support of synteny studies in tomato, potato and pepper. Plant J 71:602–614. https://doi.org/10.1111/j.1365-313X.2012.05012.x

Peters SA, Datema E, Szinay D et al (2009) *Solanum lycopersicum* cv. Heinz 1706 chromosome 6: distribution and abundance of genes and retrotransposable elements. Plant J 58:857–869. https://doi.org/10.1111/j.1365-313X.2009.03822.x

Pham GM, Braz GT, Conway M et al (2019) Genome-wide inference of somatic translocation events during potato dihaploid production. The Plant Genome 12(1–9):180079. https://doi.org/10.3835/plantgenome2018.10.0079

Plohl M, Meštrović N, Mravinac B (2014) Centromere identity from the DNA point of view. Chromosoma 123:313–325. https://doi.org/10.1007/s00412-014-0462-0

Preiszner J, Takács I, Bilgin M et al (1994) Organization of a *Solanum brevidens* repetitive sequence related to the TGRI subtelomeric repeats of *Lycopersicon esculentum*. Theor Appl Genet 89:1–8. https://doi.org/10.1007/BF00226974

Presting GG, Frary A, Pillen K, Tanksley SD (1996) Telomere-homologous sequences occur near the centromeres of many tomato chromosomes. Mol Gen Genet 251:526–531. https://doi.org/10.1007/s004380050198

Rakosy-Tican E, Thieme R, König J et al (2020) Introgression of two broad-spectrum late blight resistance genes, *Rpi-Blb1* and *Rpi-Blb3*, from *Solanum bulbocastanum* Dun plus race-specific *R* genes into potato pre-breeding lines. Front Plant Sci 11:699. https://doi.org/10.3389/fpls.2020.00699

Ramanna M, Wagenvoort M (1976) Identification of the trisomic series in diploid *Solanum tuberosum* L., group Tuberosum. I. Chromosome Identification. Euphytica 25:233–240

Ramanna MS, Prakken R (1967) Structure of and homology between pachytene and somatic metaphase chromosomes of the tomato. Genetica 38:115–133. https://doi.org/10.1007/BF01507452

Rick CM (1960) Hybridization between *Lycopersicon esculentum* and *Solanum pennellii:* phylogenetic and cytogenetic significance. Proc Natl Acad Sci U S A 46:78–82

Rodríguez F, Spooner DM (2009) Nitrate reductase phylogeny of potato (*Solanum* sect. Petota) genomes with emphasis on the origins of the polyploid species. Syst Bot 34:207–219. https://doi.org/10.1600/036364409787602195

Sawant AC (1958) Cytogenetics of interspecific hybrids, *Lycopersicon esculentum* Mill. x *L. hirsutum* Humb. and Bonpl. Genetics 43:502–514

Schweizer G, Borisjuk N, Borisjuk L et al (1993) Molecular analysis of highly repeated genome fractions in *Solanum* and their use as markers for the characterization of species and cultivars. Theor Appl Genet 85:801–808. https://doi.org/10.1007/BF00225022

Seah S, Yaghoobi J, Rossi M et al (2004) The nematode-resistance gene, *Mi-1*, is associated with an inverted chromosomal segment in susceptible compared to resistant tomato. Theor Appl Genet 108:1635–1642. https://doi.org/10.1007/s00122-004-1594-z

Song J, Dong F, Jiang J (2000) Construction of a bacterial artificial chromosome (BAC) library for potato molecular cytogenetics research. Genome 43:199–204

Spooner DM, Ghislain M, Simon R et al (2014) Systematics, diversity, genetics, and evolution of wild and cultivated potatoes. Bot Rev 80:283–383. https://doi.org/10.1007/s12229-014-9146-y

Spooner DM, Rodríguez F, Polgár Z et al (2008) Genomic origins of potato polyploids: GBSSI gene sequencing data. Crop Sci 48:27–36. https://doi.org/10.2135/cropsci2007.09.0504tpg

Stadler M, Stelzer T, Borisjuk N et al (1995) Distribution of novel and known repeated elements of *Solanum* and application for the identification of somatic hybrids among *Solanum* species. Theor Appl Genet 91:1271–1278. https://doi.org/10.1007/BF00220940

Stupar R, Song J, Tek A et al (2002) Highly condensed potato pericentromeric heterochromatin contains rDNA-related tandem repeats. Genetics 162:1435–1444

Szinay D, Bai Y, Visser R, De Jong H (2010) FISH applications for genomics and plant breeding strategies in tomato and other solanaceous crops. Cytogenet Genome Res 129:199–210. https://doi.org/10.1159/000313502

Szinay D, Wijnker E, van den Berg R et al (2012) Chromosome evolution in *Solanum* traced by cross-species BAC-FISH. New Phytol 195:688–698. https://doi.org/10.1111/j.1469-8137.2012.04195.x

Tang X, Datema E, Guzman M et al (2014) Chromosomal organizations of major repeat families on potato (*Solanum tuberosum*) and further exploring in its sequenced genome. Mol Genet Genomics 289:1307–1319. https://doi.org/10.1007/s00438-014-0891-8

Tang X, De Boer JM, Van Eck HJ et al (2009) Assignment of genetic linkage maps to diploid *Solanum tuberosum* pachytene chromosomes by BAC-FISH technology. Chromosome Res 17:899–915. https://doi.org/10.1007/s10577-009-9077-3

Tang X, Szinay D, Lang C et al (2008) Cross-species bacterial artificial chromosome-fluorescence *in situ*

hybridization painting of the tomato and potato chromosome 6 reveals undescribed chromosomal rearrangements. Genetics 180:1319–1328. https://doi.org/10.1534/genetics.108.093211
Tanksley SD, Ganal MW, Price JP et al (1992) High density molecular linkage maps of tomato and potato genomes; biological inferences and practical application. Genetics 132:1141–1160
Tek AL, Jiang J (2004) The centromeric regions of potato chromosomes contain megabase-sized tandem arrays of telomere-similar sequence. Chromosoma 113:77–83. https://doi.org/10.1007/s00412-004-0297-1
Tek AL, Song J, Macas J, Jiang J (2005) *Sobo*, a recently amplified satellite repeat of potato, and its implications for the origin of tandemly repeated sequences. Genetics 170:1231–1238. https://doi.org/10.1534/genetics.105.041087
The Potato Genome Sequencing Consortium (2011) Genome sequence and analysis of the tuber crop potato. Nature 475:189–195. https://doi.org/10.1038/nature10158
The Tomato Genome Consortium (2012) The tomato genome sequence provides insights into fleshy fruit evolution. Nature 485:635–641. https://doi.org/10.1038/nature11119
Torres GA, Gong Z, Iovene M, et al. (2011) Organization and evolution of subtelomeric satellite repeats in the potato genome. G3 (Bethesda) 1:85–92. https://doi.org/10.1534/g3.111.000125
Valkonen J, Watanabe K, Pehu E (1994) Analysis of correlation between nuclear DNA content, chromosome number, and flowering capacity of asymmetric somatic hybrids of diploid *Solanum brevidens* and (di) haploid *S. tuberosum*. Japanese J Genet 69:525–536. https://doi.org/10.1266/jjg.69.525
van der Knaap E, Sanyal A, Jackson S, Tanksley S (2004) High-resolution fine mapping and fluorescence *in Situ* hybridization analysis of *sun*, a locus controlling tomato fruit shape, reveals a region of the tomato genome prone to DNA rearrangements. Genetics 168:2127–2140. https://doi.org/10.1534/genetics.104.031013
Verlaan M, Szinay D, Hutton S et al (2011) Chromosomal rearrangements between tomato and Solanum chilense hamper mapping and breeding of the TYLCV resistance gene *Ty-1*. Plant J 68:1093–1103. https://doi.org/10.1111/j.1365-313X.2011.04762.x
Visser RGF, Hoekstra R (1988) *In situ* hybridization to somatic metaphase chromosomes of potato. Theor Appl Genet 76:420–424
Wagenvoort M (1988) Spontaneous structural rearrangements in *Solanum* 1. Chromosome Identification at Pachytene Stage. Euphytica 9:159–167
Wang L, Zeng Z, Zhang W, Jiang J (2014) Three potato centromeres are associated with distinct haplotypes with or without megabase-sized satellite repeat arrays. Genetics 196:397–401. https://doi.org/10.1534/genetics.113.160135
Wolters A-M, Caro M, Dong S et al (2015) Detection of an inversion in the *Ty-2* region between *S. lycopersicum* and *S. habrochaites* by a combination of *de novo* genome assembly and BAC cloning. Theor Appl Genet 128:1987–1997. https://doi.org/10.1007/s00122-015-2561-6
Xu J, Earle ED (1996) High resolution physical mapping of 45S (5.8S, 18S and 25S) rDNA gene loci in the tomato genome using a combination of karyotyping and FISH of pachytene chromosomes. Chromosoma 104:545–550. https://doi.org/10.1007/BF00352294
Yang Y, Yang X, Li C et al (2015) Construction and characterization of a bacterial artificial chromosome library of potato cultivar C88. Acta Horticulturae Sinica 42:361–366
Yeh B, Peloquin S (1965) Pachytene chromosomes of the potato (*Solanum tuberosum,* Group Andigena). Am J Bot 52:1014–1020
Zhang H, Kobli Kova A, Wang K et al (2014) Boom-Bust turnovers of megabase-sized centromeric dna in solanum species: rapid evolution of DNA sequences associated with centromeres. Plant Cell 26:1436–1447. https://doi.org/10.1105/tpc.114.123877
Zhong XB, Fransz PF, Van EJW et al (1998) FISH studies reveal the molecular and chromosomal organization of individual telomere domains in tomato. Plant J 13:507–517. https://doi.org/10.1046/j.1365-313X.1998.00055.x

The Wild Genetic Resources of Tomato: A Reservoir of Useful Genes for the Future

3

Maria Raffaella Ercolano, Antonio Di Matteo, Giuseppe Andolfo, and Luigi Frusciante

Abstract

Tomato wild relatives originated from Andeans mountain ranges to Galapagos archipelago offer a wide range of genetic diversity that can be exploited for different purposes. Besides their large phenotypic diversification, all wild tomato species can be intercrossed since they are diploid (2n = 24) and have a high degree of genomic synteny. However, unilateral incompatibility is present between the self-incompatible green-fruited species and self-compatible red-fruited species and the self-compatible green-fruited species *S. neorickii*. In order to promote in situ and ex situ conservation of wild tomato relatives, recent efforts were undertaken to build a comprehensive inventory for researchers and breeders working worldwide. In cultivated tomato, several agronomic traits have been lost due to the domestication bottleneck, which led to a drastic reduction of genetic variability. The exploration and utilization of wild species genetic diversity allowed the identification and introgression of important quality, abiotic, and biotic resistance stress traits. Looking for novel exotic traits remains a major goal since new breeding targets became established. Wild relatives will undoubtedly serve as a valuable reservoir of germplasm to meet future needs. Genomics research can offer the potential to accelerate the introgression of favorable wild genes in tomato crop. Insight into sequencing will greatly help to estimate the wealth of wild germplasm and enable tomato resources to be preserved and utilized efficiently.

M. R. Ercolano (✉) · A. Di Matteo · G. Andolfo · L. Frusciante
Department of Agricultural Sciences, University of Naples Federico II, Via Università 100, 80055 Portici (Naples), Italy
e-mail: ercolano@unina.it

3.1 Introduction

3.1.1 Relationships Among Tomato Closed Wild Species

Tomato (*Solanum lycopersicum*) and its wild relatives *S. arcanum*, *S. cheesmaniae*, *S. chilense*, *S. chmielewskii*, *S. corneliomulleri*, *S. galapagense*, *S. habrochaites*, *S. huaylasense*, *S. neorickii*, *S. pennellii*, *S. peruvianum*, and *S. pimpinellifolium* have evolved in extremely heterogeneous environments from Andeans Highlands to the coast of Galapagos Islands (Peralta 2008). The wide diversification among species of genus *Solanum* section *Lycopersicon* (Fig. 3.1) at the morphological, physiological, sexual, and molecular levels (Peralta and Spooner 2005, 2007) reflects the large range of ecological habitat they originated from the clade of the tomato species has distinctive morphological traits such as yellow corollas, pedicels articulated above the base,

© Springer Nature Switzerland AG 2021

D. Carputo et al. (eds.), *The Wild Solanums Genomes*, Compendium of Plant Genomes,
https://doi.org/10.1007/978-3-030-30343-3_3

pinnately segmented non-prickly leaves (Correll 1962; Rick 1988). Tomato wild relative species carry green fruits, except for the two species from the Galapagos (with yellow and orange fruits) and *S. pimpinellifolium*, which is the only species of sect. *Lycopersicon* with red fruits (Paran and van der Knaap 2007). All wild relative species are diploid (2n = 24) (Clint Nesbitt and Tanksley 2002) and are characterized by a high degree of genomic synteny (Sato et al. 2012).

In addition, *Solanum* sect. *Juglandifolia* contains two woody tomato-like nightshades *S. ochranthum* and *S. juglandifolium*, partially sympatric and morphologically similar, both being woody perennials with rampant, liana-like stems up to 30 m in length (Rick 1987; Peralta and Spooner 2005; Correll 2006; Peralta 2008). Based on evidence from molecular sequence data (Peralta 2008), sect. *Juglandifolia* is the sister group of the *Solanum* sect. *Lycopersicoides*, comprising the allopatric sister species *S. lycopersicoides* and *S. sitiens*. These four tomato-like nightshade species have in common several morphological features that make them intermediate between tomato and potato (Rick 1987; Smith and Peralta 2002; Stommel 2019). The tomato (*S. lycopersicum*) and its relatives diverged from the potatoes (section *Petota*) at ca. 8 Mya (6.6–9/9 Mya) (Särkinen et al. 2013). The diversification within the tomato clade in the strict sense (excluding sections *Juglandifolia* and *Lycopersicoides*) is estimated to be occurred ca. 2 Mya (1.2–2.6 Mya).

Fig. 3.1 Geographic distribution of wild species in *Solanum* section *lycopersicon*

3.1.2 Genetic Structure and Interspecific Crossability

Species divergence and hybridization within the tomato clade species (Nakazato and Housworth 2011) mainly shaped by the ecological complexity of the natural habitats of origin. The thirteen wild relative species, together with four closely allied *Solanum* species, harbor a wide diversity of mating systems and associated reproductive traits (Rick 1987). The cultivated tomato *S. lycopersicum* and several wild relative species are self-compatible (SC with no exert stigma) and, thus, autogamous. The allied species *S. juglandifolium*, *S. ochranthum*, *S. lycopersicoides,* and *S. sitiens* are self-incompatible (SI) with large, highly divided inflorescences and exerted stigmas, pollen shedding occur via terminal anther pores instead of through longitudinal slits, anthers are separated rather than fused, and flowers are highly scented (Chetelat et al. 2009). Two intermediate groups of species have facultative mating systems. As members of the first group, *S. pimpinellifolium* and *S. chmielewskii* are SC species with floral structures that foster outcrossing. The second group includes *S. pennellii* and *S. peruvianum* (Rick and Tanksley 1981; Graham et al. 2003), two facultative outcrossers that are mainly SI with some SC populations. All populations of *S. habrochaites* from north to central Peru are SI; whereas, populations from extreme margins of this species' distribution (on the north in Ecuador and on the south in Peru) are SC (Martin 1964; Rick et al. 1979). *S. habrochaites* populations characterized by loss of SI also harbor unilateral intraspecific incompatibility (Rick and Chetelat 1991).

Interspecific crosses have been performed among all the members of the tomato clade (Bedinger et al. 2011). Unilateral incompatibility has been shown to occur between the SI green-fruited species and SC red-fruited species. In addition, crosses between the SC green-fruited species *S. neorickii* as female parent with SI species as pollen donors are successful but not the reciprocal crosses.

Molecular structure analysis of a large set of wild *S. pimpinellifolium* accessions, cherry tomato, and tomato cultivated accessions showed that domesticated and wild tomatoes have evolved as a species complex with extensive hybridization (Labate et al. 2009; Ranc et al. 2012). This evidence suggested that spreading the tomato allelic variation has been accompanied by crossing tomato with wild relative species and distribution among different environments across the world (Labate et al. 2011). *S. lycopersicum* has likely experienced a severe population bottleneck during the colonization of the eastern Andes, followed by a rapid geographical population expansion (Caicedo and Schaal 2004; Caicedo 2008). The ancestral form of the cultivated tomato *S. lycopersicum* var. *cerasiforme* is present in both Mexico and Peru. On the contrary, *S. pimpinellifolium* is absent from Mexico. Selection for self-pollinating, as well as shortening of the stigma compared to close wild relatives, has promoted changes in *S. cerasiforme* traits such as growth habit, fruit morphology, and yield (Rick 1977).

S. cerasiforme is indicated as the direct ancestor of tomato because of its widespread occurrence in central America and its close genetic relationship with cultivated tomato (Rick and Chetelat 1995). The first hypothesis supports Peru as tomato center of origin and domestication (Candolle 1886). A second hypothesis proposed that domestication occurred primarily in Mexico in the Vera Cruz Puebla area (Jenkins 1948) as there is no evidence for the pre-Colombian cultivation of tomato in South America. Because of the genetic erosion taken place in Europe 400 years ago, the level of genetic diversity in the tomato cultivated gene pool (GP) represents a narrow part of its original variability.

3.1.3 Effort to Improve the Preservation of Tomato Genetic Resources

Various initiatives to collect large number of wild species accessions have been undertaken. However, most of the wild plant species collected primarily have been used for research rather than for long-term conservation since ex situ conservation of tomato accessions is laborious and expensive. Collections should build up to represent the full range of geographic and ecological variation in their native distributions. Recent project efforts for prioritizing conservation actions in many geographic regions were launched (TomGen, GP2 SOL, wild NSF). Higher priorities should be given to the wild relatives falling into primary or secondary GP over other more distant ones, as it is relatively easy to transfer traits between species within the primary GP. Initiative to promote in situ conservation of wild tomato relatives to take into account the importance of environmental factors and the possibility to evolve under natural conditions through crossings with wild or weedy relatives are also important (Tripp and Heide 1996; Corrado et al. 2014).

The most important bank of tomato germplasm is reported in Table 3.1 and includes facilities at the CM Rick Tomato Genetics Resource Center (TGRC, USA), at the World Vegetable Center (WVC, Taiwan), at the United States Department of Agriculture (USDA, USA), as well as at the Leibniz Institute of Plant Genetics and Crop Plant Research (IPK, Germany) and at the Centre-for-Genetic Resources (CGR, Netherlands). In the Netherlands, the Botanical and Experimental Garden (BGARD) also maintains the most extensive ex situ plant collections of non-tuberous Solanaceae species (http://www.bgard.science.ru.nl). Important tomato gene banks were established in Mexico (CNRG), Peru (INIA), Cuba (INIFAT), and Brazil (Gonçalves et al. 2008; Arizaga et al. 2016). Large collections of tomato germplasm are also conserved in Russia (VIR), Japan (NIAS). Other countries with great numbers of stored tomato accessions are Bulgaria, Canada, China, Colombia, France, Italy, Hungary, the Philippines, and Spain. The European Cooperative Programme for Plant Genetic Resources established a network to maintain more than 20,000 tomato species accessions provided with information related to taxonomic origins, regional priority categories. It proposed conservation strategies into electronic formats ECPRG (http://www.ecpgr.cgiar.org/).

The Svalbard Global Seed Vault in Norway stores several collection duplicates consistently with relevant national and international law. Each country or institution will still own and control access to the seeds they have deposited according to the black box system implying that the depositor is the only one that can withdraw the seeds and open the boxes.

The ultimate goal should be to build a comprehensive inventory for tomato wild relatives aimed at integrating the information dispersed among individual agencies or countries and to make these data available globally to researchers and breeders for efficient conservation management and usage. Consequently, several repositories hold websites or digital catalogs illustrating the collection stored (Table 3.1).

Table 3.1 List of main tomato germplasm gene banks

Repository	Acronyms	Website	Location
CM rick tomato genetics resource center	TGRC	https://tgrc.ucdavis.edu	USA
World vegetable center	WVC	https://avrdc.org	Taiwan
Germplasm resources information network (USDA)	GRIN	https://www.ars-grin.gov	USA
The European cooperative programme for plant genetic resources	ECPRG	www.ecpgr.cgiar.org/resources/germplasm-databases	Europe
The genetic resources center, (National agriculture and food research organization)	NARO	https://www.gene.affrc.go.jp	Japan
The Leibniz institute of plant genetics and crop plant research	IPK	https://www.ipkgatersleben.de/en/genebank/	Germany
Embrapa genetic resources & biotechnology, national genetic resources network	CENARGEN (RENARGEN)	https://www.embrapa.br/en/international	Brazil
National genetic resources center	CNRG	www.inifap.gob.mx/SitePages/centros/cnrg.asp	Mexico
Centre-for-genetic-resources-the-Netherlands	CGR	https://www.wur.nl/en/	Netherlands
Instituto nacional de innovacion Agraria	INIA	http://www.inia.gob.pe/aip/	Perù
Fundamental research institute on tropical agriculture	INIFAT	https://www.gfar.net/organizations/fundamental-research-institute-tropical-agriculture-alejandro-von-humboldt	Cuba
The botanical and experimental garden	BGARD	www.bgard.science.ru.nl	Netherlands
Australian tropical field crops genetic resource centre	ATFGRC	http://www.dpi.qld.gov.au/auspgris	Australia
INRA genetics and breeding of fruit and vegetables	GAFL	https://www6.paca.inra.fr/gafl_eng/Vegetables-GRC/Our-Collections/Tomato-Collection/The-Solanaceae-Genetic-Resources-Network	France
Institute of conservation and improvement of the agrodiversity	COMAV	https://www.comav.upv.es	Spain
Mediterranean germplasm database	IBBR	http://ibbr.cnr.it/mgd/	Italy
National programme on conservation and utilization of plant genetic resources and agro-biodiversity		http//genebank.vurv.cz/genetic/resources	Cezch Republic
Institute of plant genetic resources "Konstantin Malkov"	IPGRBG	http://ipgrbg.com	Bulgaria
Vavilov institute of general genetics Russian academy of sciences	VIGG	http://www.vigg.ru	Russia
Plant gene resources of Canada	PGRC	http://pgrc3.agr.gc.ca	Canada
Institute of crop germplasm resources	ICGR	http://www.cgris.net/icgr/icgr	China
National plant genetic resources laboratory	NPGRL	https://cafs.uplb.edu.ph/	Philippines
Svalbard global seed vault		https://www.seedvault.no	Norway

3.1.4 Strategies for Discovery Wild Species Donor Genes

The exploration and utilization of the natural genetic diversity stored in tomato wild relatives allowed to achieve important insights in the last century. Sources for genetic tolerance to pathogen and quality traits were successfully introgressed in tomato crop by conventional breeding. However, several agronomically important traits have been lost in cultivated tomato due to domestication bottlenecking of genetic variability.

Molecular-based approaches, developed in the last few decades, highly improved the exploitation of tomato wild accessions genetic potential. High-density genetic maps based on interspecific crosses between cultivated tomato and related wild species from the *lycopersicon* group (Grandillo et al. 2011), as well as from *lycopersicoides* (Pertuzé et al. 2003) and *juglandifolia* groups (Albrecht and Chetelat 2009), supported the analysis of important loci. The characterization, the conservation, and managing of Genebank wild accessions were promoted by fingerprinting analysis (Robertson and Labate 2007). Phylogenetic studies allowed to establish important relationships among and within wild species (Spooner et al. 2005; Zuriaga et al. 2009; Albrecht and Chetelat 2009). Rapidly increasing of high throughput DNA technologies allowed the construction of oligonucleotide-based arrays used for comparing DNA sequence polymorphisms among different tomato wild species accessions and cultivated tomato (Sim et al. 2009). Next-generation sequencing (NGS) supported the identification of thousands of genomic variants in hundreds of wild populations (Aflitos et al. 2014; Lin et al. 2014; Sahu and Chattopadhyay 2017). The integration of "omics" data with additional phenotypic and genetic information and ad hoc developed genomic resources aided the discovering and characterizing a myriad of quantitative trait loci (QTLs).

For instance, exotic introgression lines (IL) resulted extremely useful for reintroducing genetic variability from wild species into tomato germplasm (Tanksley and McCouch 1997; Zamir and Geiger 2001; Gur and Zamir 2004). Introgression populations were obtained using several species, including *S. pennellii* (Eshed and Zamir 1994), *S. pimpinellifolium* (Doganlar et al. 2002), *S. habrochaites* (Monforte and Tanksley 2000; Francis et al. 2001; Finkers et al. 2007; Foolad and Panthee 2012; Kinkade and Foolad 2013; Valverde-Barrantes and Blackwood 2016), and *S. lycopersicoides* (Chetelat et al. 1998; Canady et al. 2006). The most widely used IL population is the *S. pennellii* LA 0716 collection, which facilitated the identification of more than 2700 traits involved in agronomically important traits (Lippmann et al. 2007).

Over the last few years, genome-wide association study (GWAS) combined with extensive mapping of various traits led to the discovery and partitioning of several traits via approaches such as QTL-seq and integration of multi-omics data (Tieman et al. 2017). Association studies reached very high levels of resolution by integrating genomic resources such as SNP data (~26 million SNPs) with fruit transcriptome (RNA-seq, ~30,000 genes) and metabolome (362 annotated metabolites) datasets from between 399 and 610 accessions of the tomato species *S. pimpinellifolium*, *S. lycopersicum* var. *cerasiforme*, and *S. lycopersicum* (Zhu et al. 2018).

3.1.5 Identification of Biotic Stress Tolerance Sources

Wild germplasm has been extensively used as a source of major pests and pathogens resistance in tomato. Breeders have introgressed wild genome regions into cultivated varieties since 1917, starting with the leaf mold *Cf* resistance gene. Subsequently, several disease resistance genes have been introgressed from wild relatives, such as *S. pimpinellifolium, S. chilense, S. peruvianum, S. habrochaites, S. pennellii* (Laterrot 2000; Foolad 2007; Scott and Gardner 2007).

The identification of new sources of resistance in wild species is still an ongoing process. A wealth of monogenic resistance genes deriving from wild tomato species has also been cloned (Table 3.2). The recent characterization of three fusarium wilt resistance genes provided effective means to control this soil-born pathogen, besides the nematode and verticillium resistance genes introgressed from *S. peruvianum*. Sources of resistance for foliar and fruit diseases have also been reported (bacterial speck, leaf mold, bacterial cancer, early blight, blackmold). Multiple genes resources for some diseases, such as late blight (caused by oomycete *Phytophthora infestans*) and powdery mildew (caused by fungus *Oidium lycopersicum*), have been identified, and both vertical and horizontal resistances have been transferred to the cultivated tomato. The tomato is also very susceptible to viruses, and several resistance genes found in wild species have proved to be effective and reliable for their control. Monogenic resistance genes to various viruses have been recognized in wild accessions, and the resistance often resulted race-specific as well as environment-dependent. Finally, *S. habrochaites* and *S. pennellii* species resulted very valuable sources of insect and pest resistance. Mostly, introgression of genes conferring resistance to insects in tomato has been carried out using *S. pennellii* accession LA716 or *S. habrochaites* LA1777 as resistance parents. Effort to exploit wild species variability for such traits is not fully deserved. Knowledge of the kind of resistance to specific vector insects and pests will facilitate the effort to utilize such variability.

The characterization, disease evaluation, and transfer of resistance genes were carried out through phenotypic selection and traditional breeding protocols. Subsequently, genetic mapping and molecular-based assisted selection (MAS) accelerated the breeding work. Modern cultivars can cumulate different disease resistance genes, all deriving from wild species, and at least 20 of them have been bred into tomato cultivars (Ji et al. 2007; Robertson and Labate 2007). Looking for resistance sources remains a major goal since new diseases, or new forms of existing pathogens and pests become established. Wild relatives will undoubtedly serve as a valuable reservoir of germplasm to meet future needs.

Table 3.2 Summary of cloned resistance genes deriving from tomato wild relatives

Plant species	Gene	Pathogen	Disease common name	References
Solanum pennellii	*Asc*	*Alternaria alternata* f. sp. *lycopersici*	Alternaria stem cancer	Brandwagt et al. (2002)
Solanum pimpinellifolium	*Cf1*	*Cladosporium fulvum*	Leaf mold	Dickinson et al. (1993)
Solanum habrochaites	*Cf2*	*C. fulvum*	Leaf mold	Dixon et al. (1996)
S. habrochaites	*Cf4*	*C. fulvum*	Leaf mold	Dixon et al. (1998)
S. pimpinellifolium	*Cf5*	*C. fulvum*	Leaf mold	Jones et al. (1994)
S. pimpinellifolium	*Cf9*	*C. fulvum*	Leaf mold	Thomas et al. (1997)
S. pimpinellifolium	*Hero*	*Globodera rostochiensis*	Yellow potato cyst nematode	Ernst et al. (2002)
S. pimpinellifolium	*I1*	*Fusarium oxysporum* fs *lycopersici*	Fusarium wilt	Catanzariti et al. (2017)
S. pimpinellifolium	*I2*	*F. oxysporum* fs *lycopersici*	Fusarium wilt	Simons et al. (1998)
S. pennellii	*I3*	*F. oxysporum* fs *lycopersici*	Fusarium wilt	Catanzariti et al. (2015)
S. pennellii	*I7*	*F. oxysporum* fs *lycopersici*	Fusarium wilt	

(continued)

Table 3.2 (continued)

Plant species	Gene	Pathogen	Disease common name	References
				Gonzalez-Cendales et al. (2016)
Solanum peruvianum	*Mi-1*	*Melaydogyne* spp., Macrosiphum euphorbiae	Root-knot	Rossi et al. (1998)
S. pimpinellifolium	*Ph3*	*Phythopthora infestans*	Late blight	Zhang et al. (2014)
S. pimpinellifolium	*Prf*	Required for *Pto Pseudomonas syringae* pv. *Tomato*	Bacterial speck	Salmeron et al. (1996)
S. pimpinellifolium	*Pto*	*P. syringae* pv. *Tomato*	Bacterial speck	Martin et al. (1993)
S. habrochaites	*Pto 1*	*P. syringae* pv. *Tomato*	Bacterial speck	Riely and Martin (2001)
S. peruvianum	*Sw5*	TSWV	Tomato spotted wilt	Brommonschenkel et al. (2000)
S. peruvianum	*Ve*	*Verticiullim dahlia*	Verticillium wilt	Kawchuk et al. (2001)
S. peruvianum	*Ve2*	*V. dahlia*	Verticillium wilt	Kawchuk et al. (2001)
S. habrochaites	*Tm1*	TMV	Tobacco mosaic	Ishibashi et al. (2007)
S. habrochaites	*Tm2*	TMV	Tobacco mosaic	Lanfermeijer et al. (2005)
S. peruvianum	*Tm2a*	TMV	Tobacco mosaic	Lanfermeijer et al. (2003)
Solanum chilense	*Ty1*	TYLCV	Tomato yellow leaf curl	Verlaan et al. (2011)
S. habrochaites	*Ty2*	TYLCV	Tomato yellow leaf curl	Yamaguchi et al. (2018)
S. chilense	*Ty3*	TYLCV; ToMoV	Tomato yellow leaf curl; tomato mottle virus	Verlaan et al. (2013)

3.1.6 Wild Germplasm Surveys for Abiotic Stress Tolerance

Sources of genetic tolerance to different abiotic stresses were found within wild species, including *S. chilense*, *S. peruvianum*, *S. pennellii*, *S. pimpinellifolium*, and *S. harbochaites*. In addition, there are a few species within *Solanum* that exhibit tolerance to environmental stresses and that may be exploited in tomato breeding for stress tolerance, such as *S. lycopersicoides*, *S. juglandifolium*, and *S. ochranthum* (Zhao et al. 2005). Five SC wild tomato species (*S. cheesmaniae*, *S. chmielewskii*, *S. lycopersicum* var. *cerasiforme*, *S. neorickii*, and *S. pimpinellifolium*) displayed a quantitative distribution of the tolerance levels to drought regardless of their shoot physiological traits (Easlon and Richards 2009). Several accessions of *S. pimpinellifolium*, *S. pennellii*, *S. chilense*, and *S. peruvianum* exhibit robust drought tolerance (Peralta et al. 2005; Moyle 2008; Chetelat et al. 2009). QTLs associated with drought tolerance traits were also identified in *S. pennellii* introgression lines (Rigano et al 2016; Bolger et al. 2014).

The genetic control of traits conferring salt stress tolerance in *S. pimpinellifolium* showed quantitative inheritance with low heritability. However, the efficiency of selection for salt tolerance deriving from *S. pimpinellifolium* was proved by Monforte et al. (1996). Interestingly, *S. pimpinellifolium* accession LA722 contains QTLs affecting cold and salt tolerance during germination (Foolad et al. 2008). The *S. pennellii* genome revealed about 100 candidate genes having differential expression under drought or salt stress conditions mapping to drought- or salt-related QTL regions (Bolger et al. 2014). Major QTLs on chromosomes 6, 7, and 11 were identified in introgression lines derived from *S. pennellii* LA716. In *S. lycopersicoides,* six major QTLs for abiotic stress tolerance were discovered, which are located on chromosomes 4, 6, 9, and 12. Co-localization of QTLs on chromosome 6 in *S. pennellii* and *S. lycopersicoides* highlighted a shared conserved locus, and this knowledge was also used to identify QTLs for salt tolerance at the seedling stage. Three genes were induced by salt stress exclusively in *S. pimpinellifolium* PI365967 (Sun et al. 2012).

Because of the natural complexity of temperature stress tolerance, it was difficult identifying key genetic components leading to the development of tomatoes tolerant to high temperatures. A successful approach could be based on the analysis of the various sub-traits contributing individually to the trait under study and combine them afterward. Recently, the screening of wild species accessions for fertility performances under long-term mild heat (LTMH) allowed to identify *S. pimpinellifolium* germplasm with local adaptation of reproductive heat tolerance (Driedonks et al. 2018). Two QTLs for heat resistance in tomato were reported on chromosomes 6 and 12 (Kadirvel et al. 2013). Several QTLs associated with shoot wilting and root ammonium uptake under chilling temperatures were identified in an *S. habrochaites* segregating population. The high chilling-tolerant *S. habrochaites* LA1777 showed a complex regulation of miRNAs targeting six genes controlling anti-stress proteins, antioxidant enzymes, and cell wall functions (Cai et al. 2014).

3.1.7 Main Achievements for Quality and Yield-Related Traits

In cultivated tomato, identification of yield and quality traits from wild relatives has become successful in the last 30 years. For example, the jointless (*j2*) recessive allele (Szymkowiak and Irish 2006; Quinet et al. 2011) was introgressed from *S. cheesmanii* into many processing varieties. Several wild species accessions, including *S. pimpinellifolium*, *S. chmielewskii,* and *S. cheesmanii,* with much higher concentrations (~9–15%) of soluble solids have been identified. QTLs for soluble solids were introgressed from. *S. chmielewskii* by Azada et al. (2005) and from *S. habrochaites* by Petreikov et al. (2006). The functional characterization of the TIV gene from wild green-fruited tomatoes revealed its vacuolar invertase activity involved in controlling sucrose partitioning (Hadas et al. 1995; Chetelat et al. 1995; Miron et al. 2002; Klann et al. 2016). The introgression of the wild TIV gene in cultivated tomatoes has been exploited in order to improve fruit soluble solids content. Two epistatic genes that increase the proportion of fructose over glucose, thus contributing to fruit sweetness, have been identified in *S. habrochaites*. The major locus controlling the Fgr trait was mapped to chromosome 4 (Levin et al. 2000), while a second locus was mapped to chromosome 6, near to loci for a fructokinase (FK2) and a hexokinase (Levin et al. 2007). QTLs for increased fruit size were introduced from the small-berried *S. pimpinellifolium* (Tanksley et al. 1996), whereas QTLs for fruit color were introgressed from green-berried *S. habrochaites* (Bernacchi et al. 1998). Genes improving the fruit content of ascorbic acid (Lima-Silva et al. 2012) were introgressed from *S. pennellii*, whereas genes for sugar, organic acids, and carotenoid fruit content from *S. pimpinellifolium* (Causse et al. 2004; Capel et al. 2015). A unique *S. pennellii* QTL was proven to be effective at the heterozygous level in improving harvest index, earliness, and sugars and amino acids content in processing tomatoes (Gur et al. 2010, 2011).

Screening of *S. pennellii* IL population was very successful in identifying QTLs for fruit traits (Causse et al. 2004), antioxidants (Rousseaux et al. 2005), soluble solids (Fridman et al. 2002; Gur and Zamir 2004; Schauer et al. 2008), vitamin C (Stevens et al. 2007; Di Matteo et al. 2010), and volatile aromas (Tadmor et al. 2002). More of 900 mQTLs for 74 molecules (Tieman et al. 2006; Stevens et al. 2007), mQTLs modifying aroma volatiles (Rambla et al. 2014) and the *Uniform ripening* locus (Powell et al. 2012) were also identified. ILs were also used to dissect the genetic basis of heterosis (Lippman et al. 2007).

However, the exploitation of wild fruit-quality QTLs in practical breeding has been very limited because of the inherent difficulties in implementing selection (Collard and Mackill 2008).

3.1.8 Development of Genomic Tools for Future Challenges

Advances in genomics offer the potential to accelerate the introgression of favorable wild genes in tomato crop. In previous paragraphs, several examples of discovered wild alleles/genes useful for improving tomato performances were illustrated. Genomic tools could strongly enhance the exploitation of wild relatives (Zamir 2008), but the difficulty of transferring the targeted allele (with favorable effect) without unfavorable ones, carried on by "linkage drag," still limits feasibility of this approach. Genome-wide genotyping using next-generation sequencing is a very valuable method to be explored for future needs. Advances into these technologies will foster discovering the wealth of wild germplasm and enable tomato resources to be preserved and utilized more efficiently.

Resequencing of wild relative genomes and mapping of reads against the tomato reference genome revealed important differences, particularly in gene content and sequence variation (Aflitos et al. 2014; Lin et al. 2014). Specific genomic regions depleted of diversity as a consequence of domestication and selection have been identified (Lin et al. 2014). Thanks to variome-guided analysis, knowledge of sweeps loci, and linkage drags will help identifying genes for favorable traits from their embedded. Also, the comprehension of genetic erosion basis and loci selection during domestication will enable redesigning the genomic foundation for future tomato breeding (Di Donato et al. 2018; Ercolano et al. 2020). Sequencing of wild species genome allows the targeted reintroduction of the diversity into the domesticated gene pool (Rodríguez-Leal et al. 2017). Interestingly, a genome engineering strategy enabled de novo domestication of wild *S. pimpinellifolium* (Zsögön et al. 2018).

Currently, the precision of genome sequencing has been improved by new technologies such as Pac Bio platform, as the DNA sequencing technologies advance the opportunities to explore genome sequence diversity in wild species increase. Sequencing of wild reference genomes (*S. pennellii, S. habrochaites, S.chilense* and *S. pimpinellifolum*) helped to characterize more complex-related traits and understand the means of plant evolution and adaptation (Bolger et al. 2014). Sequencing of maternal genomes, especially chloroplast genomes (Nock et al. 2011), supported the exploration of genetic diversity in tomato wild populations. In addition, the sequencing of transcriptomes allowed the rapid identification of novel alleles or expression variants for crop improvement. Transcriptome analysis of wild relatives can reveal expression and gene sequence divergence resulting from the selection of environmental responsive and stress tolerance genes (Chitwood and Kumar 2013) induced by domestication/bottleneck events. Several genomic resources (DNA sequencing data, RNA-seq data, and marker data) developed from wild tomato species by individual laboratories and by community initiatives are stored in repositories such as SGN (http://solgenomics.net/search/search=library) and will facilitate future efforts.

References

Aflitos S, Schijlen E, De Jong H et al (2014) Exploring genetic variation in the tomato (*Solanum* section *Lycopersicon*) clade by whole-genome sequencing. Plant J 80:136–148. https://doi.org/10.1111/tpj.12616

Albrecht E, Chetelat RT (2009) Comparative genetic linkage map of *Solanum* sect. *Juglandifolia*: evidence of chromosomal rearrangements and overall synteny with the tomatoes and related nightshades. Theor Appl Genet 118:831–847. https://doi.org/10.1007/s00122-008-0943-8

Bedinger PA, Chetelat RT, McClure B et al (2011) Interspecific reproductive barriers in the tomato clade: opportunities to decipher mechanisms of reproductive isolation. Sex Plant Reprod 24(3):171–187. https://doi.org/10.1007/s00497-010-0155-7

Bernacchi D, Beck-Bunn T, Emmatty D et al (1998) Advanced backcross QTL analysis of tomato. II. Evaluation of near-isogenic lines carrying single-donor introgressions for desirable wild QTL-alleles derived from *Lycopersicon hirsutum* and *L. pimpinellifolium*. Theor Appl Genet 97:170–180. https://doi.org/10.1007/s001220050882

Bolger A, Scossa F, Bolger ME et al (2014) The genome of the stress-tolerant wild tomato species *Solanum pennellii*. Nat Genet 46:1034–1038. https://doi.org/10.1038/ng.3046

Brandwagt BF, Kneppers TJA, Nijkamp HJJ et al (2002) Overexpression of the tomato Asc-1 gene mediates high insensitivity to AAL toxins and fumonisin B1 in tomato hairy roots and confers resistance to Alternaria alternata f. sp. lycopersici in Nicotiana umbratica plants. Mol Plant–Microbe Interact 15:35–42

Brommonschenkel SH, Frary A, Frary A et al (2000) The broad-spectrum tospovirus resistance gene Sw-5 of tomato Is a homolog of the root-knot nematode resistance gene Mi. Mol Plant-Microbe Interact 13:1130–1138

Cai X, Ye J, Hu T et al (2014) Genome-wide classification and expression analysis of nucleobase–ascorbate transporter (NAT) gene family in tomato. Plant Growth Regul 73:19–30. https://doi.org/10.1007/s10725-013-9864-x

Caicedo AL (2008) Geographic diversity cline of R gene homologs in wild populations of *Solanum pimpinellifolium* (Solanaceae). Am J Bot 95:393–398. https://doi.org/10.3732/ajb.95.3.393

Caicedo AL, Schaal BA (2004) Population structure and phylogeography of *Solanum pimpinellifolium* inferred from a nuclear gene. Mol Ecol 13:1871–1882. https://doi.org/10.1111/j.1365-294X.2004.02191.x

Canady MA, Ji Y, Chetelat RT (2006) Homeologous recombination in Solanum lycopersicoides introgression lines of cultivated tomato. Genetics 174(4):1775–1788. https://doi.org/10.1534/genetics.106.065144

de Candolle A (1886) Origin of cultivated plants. D. Appleton, New York

Catanzariti A-M, Do HTT, Bru P et al (2017) The tomato I gene for Fusarium wilt resistance encodes an atypical leucine-rich repeat receptor-like protein whose function is nevertheless dependent on SOBIR1 and SERK3/BAK1. Plant J 89:1195–1209

Catanzariti A-M, Lim GTT, Jones DA (2015) The tomato I-3 gene: a novel gene for resistance to Fusarium wilt disease. New Phytologist 207(1):106–118. https://doi.org/10.1111/nph.13348

Capel C, del Carmen AF, Alba JM et al (2015) Wide-genome QTL mapping of fruit quality traits in a tomato RIL population derived from the wild-relative species *Solanum pimpinellifolium* L. Theor Appl Genet 128:2019–2035. https://doi.org/10.1007/s00122-015-2563-4

Causse M, Duffe P, Gomez MC et al (2004) A genetic map of candidate genes and QTLs involved in tomato fruit size and composition. J Exp Bot 55:1671–1685. https://doi.org/10.1093/jxb/erh207

Chetelat RT, DeVerna JW, Bennett AB (1995) Introgression into tomato (*Lycopersicon esculentum*) of the *L. chmielewskii* sucrose accumulator gene (sucr) controlling fruit sugar composition. Theor Appl Genet 91:327–333. https://doi.org/10.1007/BF00220895

Chetelat RT, Pertuzé RA, Faúndez L et al (2009) Distribution, ecology and reproductive biology of wild tomatoes and related nightshades from the Atacama desert region of Northern Chile. Euphytica 167:77–93. https://doi.org/10.1007/s10681-008-9863-6

Chitwood DH, Kumar R (2013) A quantitative genetic basis for leaf morphology is revealed in a set of precisely defined tomato introgression lines. Plant Cell 25:2379–2379. https://doi.org/10.1105/tpc.113.250710

Clint Nesbitt T, Tanksley SD (2002) Comparative sequencing in the genus lycopersicon: implications for the evolution of fruit size in the domestication of cultivated tomatoes. Genetics 162:365–379

Collard BCY, Mackill DJ (2008) Marker-assisted selection: an approach for precision plant breeding in the twenty-first century. Philos Trans R Soc B Biol Sci 363:557–572

Corrado G, Caramante M, Piffanelli P, Rao R (2014) Genetic diversity in Italian tomato landraces: implications for the development of a core collection. Sci Hortic (amsterdam) 168:138–144. https://doi.org/10.1016/j.scienta.2014.01.027

Correll DS (1962) The potato and its wild relatives. Contr Texas Res Found Bot Stud 4:1–606

Correll DS (2006) The potato and its wild relatives. Taxon 12:38. https://doi.org/10.2307/1216687

Di Donato A, Filippone E, Ercolano MR et al (2018) Genome sequencing of ancient plant remains: findings, uses and potential applications for the study and improvement of modern crops. Front Plant Sci 9:441

Di Matteo A, Sacco A, Anacleria M et al (2010) The ascorbic acid content of tomato fruits is associated with the expression of genes involved in pectin degradation. BMC Plant Biol 10:163. https://doi.org/10.1186/1471-2229-10-163

Di Matteo A, Sacco A, De Stefano R et al (2012) Comparative transcriptomic profiling of two tomato lines with different ascorbate content in the fruit. Biochem Genet 50:908–921. https://doi.org/10.1007/s10528-012-9531-3

Doganlar S, Frary A, Daunay MC et al (2002) A comparative genetic linkage map of eggplant (*Solanum melongena*) and its implications for genome evolution in the Solanaceae. Genetics 161:1697–1711

Driedonks N, Wolters-Arts M, Huber H et al (2018) Exploring the natural variation for reproductive thermotolerance in wild tomato species. Euphytica 214. https://doi.org/10.1007/s10681-018-2150-2

Easlon HM, Richards JH (2009) Drought response in self-compatible species of tomato (Solanaceae). Am J Bot 96:605–611. https://doi.org/10.3732/ajb.0800189

Ercolano MR, Di Donato A, Sanseverino W et al (2020) Complex migration history is revealed by genetic diversity of tomato samples collected in Italy during the eighteenth and nineteenth centuries. Hortic Res 7:100. https://doi.org/10.1038/s41438-020-0322-4

Ernst K, Kumar A, Kriseleit D at al (2002) The broad-spectrum potato cyst nematode resistance gene (Hero) from tomato is the only member of a large gene family of NBS-LRR genes with an unusual amino acid repeat in the LRR region. Plant J 31:127–136

Eshed Y, Zamir D (1994) A genomic library of *Lycopersicon pennellii* in L. esculentum: a tool for fine mapping of genes. Euphytica 79:175–179. https://doi.org/10.1007/BF00022516

Finkers R, Van Heusden AW, Meijer-Dekens F et al (2007) The construction of a *Solanum habrochaites* LYC4 introgression line population and the identification of QTLs for resistance to *Botrytis cinerea*. Theor Appl Genet 114:1071–1080. https://doi.org/10.1007/s00122-006-0500-2

Foolad MR (2007) Genome mapping and molecular breeding of tomato. Int J Plant Genomics 2007. https://doi.org/10.1155/2007/64358

Foolad MR, Merk HL, Ashrafi H (2008) Genetics, genomics and breeding of late blight and early blight resistance in tomato. CRC Crit Rev Plant Sci 27:75–107

Foolad MR, Panthee DR (2012) Marker-assisted selection in tomato breeding. CRC Crit Rev Plant Sci 31:93–123

Francis NJ, Saurin AJ, Shao Z, Kingston RE (2001) Reconstitution of a functional core polycomb repressive complex. Mol Cell 8:545–556. https://doi.org/10.1016/S1097-2765(01)00316-1

Fridman E, Pleban T, Zamir D (2002) A recombination hotspot delimits a wild-species quantitative trait locus for tomato sugar content to 484 bp within an invertase gene. Proc Natl Acad Sci 97:4718–4723. https://doi.org/10.1073/pnas.97.9.4718

Gonçalves LSA, Rodrigues R, Sudré CP et al (2008) Divergência genética em tomate estimada pormarcadores RAPD em comparação com descritores multicategóricos. Hortic Bras 26:364–370. http://dx.doi.org/10.1590/S0102-05362008000300014

Gonzalez-Cendales Y, Catanzariti AM, Baker B et al (2016) Identification of I-7 expands the repertoire of genes for resistance to Fusarium wilt in tomato to three resistance gene classes. Mol Plant Pathol 17 (3):448–463

Graham EB, Shannon SM, Petersen JP, Chetelat RT (2003) A self-compatible population of *Lycopersicon peruvianum* collected from N. Chile

Grandillo S, Chetelat R, Knapp S et al (2011) Solanum sect. Lycopersicon. In: Wild crop relatives: genomic and breeding resources. Springer, Berlin, Heidelberg, pp 129–215

Gur A, Osorio S, Fridman E et al (2010) hi2-1, A QTL which improves harvest index, earliness and alters metabolite accumulation of processing tomatoes. Theor Appl Genet 121:1587–1599. https://doi.org/10.1007/s00122-010-1412-8

Gur A, Semel Y, Osorio S et al (2011) Yield quantitative trait loci from wild tomato are predominately expressed by the shoot. Theor Appl Genet 122:405–420. https://doi.org/10.1007/s00122-010-1456-9

Gur A, Zamir D (2004) Unused natural variation can lift yield barriers in plant breeding. PLoS Biol 2. https://doi.org/10.1371/journal.pbio.0020245

Hadas R, Schaffer A, Miron D et al (1995) PCR-generated molecular markers for the invertase gene and sucrose accumulation in tomato. Theor Appl Genet 90:1142–1148. https://doi.org/10.1007/BF00222935

Ishibashi K, Masuda K, Naito S et al (2007) An inhibitor of viral RNA replication is encoded by a plant resistance gene. Proc Natl Acad Sci USA 104:13833

Ji Y, Schuster DJ, Scott JW (2007) Ty-3, a begomovirus resistance locus near the Tomato yellow leaf curl virus resistance locus Ty-1 on chromosome 6 of tomato. Mol Breeding 20:271–284. https://doi.org/10.1007/s11032-007-9089-7

Jenkins JA (1948) The origin of the cultivated tomato

Kadirvel P, de la Peña R, Schafleitner R et al (2013) Mapping of QTLs in tomato line FLA456 associated with resistance to a virus causing tomato yellow leaf curl disease. Euphytica 190:297–308. https://doi.org/10.1007/s10681-012-0848-0

Kawchuk LM, Hachey J, Lynch DR et al (2001) Tomato Ve disease resistance genes encode cell surface-like receptors. Proc Natl Acad Sci USA 98(11):6511–6515. https://doi.org/10.1073/pnas.091114198

Kinkade MP, Foolad MR (2013) Validation and fine mapping of lyc12.1, a QTL for increased tomato fruit lycopene content. Theor Appl Genet 126:2163–2175. https://doi.org/10.1007/s00122-013-2126-5

Klann EM, Chetelat RT, Bennett AB (2016) Expression of acid invertase gene controls sugar composition in tomato (Lycopersicon) fruit. Plant Physiol 103:863–870. https://doi.org/10.1104/pp.103.3.863

Labate J, Robertson L, Baldo A (2009) Multilocus sequence data reveal extensive departures from equilibrium in domesticated tomato (Solanum lycopersicum L.). Heredity 103:257–267. https://doi.org/10.1038/hdy.2009.58

Labate JA, Sheffer SM et al (2011) Diversity and population structure in a geographic sample of tomato accessions. Crop Sci 51(3):1068–1079. https://doi.org/10.2135/cropsci2010.05.0305

Lanfermeijer FC, Warmink J, Hille J (2005) The products of the broken Tm-2 and the durable Tm-22 resistance genes from tomato differ in four amino acids. J Exp Bot 56:2925–2933

Laterrot H (2000) Disease resistance in tomato: practical situation. Acta Physiol Plant 22:328–331. https://doi.org/10.1007/s11738-000-0048-8

Levin I, Gilboa N, Yeselson E et al (2000) Fgr, a major locus that modulates the fructose to glucose ratio in mature tomato fruits. Theor Appl Genet 100:256–262. https://doi.org/10.1007/s001220050034

Levin I, Gilboa N, Cincarevsky F et al (2007) Epistatic interaction between the Fgr and FK2 genes determines the fructose to glucose ratio in mature tomato fruit. Isr J Plant Sci 54:215–223

Lima-Silva V, Rosado A, Amorim-Silva V, et al (2012) Genetic and genome-wide transcriptomic analyses identify co-regulation of oxidative response and hormone transcript abundance with vitamin C content in tomato fruit. BMC Genomics 13. https://doi.org/10.1186/1471-2164-13-187

Lin T, Zhu G, Zhang J et al (2014) Genomic analyses provide insights into the history of tomato breeding. Nat Genet 46:1220–1226

Lippman ZB, Semel Y, Zamir D (2007) An integrated view of quantitative trait variation using tomato interspecific introgression lines. Curr Opin Genet Dev 17:545–552

Lippmann M, Bress A, Nemeroff CB et al (2007) Long-term behavioural and molecular alterations associated with maternal separation in rats. Eur J Neurosci 25:3091–3098. https://doi.org/10.1111/j.1460-9568.2007.05522.x

Martin FW (1964) The understanding of unilateral incompatibility and its relation to self-income. Fruits

Miron D, Petreikov M, Carmi N et al (2002) Sucrose uptake, invertase localization and gene expression in developing fruit of *Lycopersicon esculentum* and the sucrose-accumulating *Lycopersicon hirsutum*. Physiol Plant 115:35–47. https://doi.org/10.1034/j.1399-3054.2002.1150104.x

Monforte AJ, Asíns MJ, Carbonell EA (1996) Salt tolerance in Lycopersicon species. IV. Efficiency of marker-assisted selection for salt tolerance improvement. Theor Appl Genet 93:765–772. https://doi.org/10.1007/BF00224074

Moyle LC (2008) Ecological and evolutionary genomics in the wild tomatoes (*Solanum* sect. *Lycopersicon*). Evolution (N.Y.) 62:2995–3013

Nakazato T, Housworth EA (2011) Spatial genetics of wild tomato species reveals roles of the Andean geography on demographic history. Am J Bot 98:88–98. https://doi.org/10.3732/ajb.1000272

Nock CJ, Waters DLE, Edwards MA et al (2011) Chloroplast genome sequences from total DNA for plant identification. Plant Biotechnol J 9:328–333. https://doi.org/10.1111/j.1467-7652.2010.00558.x

Paran I, van der Knaap E (2007) Genetic and molecular regulation of fruit and plant domestication traits in tomato and pepper. J Exp Bot 58:3841–3852

Peralta I, Spooner D (2005) Morphological characterization and relationships of wild tomatoes (*Solanum* L. sect. *Lycopersicon*). Monogr Syst Bot 104:227–257

Peralta I, Spooner D (2007) History, origin and early cultivation of tomato (Solanaceae). In: Genetic improvement solanaceous crops. pp 1–27

Peralta IE, Spooner DM, Knapp S (2008) Taxonomy of wild tomatoes and their relatives (Solanum sect. *Lycopersicoides*, sect. Author(s): Peralta IE, Spooner DM, Knapp S Source: systematic botany monographs, taxonomy of wild tomatoes and their relatives. Am Soc Plant Taxon 84:1–186

Peralta IE, Knapp S, Spooner DM (2005) New species of wild tomatoes (*Solanum* section *Lycopersicon*: Solanaceae) from Northern Peru. Syst Bot 30:424–434. https://doi.org/10.1600/0363644054223657

Pertuzé RA, Ji Y, Chetelat RT (2003) Comparative linkage map of the *Solanum lycopersicoides* and *S. sitiens* genomes and their differentiation from tomato. Genome 45:1003–1012. https://doi.org/10.1139/g02-066

Petreikov M, Shen S, Yeselson Y et al (2006) Temporally extended gene expression of the ADP-Glc pyrophosphorylase large subunit (AgpL1) leads to increased enzyme activity in developing tomato fruit. Planta 224:1465–1479. https://doi.org/10.1007/s00425-006-0316-y

Powell ALT, Nguyen CV, Hill T et al (2012) Uniform ripening encodes a golden 2-like transcription factor regulating tomato fruit chloroplast development. Science (80) 336:1711–1715. https://doi.org/10.1126/science.1222218

Quinet M, Kinet J M and Lutts S (2011) Flowering response of the uniflora: blind:self-pruning and jointless:uniflora:self-pruning tomato (Solanum lycopersicum) triple mutants. Physiol Plant 141(2):166–176. https://doi.org/10.1111/j.1399-3054.2010.01426.x

Rambla JL, Tikunov YM, Monforte AJ et al (2014) The expanded tomato fruit volatile landscape. J Exp Bot 65:4613–4623

Ranc N, Muños S, Xu J et al (2012) Genome-wide association mapping in tomato (*Solanum lycopersicum*) is possible using genome admixture of *solanum lycopersicum* var. *cerasiforme*. G3: Genes|Genomes|Genetics 2:853–864. https://doi.org/10.1534/g3.112.002667

Rigano MM, Arena C, Di Matteo A, Sellitto S, Frusciante L, Barone A (2016) Eco-physiological response to water stress of drought-tolerant and drought-sensitive tomato genotypes. Plant Biosyst 150(4):682–691. https://doi.org/10.1080/11263504.2014.989286

Rick CM (1979) Biosystematic studies in Lycopersicon and closely related species of Solanum. In: Hawkes JG, Lester RN, Skelding AD (eds) The biology and taxonomy of the Solanaceae. Academic Press, New York, pp 667–678

Rick CM (1987) Tomato-like nightshades: affinities, autoecology, and breeders' opportunities. Econ Bot

Rick CM (1988) Tomato-like nightshades: affinities, autoecology, and breeders' opportunities. Econ Bot 42:145–154

Rick CM, Chetelat RT (1995) Utilization of related wild species for tomato improvement. Acta Hortic 21–38. https://doi.org/10.17660/ActaHortic.1995.412.1

Rick CM, Tanksley SD (1981) Genetic variation in *Solanum pennellii*: comparisons with two other sympatric tomato species. Plant Syst Evol 139:11–45. https://doi.org/10.1007/BF00983920

Robertson LD, Labate JA (2007) Genetic resources of tomato (*Lycopersicon esculentum* var. esculentum) and wild relatives. In: Genetic improvement of Solanaceous crops: tomato, vol 2. pp 25–75

Rodríguez-Leal D, Lemmon ZH, Man J et al (2017) Engineering quantitative trait variation for crop improvement by genome editing. Cell 171:470-480. e8. https://doi.org/10.1016/j.cell.2017.08.030

Rousseaux MC, Jones CM, Adams D et al (2005) QTL analysis of fruit antioxidants in tomato using *Lycopersicon pennellii* introgression lines. Theor Appl Genet 111:1396–1408. https://doi.org/10.1007/s00122-005-0071-7

Sahu KK, Chattopadhyay D (2017) Genome-wide sequence variations between wild and cultivated tomato species revisited by whole genome sequence mapping. BMC Genomics 18. https://doi.org/10.1186/s12864-017-3822-3

Särkinen T, Bohs L, Olmstead RG, Knapp S (2013) A phylogenetic framework for evolutionary study of the nightshades (Solanaceae): a dated 1000-tip tree. BMC Evol Biol 13. https://doi.org/10.1186/1471-2148-13-214

Sato S, Tabata S, Hirakawa H et al (2012) The tomato genome sequence provides insights into fleshy fruit evolution. Nature 485:635–641. https://doi.org/10.1038/nature11119

Schauer N, Semel Y, Balbo I et al (2008) Mode of inheritance of primary metabolic traits in tomato. Plant Cell Online 20:509–523. https://doi.org/10.1105/tpc.107.056523

Sim SC, Robbins MD, Chilcott C et al (2009) Oligonucleotide array discovery of polymorphisms in cultivated tomato (*Solanum lycopersicum* L.) reveals patterns of SNP variation associated with breeding. BMC Genomics 10:466. https://doi.org/10.1186/1471-2164-10-466

Smith SD, Peralta IE (2002) Ecogeographic surveys as tools for analyzing potential reproductive isolating mechanisms: an example using *Solanum juglandifolium* Dunal, S. ochranthum. 51:341–349

Spooner DM, McLean K, Ramsay G et al (2005) A single domestication for potato based on multilocus amplified fragment length polymorphism genotyping. Proc Natl Acad Sci 102:14694–14699. https://doi.org/10.1073/pnas.0507400102

Stevens R, Buret M, Duffe P et al (2007) Candidate genes and quantitative trait loci affecting fruit ascorbic acid content in three tomato populations. Plant Physiol 143:1943–1953. https://doi.org/10.1104/pp.106.091413

Stommel JR (2019) Barriers for introgression of *Solanum ochranthum* into tomato via somatic hybrids. J Am Soc Hortic Sci 126:587–592. https://doi.org/10.21273/jashs.126.5.587

Sun YD, Liang Y, Wu JM et al (2012) Dynamic QTL analysis for fruit lycopene content and total soluble solid content in a *Solanum lycopersicum* x *S. pimpinellifolium* cross. Genet Mol Res 11:3696–3710. https://doi.org/10.4238/2012.August.17.8

Szymkowiak EJ, Irish EE (2006) JOINTLESS suppresses sympodial identity in inflorescence meristems of tomato. Planta 223:646–658. https://doi.org/10.1007/s00425-005-0115-x

Tadmor Y, Fridman E, Gur A et al (2002) Identification of malodorous, a wild species allele affecting tomato aroma that was selected against during domestication. J Agric Food Chem 50:2005–2009. https://doi.org/10.1021/jf011237x

Tanksley SD, Grandillo S, Fulton TM et al (1996) Advanced backcross QTL analysis in a cross between an elite processing line of tomato and its wild relative *L. pimpinellifolium*. Theor Appl Genet 92:213–224. https://doi.org/10.1007/BF00223378

Tanksley SD, McCouch SR (1997) Seed banks and molecular maps: unlocking genetic potential from the wild. Science (80-) 277:1063–1066

Tieman D, Zhu G, Resende MFR et al (2017) Plant science a chemical genetic roadmap to improved tomato flavor downloaded from. Science (80) 355:27

Tieman DM, Zeigler M, Schmelz EA et al (2006) Identification of loci affecting flavour volatile emissions in tomato fruits. J Exp Bot 57:887–896. https://doi.org/10.1093/jxb/erj074

Tripp R, Van der Heide W (1996) The erosion of crop genetic diversity: challenges, strategies and uncertainties. Nat Resour Perspect 7:10s

Valverde-Barrantes OJ, Blackwood CB (2016) Root traits are multidimensional: specific root length is independent from root tissue density and the plant economic spectrum. J Ecol 104:1299–1310. https://doi.org/10.1111/1365-2745.12562

Verlaan MG, Szinay D, Hutton SF et al (2011) Chromosomal rearrangements between tomato and Solanum chilense hamper mapping and breeding of the TYLCV resistance gene Ty-1. The Plant Journal 68:1093–1103

Verlaan MG, Hutton SF, Ibrahem RM et al (2013) The Tomato Yellow Leaf Curl Virus resistance genes Ty-1 and Ty-3 are allelic and code for DFDGD-class RNA-dependent RNA polymerases. PLoS Genet 9(3): e1003399

Yamaguchi H, Ohnishi J, Saito A et al (2018) An NB-LRR gene, TYNBS1, is responsible for resistance

mediated by the Ty-2 Begomovirus resistance locus of tomato. Theor Appl Genet 131:1345–1362
Zamir D (2008) Plant breeders go back to nature. Nat Genet 40:269–270
Zamir E, Geiger B (2001) Molecular complexity and dynamics of cell-matrix adhesions. J Cell Sci 114:3583–3590
Zhao L, Qiu C, Li J et al (2005) Investigation of disease resistance and cold tolerance of *Solanum lycopersicoides* for tomato improvement. HortScience 40:43–46
Zhang C, Liu L, Wang X et al (2014) The Ph-3 gene from Solanum pimpinellifolium encodes CC-NBS-LRR protein conferring resistance to Phytophthora infestans. Theor Appl Genet 127:1353–1364
Zhu G, Wang S, Huang Z et al (2018) Rewiring of the fruit metabolome in tomato breeding. Cell 172:249-261.e12. https://doi.org/10.1016/j.cell.2017.12.019
Zsögön A, Čermák T, Naves ER et al (2018) De novo domestication of wild tomato using genome editing. Nat Biotechnol 36:1211–1216
Zuriaga E, Blanca JM, Cordero L et al (2009) Genetic and bioclimatic variation in *Solanum pimpinellifolium*. Genet Resour Crop Evol 56:39–51. https://doi.org/10.1007/s10722-008-9340-z

4 Wild and Cultivated Potato Species Diversity, Taxonomy, and Conservation

Iris Edith Peralta, Andrea Martina Clausen, Cinthya Zorrilla, Mercedes Ames, Ariana Digilio, and Flor Rodriguez

Abstract

In the present chapter, we summarize the knowledge of wild and cultivated potato species, diversity, a taxonomic update including group concept classification and description, species valid names, and a complete synonymy, distribution, and habitat. Likewise, the importance of reproductive characters, breeding barriers, interspecific hybridization, and gene flow, introgression, polyploidy in potato evolution and ecological adaptation, and conservation strategies is explained. Also a comprehensive taxonomy of all wild and cultivated potatoes, based on the integration of multiple evidences and phylogenetic relationships between taxa is discussed, providing a framework for further investigation of complex groups as well as rare endemic species. Hypothesis regarding patterns of species diversity and distributions, and adaptive mechanisms to different extreme environments are proposed. More recent genomic approaches are promissory not only to investigate wild potato genome evolution but also to detect alleles related to important agronomic traits. Germplasm of more wild species or potato landraces can be explored considering hypothesis of relationships. A taxonomic framework could be useful for harmonizing names and classification of potato collection among genebanks. The knowledge of species diversity and distribution patterns is fundamental for collecting strategies and the establishment of natural protected areas as well as agrobiodiversity zones, and for management and sustainable use of potato genetic resources.

I. E. Peralta (✉)
Facultad de Ciencias Agrarias, Universidad Nacional de Cuyo & IADIZA CCT CONICET Mendoza, Alte. Brown 500 (5505) Luján de Cuyo, Mendoza, Argentina
e-mail: iperalta@fca.uncu.edu.ar

A. M. Clausen · A. Digilio
Estación Experimental Agropecuaria Balcarce, Instituto Nacional de Tecnología Agropecuaria (INTA) C.C. 276, 7620 Balcarce, Bs. As., Argentina

C. Zorrilla · F. Rodriguez
Universidad Nacional Agraria La Molina, La Molina, 15024 Lima, Perú

M. Ames
6348 Goose Lake Verona, Madison, WI 53593, USA

4.1 Introduction

Ancient American farmers domesticated potato species on the high plateaus of Andean Punas, and also in the lowlands of Southern Chile. The temperament and capacity of these communities, their knowledge of natural diversity and environment, led to the development of a rich culture, reflected in the diversity of their food resources, farming practices, and traditions (Zhang and Rodríguez 2015; De Haan et al. 2019). Andean

© Springer Nature Switzerland AG 2021

D. Carputo et al. (eds.), *The Wild Solanums Genomes*, Compendium of Plant Genomes,
https://doi.org/10.1007/978-3-030-30343-3_4

farmers not only domesticated potatoes, but also more than 40 species for their subsistence (Parodi 1966; Popenoe et al. 1990) and generated ingenious cultivation methods on the slopes of the high mountains, which still last to this day. This legacy is one of America's great treasures bequeathed to the world, and potatoes are essentials for human food subsistence.

Potatoes were introduced to Europe in the mid-sixteenth century and then its cultivation spread to the whole world (Hawkes 1990; Ames and Spooner 2008). Nowadays, potato is one of the most important crops for human nutrition and health (Burgos et al. 2020), and it is the first tuberous species cultivated worldwide, with an average production of 378,201,964 Ton and a harvested area of 19,062,653 Ha (http://www.fao.org/faostat).

Cultivated potatoes were domesticated from its indigenous relatives. Wild species are native of America distributed from Southwestern United States (latitude 38 °N), Central and South America to Argentina, and adjacent mainland Chile (latitude 41 °S) (Fig. 4.1), with greatest species richness at latitude 21 °S in South America and a secondary center of speciation around 20 °N in the Central Mexican highlands (Hijmans and Spooner 2001; Hijmans et al. 2002). Landraces of cultivated potatoes are grown throughout mid to high (about 3000–3500 m) elevations in the Andes from Northern South America to Northwestern Argentina, and in lowland South-Central Chile, concentrated in the Chonos Archipelago (Fig. 4.2).

Traditionally, potatoes have been included in the genus *Solanum* section *Petota* Dumort., which comprises all wild tuber bearing species and the cultivated potato (*S. tuberosum* L.). A closely related group, *Solanum* section *Etuberosum* (Bukazov & Kameraz.) A. Child, includes three wild non-tuber bearing species but morphologically similar to potatoes (Contreras and Spooner 1999; Spooner et al. 2016).

The taxonomy of potatoes have been difficult to elucidate due to their great diversity along a wide geographic range, ecological adaptation to different habitat, great phenotypic variation that made difficult the interpretation of morphological characters, and also complicated by reproductive features such as interspecific hybridization, introgression, allopolyploidy, prezygotic and postzygotic mechanisms, also a unique mixture of sexual and asexual reproduction, and possible recent species divergence (Ghislain et al. 2006; Rodríguez et al. 2009; Spooner et al. 2004, 2014, 2016, 2019).

Besides the biological processes involved in potatoes evolution and ecological adaptations, early taxonomists who have worked on section *Petota* applied different concepts in their treatments, fundamentally in the criteria to delimitate taxa and number of species, hypotheses of their interrelationships, and interpretations of the hybrid origins of various taxa (Spooner and Van den Berg 1992; Spooner and Salas 2006; Rodríguez et al. 2010; Spooner et al. 2014, 2016, 2019).

More recently comprehensive monographs of section *Petota* and section *Etuberosum* (Spooner et al. 2004, 2016, 2019; Ovchinnikova et al. 2011) revised initial taxonomic treatments, incorporating studies of numerous herbarium specimens, including types, and cultivated representatives of all recognized species. In these comprehensive monographs, it is important to highlight the application of common taxonomic concepts based on phylogeny to elucidate potato diversity. Furthermore, the results of recent morphological and molecular phylogenetic studies have driven us to continuously reduce the number of potato species relative to early taxonomic treatments (Spooner et al. 2004, 2016, 2019; Ovchinnikova et al. 2011).

In the present chapter, we summarize the knowledge of potato species diversity, a taxonomic update including group concept classification and description, species valid names and a complete synonymy, distribution and habitat, as well as the importance of reproductive characters, breeding barriers, interspecific hybridization, gene flow, introgression, polyploidy in potato evolution and ecological adaptation. Additionally, a discussion is included about difficult groups for further taxonomical studies, and possible approaches to clarify and solve taxonomical controversies. The methods and issues

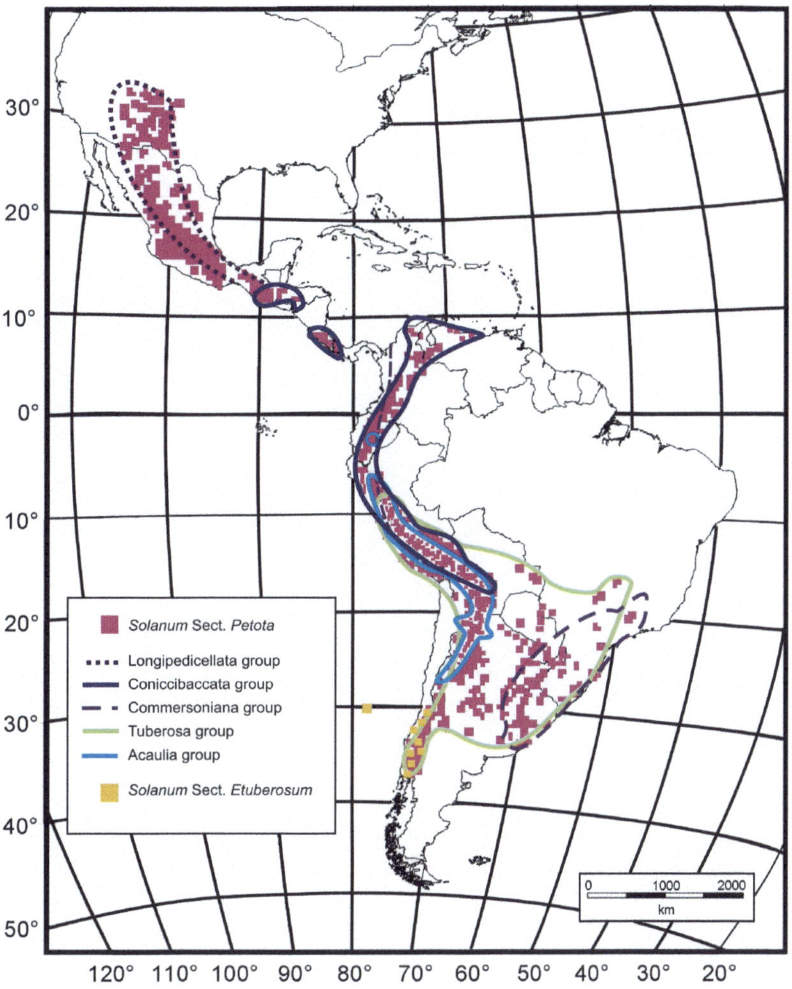

Fig. 4.1 Distribution of wild species of sect. *Etuberosum* and sect. *Petota*, delimited areas were drawn in five most widespread and diverse non-formal groups: Longipedicellata, Conicibaccata, Commersoniana, Tuberosa, and Acaulia

for ex-situ conservation of potato genetic resources are also considered. Finally, the strategies for in situ conservation in natural and agroecological areas are discussed.

4.2 Potato Reproductive Characteristics

The reproductive characteristics of wild potato species and the evidences of natural hybridization phenomena have led botanists to have different taxonomical interpretations that are sometimes conflicting (Spooner and Van den Berg 1992; Spooner and Salas 2006; Spooner et al. 2016, 2019).

All species of section *Etuberosum* and *Petota* have the same basic chromosome number of x = 12. Ploidy refers to the number of chromosome sets in the genome, and Rybin (1929, 1933) first described the polyploid series in wild potatoes (2*x*, 3*x*, 4*x*, 5*x*, 6*x*). Ploidy assessment, summarized in Table 4.1, revealed that the majority (66%) of wild species are diploids (2*x* = 24), but there is also variation in species ploidy (Gavrilenco 2011). Hijmans et al. (2007) determined the geographical and environmental correlations of ploidy for the wild taxa of *Solanum* sect. *Petota*, documented multiple cytotypes in 21 wild species, and found that diploids occupy a larger geographical area, at the northern and southern edges of distribution, than

Fig. 4.2 Distribution of cultivated species: *S. tuberosum* L. with two cultivar groups, the 'Andigenum Group' of upland Andean genotypes with diploid, triploid, and tetraploids, and the 'Chilotanum Group' of lowland tetraploid Chilean landraces, and other three Andean cultivated species *S. ajanhuiri* (diploid), *S. juzepzukii* (triploid) and *S. curtilobum* (pentaploid)

polyploids that most frequently occur in small areas at ecological extremes where higher-level polyploids species occur in colder habitats and triploids in warmer and drier sites than diploid. In Table 4.1, ten diploid species have additional triploid populations with 36 chromosomes ($3x$), and one diploid species also present tetraploid populations with 48 chromosomes ($4x$); two of hybrid origin were found exclusively triploids ($3x = 36$); eleven exclusively tetraploids ($4x = 48$); one exclusively pentaploid ($5x = 60$) and six exclusively hexaploids ($6x = 72$). In few species, populations of diploids and hexaploids, triploids and tetraploids and tetraploids and hexaploids have been detected. Four species have populations with more than one even ploidy level (S. *colombianum, S. andreanum, S. brevicaule* and *S. candolleanum*) (Table 4.1). Triploid and pentaploid populations are generally highly sterile. Ploidy in cultivated potatoes has been also investigated in the *S. tuberosum* Andigenum Group, where no ecogeographical association for the ploidy variants and different habitat was found, while in *S. tuberosum* Chilotanum Group ploidy was related with extreme northern and southern distribution (Spooner et al. 2010).

Genome structure has been analyzed by Matsubayashi (1991) and Gavrilenko (2007, 2011) through various cytological techniques (see also Chap. 2). Additional approaches, like the analysis of orthologous GBSSI genes sequences, showed the first molecular evidence of allopolyploidy in potato (Spooner et al. 2008). Analysis of single-copy genes have been useful to understand genomic complexity, revealing patterns of hybrid origins and allele losses in potato polyploids (Rodriguez and Spooner 2009; Cai et al. 2012). Genome rearrangements in *S. bulbocastanum*, a wild potato species with B genome, were uncovered for the first time when its linkage map was compared with potato and tomato physical maps, and provided a promissory approach for investigation of genome-specific structural chromosome rearrangements between *Solanum* A and B genomes as well as for mapping of agronomical traits (Iorizzo et al. 2014; Mann et al. 2011; Aversano et al. 2015).

Wild potatoes are distributed along a wide geographic range in America, where physical and ecological barriers can prevent gene flow among species. Nevertheless, Camadro et al. (2004) argued that these external factors are not sufficient to explain maintenance of potato species integrity. These authors, based on the evidences of little genome differentiation in potatoes, and also taking into account the lack of interspecific crossing in several sympatric populations, proposed that internal barriers to hybridization may have played a fundamental role in wild potatoes evolution. Interspecific pollen-pistil incompatibility, nuclear-cytoplasmic male sterility, and seed endosperm development are major forces that strengthen geographic and ecological barriers, even though a certain amount of gene exchange could be possible, species remain separate in an evolutionary context (Camadro et al. 2004).

Most diploid potato species are self-incompatible due to a multiallelic S-locus with gametophytic expression (Cipar et al. 1964; Goldberg et al. 2010). The style produces a ribonuclease codified by the S-locus that prevents the normal growth of genetically matching pollen tubes (Dodds et al. 1996; Luu et al. 2000). The S locus has been mapped to chromosome 1 in potato germplasm (Gebhardt et al. 1991) and S-RNase genes have also been mapped in the same chromosome (Rivard et al. 1996). A dominant self-incompatibility inhibitor gene that allows self-pollination has been reported in wild diploid species and mapped to the distal end of chromosome 12 (Hosaka and Hanneman 1998). Interspecific pollen-pistil interaction has been explained by a genetic system of cross incompatibility or incongruity (CI), in which genes interact on a one-to-one basis to allow or prevent hybridization (Hogenboom 1973, 1979). Camadro and Peloquin (1981) proposed a genetic model to explain the isolation in tuber-bearing *Solanum* species and the maintenance of their genetic integrity, with dominant CI genes in styles that prevent fertilization by pollen carrying specific dominant complementary genes. These genetic systems developed during the evolution

Table 4.1 Three species of sect. *Etuberosum*, and 108 epecies of sect. *Petota* are included in non-formal groups, information about countries and regions of occurrence, as well as altitude, ploidy levels, endosperm balance numbers (EBN), and phenology are indicated. Abbreviations are used for the countries: ARG Argentina, BOL Bolivia, BRA Brazil, CHI Chile, COL Colombia, CRI Costa Rica, ECU Ecuador, GUA Guatemala, HON Honduras, MEX México, PAN Panmá, PAR Paraguay, PER Perú, URU Uruguay, USA United States of America, VEN Venezuela, and numbers indicated species quantity

Section		Wild Species	Distribution (Countries)	Distribution (Regions)	Altitude (m)	Ploidy (EBN)	Phenology (Flowering & Fruiting)
ETUBEROSUM	1	*S. etuberosum* Lindl.	CHI	V, VI, VII, VIII, IX Regions	430–2500	2*X* (1)	December–March
	2	*S. palustre* Schltdl.	ARG CHI	ARG (Neuquén) CHI (V, VIII, IX, X)	40–1170	2*X* (1)	January–March
	3	*S. fernandezianum* Phil.	CHI	Masatierra Island	100–610	2*X* (1)	January–March
PETOTA	ACAULIA						
	1	*S. acaule* Bitter	ARG BOL CHI PER	ARG (Catamarca, Jujuy, La Rioja, Salta, San Juan, Tucumán) BOL (Chuquisaca, Cochabamba, La Paz, Oruro, Potosí, Tarija) CHI (II Región) PER (Cajamarca)	2000–4700	4*X* (2)	December–April
	2	*S. albicans* (Ochoa) Ochoa	ECU PER	ECU (Chimborazo) PER (Cajamarca)	3340–4800	6*X* (4)	January–May
	3	*S. demissum* Lindl.	GUA MEX	GUA (Huehuetenango, Sacatepequez, Totonicapán), MEX (Aguascalientes, Chiguagua, Distrito Federal, Durango, Hidalgo, México, Michoacán, Morelos, Oaxaca, Puebla, Sinaloa, Sonora, Tlaxcala, Veracruz)	[1900] 2100–3700	6*X* (4)	August–October
	4	*S.* × *aemulans* Bitter & Wittm.	ARG	Jujuy, La Rioja, Salta, San Juan, Tucumán	2690–4020	3*X* 4*X* (2)	January–April
	5	*S.* × *brucheri* Correll	ARG	Jujuy	3100–4000	3*X*	
	6	*S.* × *edinense* Berthault	MEX	Distrito Federal, Guanajuato, Hidalgo, México, Michoacán, Puebla, Tlaxcala, Veracruz	2050–3560	5*X*	August–October
	BULBOCASTANA						
	7	*S. bulbocastanum* Dunal	GUA, HON, MEX	GUA (Baja Verapaz, Guatemala, Huehuetenango, Quezaltenango, Sacatepequez, Sololá) HON (La Paz) MEX (Chiapas, Colima, Distrito Federal, Durango, Guanajuato, Hidalgo, Jalisco, México, Michoacán, Morelos, Nayarit, Oaxaca, Puebla, Querétaro, San Luis Potosí, Sinaloa, Tlaxcala, Veracruz)	1200–2300	2*X* (1) 3*X*	July–November
	8	*S. cardiophyllum* Lindl.	MEX	Aguascalientes, Chihuahua, Distrito Federal, Durango, Guerrero, Hidalgo, Jalisco, México, Michoacán, Morelos, Oaxaca, Puebla, Querétaro, Sinaloa, Tlaxcala, Zacatecas	1320–2800	2*X* (1) 3*X*	July–October
	COMMERSONIANA						
	9	*S. commersonii* Dunal	ARG BRA URU	ARG (Buenos Aires, Chaco, Corrientes, Entre Ríos, Misiones) BRA (Rio Grande do Sul, Santa Catarina) URU (Artigas, Canelones, Cerro Largo, Florida, Lavalleja,	0–400	2*X* (1) 3*X*	October–July

(continued)

Table 4.1 (continued)

Section		Wild Species	Distribution (Countries)	Distribution (Regions)	Altitude (m)	Ploidy (EBN)	Phenology (Flowering & Fruiting)
				Maldonado, Montevideo, Paysandú, Río Negro, Rocha, Salto, San José, Soriano, Tacuarembó)			
PETOTA	10	*S. malmeanum* Bitter	ARG BRA PAR URU	ARG (Buenos Aires, Chaco, Corrientes, Entre Ríos, Formosa, Misiones, Santa Fé) BRA (Rio Grande do Sul) PAR (Boquerón, Central, Cordillera, Itapuá, Presidente Hayes) URU (Colonia, Maldonado, Soriano)	0–330	2*X* (1)	October–July
	CONICIBACCATA						
	11	*S. agrimonifolium* Rydb.	GUA HON MEX	GUA (Chimaltenango, El Progreso, Huehuetenango, Quezaltenango, San Marcos, Sololá, Totonicapán) HON (Morazán) MEX (Chiapas)	1800–3400 [3800]	4*X* (2)	June–March
	12	*S. ayacuchense* Ochoa	PER	Ayacucho	2000–3200	2*X* (2)	February–April
	13	*S. bombycinum* Ochoa	BOL	La Paz	2000–2870	4*X*	February
	14	*S. buesii* Vargas	PER	Cuzco	2400–3700	2*X* (2)	February–September
	15	*S. burkartii* Ochoa	PER	Amazonas, Cajamarca	2600–3350	2*X*	March–July
	16	*S. colombianum* Dunal	COL ECU PER VEN	COL (Antioquia, Boyacá, Caldas, Caquetá, Cauca, Cesar, Cundinamarca, Huila, Nariño, Norte de Santander, Putumayo, Quindío, Risaralda, Santander, Tolima, Valle) ECU (Azuay, Cañar, Carchi, Chimborazo, Cotopaxi, El oro, Loja, Morona-Santiago, Napo, Orellana, Pichincha, Tungurahua, Zamora Chinchipe) PER (Piura) VEN (Apure, Carabobo, Falcón, Mérida, Portuguesa, Táchira, Trujillo)	[1200] 2000–3950	4*X* (2) 6*X* (4)	All year
	17	*S. flahaultii* Bitter	COL	Cundinamarca, Boyacá, Meta, Santander, Cauca	[2500] 3150–3610 [4310]	4*X*	April–August
	18	*S. garcia-barrigae* Ochoa	COL	Santander	3010–3900	4*X*	July–August
	19	*S. laxissimum* Bitter	PER	Ayacucho, Cuzco, Huánuco, Junín, Pasco	[670, 1200] 1700–3580	2*X* (2)	All year
	20	*S. limbaniense* Ochoa	PER	Puno	2900–3750	2*X* (2)	January–April
	21	*S. lobbianum* Bitter	COL	Caldas	3000–3570	4*X* (2)	July
	22	*S. longiconicum* Bitter	CRI, PAN	CRI (Alajuela, Cartago, Guanacaste, Heredia, Limón, Puntarenas, San José), PAN (Chiriquí, Bocas del Toro)	[1050] 1400–3300	4*X*	All year
	23	*S. nubicola* Ochoa	PER	Huánuco, La Libertad	3260–3600	4*X* (2)	April–May
	24	*S. oxycarpum* Schiede	MEX	Hidalgo, Oaxaca, Puebla, Veracruz		4*X* (2)	

(continued)

Table 4.1 (continued)

Section		Wild Species	Distribution (Countries)	Distribution (Regions)	Altitude (m)	Ploidy (EBN)	Phenology (Flowering & Fruiting)
					[1520] 1870–2870		July–November
	25	*S. pillahuatense* Vargas	PER	Apurímac, Cuzco	2700–3700	2*X* (2)	January–March
PETOTA	26	*S. rhomboideilanceolatum* Ochoa	PER	Ayacucho, Junín	2100–3100	2*X* (2)	January–April
	27	*S. salasianum* Ochoa	PER	Huánuco	2600–3000	2*X*	March–May
	28	*S. violaceimarmoratum* Bitter	BOL PER	BOL (Cochabamba, La Paz) PER (Cuzco)	1800–3800	2*X* (2)	January–October
	29	*S. woodsonii* Correll	PAN	Chiriquí	3000–3500		All year
	IOPETALA						
	30	*S. guerreroense* Correll	MEX	Guerrero	2800–3000	6*X* (4)	July–December
	31	*S. hougasii* Correll	MEX	Colima, Guerrero, Jalisco, Michoacan	1600–3135	6*X* (4)	August–December
	32	*S. iopetalum* (Bitter) Hawkes	MEX	Distrito Federal, Guanajuato, Guerrero, Hidalgo, Jalisco, México, Michoacán, Morelos, Oaxaca, Puebla, Querétaro, Tlaxcala, Veracruz	1700–3350	6*X* (4)	July–November
	33	*S. schenckii* Bitter	MEX	Oaxaca, Puebla, Querétaro, Veracruz	[1780] 2420–2900 [3700]	6*X* (4)	August–October
	LONGIPEDICELLATA						
	34	*S. hjertingii* Hawkes	MEX	Coahuila, Nuevo León, San Luis Potosí, Tamaulipas	[1230] 1650–3210	4*X* (2)	July–October
	35	*S. stoloniferum* Schltdl.	MEX USA	MEX (Aguascalientes, Baja California Sur, Chihuahua, Coahuila, Distrito Federal, Durango, Guanajuato, Hidalgo, Jalisco, México, Michoacán, Morelos, Nayarit, Nuevo León, Oaxaca, Puebla, Querétaro, San Luis Potosí, Sinaloa, Sonora, Tlaxcala, Veracruz) USA (Arizona, Colorado, New Mexico, Texas)	[1040] 1440–3400 [3700]	4*X* (2)	USA: July–September MEX: July–November
	36	*S.* × *vallis-mexici* Juz.	MEX	Distrito Federal	2280–3000	3*X*	July–October
	MEGISTACROLOBA						
	37	*S. boliviense* Dunal	ARG BOL PER	ARG (Catamarca, Jujuy, Salta, Tucumán) BOL (Chuquisaca, Cochabamba, La Paz, Oruro, Potosí, Santa Cruz, Tarija) PER (Apurímac, Cusco)	1600–4270	2*X* (2)	
	38	*S. hastiforme* Correll	PER	Amazonas, La Libertad, Ancash	[2850] 3200–3900 [4700]	2*X* (2)	March–May

(continued)

Table 4.1 (continued)

Section		Wild Species	Distribution (Countries)	Distribution (Regions)	Altitude (m)	Ploidy (EBN)	Phenology (Flowering & Fruiting)
	39	*S. raphanifolium* Cárdenas & Hawkes	PER	Cuzco, Apurimac, Puno	[2000] 2700–4200 [4500]	2*X* (2)	January–April
	40	*S. sogarandinum* Ochoa	PER	Cajamarca, Lima	2800–4100	2*X* (2) 3*X*	January–March
PETOTA	*MORELLIFORME*						
	41	*S. clarum* Correll	GUA, MEX	GUA (Huehuetenango, Quezaltenango, Sacatepequez, San Marcos, Sololá, Totonicapán) MEX (Chiapas)	2740–3800	2*X*	July–November
	42	*S. morelliforme* Bitter & Muench	BOL GUA HON MEX	BOL (La Paz) GUA (Chimaltenango, Huehuetenalgo, Quetzaltenango, Quiché, San Marcos, Totonicapán) HON (Morazán) MEX (Chiapas, Guerrero, Hidalgo, México, Michoacán, Oaxaca, Puebla, Querétaro, Veracruz)	1600–3050	2*X*	July–October
	PINNATISECTA						
	43	*S. jamesii* Torr.	MEX, USA	MEX (Chihuahua, Querétaro, San Luis Potosí, Sonora) USA (Arizona, Colorado, Nebraska, New Mexico, Texas, Utah)	1370–2870	2*X* (1)	June–October
	44	*S. pinnatisectum* Dunal	MEX	Guanajuato, Jalisco, Michoacán, Querétaro	1500–2200	2*X* (1)	July–September
	45	*S.* × *michoacanum* (Bitter) Rydb.	MEX	Michoacán	1900–2100	2*X*	July–September
	46	*S.* × *sambucinum* Rydb.	MEX	Guanajuato, Querétaro	1720–2200	2*X*	August–October
	PIURANA						
	47	*S. acroglossum* Juz.	PER	Huánuco, Pasco	2025–3800	2*X* (2)	January–May
	48	*S. acroscopicum* Ochoa	PER	Arequipa, Ayacucho, Cajamarca, Tacna	2350–3900	2*X*	February–June
	49	*S. albornozii* Correll	ECU	Azuay, Loja	2350–3400	2*X* (2)	March–July
	50	*S. anamatophilum* Ochoa	PER	Ancash	1720–2800	2*X* (2)	March–May
	51	*S. andreanum* Baker	COL ECU	COL (Nariño, Putumayo) ECU (Azuay, Bolivar, Cañar, Carchi, Cayambe, Chimborazo, Cotopaxi, Imbabura, Loja, Morona Santiago, Napo, Orellana, Pichincha, Santiago-Zamora, Sucumbíos, Tungurahua)	Diploid populations: 1900–3700 Polyploid populations: 2200–4000	2*X* (2) 4*X* (4)	Mainly April–July but all year
	52	*S. augustii* Ochoa	PER	Ancash	3200–3800	2*X* (1)	April–May
	53	*S. cajamarquense* Ochoa	PER	Cajamarca	2200–3000	2*X* (1)	March–May

(continued)

Table 4.1 (continued)

Section		Wild Species	Distribution (Countries)	Distribution (Regions)	Altitude (m)	Ploidy (EBN)	Phenology (Flowering & Fruiting)
	54	*S. cantense* Ochoa	PER	Ancash, Lima	2350–3400	2*X* (2)	March–May
	55	*S. chilliasense* Ochoa	ECU	El Oro	3200–3450	2*X* (2)	April–May
	56	*S. chiquidenum* Ochoa	PER	Ancash, Cajamarca, Huánuco, La Libertad	[420] 2000–3500 [3700]	2*X* (2)	March–June
PETOTA	57	*S. chomatophilum* Bitter	ECU PER	ECU (Azuay, Pichincha) PER (Amazonas, Ancash, Cajamarca, Huánuco, La Libertad, Lima, Junín, Pasco, San Martín)	1950–4500 [4800]	2*X* (2)	Mainly March–June but all year
	58	*S. contumazaense* Ochoa	PER	Cajamarca	2150–2900	2*X* (2)	April–June
	59	*S. dolichocremastrum* Bitter	PER	Ancash, Huánuco	3400–4400	2*X* (1)	February–May
	60	*S. huancabambense* Ochoa	PER	Cajamarca, Lambayeque, Piura	1650–3000 [3460]	2*X* (2)	February–June
	61	*S. humectophilum* Ochoa	PER	Amazonas	2800–3200	2*X* (1)	March–May
	62	*S. hypacrarthrum* Bitter	PER	Ancash, Cajamarca, Lima	1800–3800	2*X* (1)	March–June
	63	*S. immite* Dunal	PER	Cajamarca, Lima	Lomas: 80–480 Mountains: 1650–3160	2*X* (1) 3*X*	Lomas: August–October Uplands: March–June
	64	*S. minutifoliolum* Correll	ECU	Napo, Cañar	1200–3400	2*X* (1)	December–August
	65	*S. mochiquense* Ochoa	PER	Piura, Lima	Lomas: 150–800 Mountains: 1170–3000	2*X* (1)	Lomas: August–October Uplands: March -May
	66	*S. multiinterruptum* Bitter	PER	Ancash, Huancavelica, Lima	2675–3900	2*X* (2)	March–May
	67	*S. olmosense* Ochoa	ECU PER	ECU (Loja) PER (Lambayeque)	1200–2650	2*X* (2)	March–May
	68	*S. paucissectum* Ochoa	PER	Cajamarca, Piura	2350–3360	2*X* (2)	February–May
	69	*S. piurae* Bitter	PER	Piura	2000–3360	2*X* (2)	April–June

(continued)

Table 4.1 (continued)

Section		Wild Species	Distribution (Countries)	Distribution (Regions)	Altitude (m)	Ploidy (EBN)	Phenology (Flowering & Fruiting)
	70	*S. raquialatum* Ochoa	PER	Piura	[1350] 1900–3100	2*X* (1)	March–June
	71	*S. scabrifolium* Ochoa	PER	Huánuco	2800–3340	2*X*	March–April
	72	*S. simplicissimum* Ochoa	PER	Lima	1600–2720	2*X* (1)	January–April
	73	*S. trinitense* Ochoa	PER	Cajamarca	2700–3450	2*X* (1)	April–June
PETOTA	74	*S. wittmackii* Bitter	PER	Lima	Lomas: 30–480 Mountains: 2200–3400	2*X* (1)	Lomas: July–October Uplands: March–May
	75	*S. xblanco-galdosii* Ochoa	PER	Ancash, Cajamarca, La Libertad	2700–3260	2*X* (2)	February–June
	POLYADENIA						
	76	*S. lesteri* Hawkes & Hjert.	MEX	Oaxaca	2100–2390	2*X*	September–October
	77	*S. polyadenium* Greenmam	MEX	Hidalgo, Jalisco, México, Michoacán, Oaxaca, Puebla, Querétaro, Veracruz	1900–2900	2*X*	August–October
	STENOPHYLLIDIA						
	78	*S. hintonii* Correll	MEX	Colima, Guanajuato, México, Querétaro	1700–2800	2*X* (1)	August–October
	79	*S. stenophyllidium* Bitter	MEX	Aguascalientes, Chihuahua, Durango, Jalisco, México, Michoacán, Nayarit, Sonora, Zacatecas	[1100] 1380–2500	2*X* (1)	July–September
	80	*S. ehrenbergii* (Bitter) Rydb.	MEX	Aguascalientes, Distrito Federal, Guanajuato, Hidalgo, Jalisco, México, Michoacán, Nayarit, Puebla, Queretaro, San Luis Potosí, Zacatecas	[800] 1450–2500	2*X* (1)	July–October
	STIPULOIDEA						
	81	*S. stipuloideum* Rusby	BOL	Chuquisaca, Cochabamba, La Paz, Santa Cruz	2000–4000	2*X* (1)	December–April
	82	*S. neocardenasii* Hawkes & Hjert.	BOL	Santa Cruz	1400–1700	2*X* (2)	January–February
	TRIFIDA						
	83	*S. tarnii* Hawkes & Hjert.	MEX	Hidalgo, Querétaro, Veracruz	2000–2600	2*X*	September–October

(continued)

Table 4.1 (continued)

Section		Wild Species	Distribution (Countries)	Distribution (Regions)	Altitude (m)	Ploidy (EBN)	Phenology (Flowering & Fruiting)
	84	*S. trifidum* Correll	MEX	Jalisco, Michoacán	[1800] 2000–2800 [3050]	2*X* (1)	July–October
	TUBEROSA						
	85	*S. amayanum* Ochoa	PER	Huancavelica	3000–3900	2*X* (2)	January–March
	86	*S. ancophilum* (Correll) Ochoa	PER	Ancash, La Libertad	[2600] 3000–3800	2*X* (2)	March–May
PETOTA	87	*S. berthaultii* Hawkes	ARG BOL	ARG (Jujuy, Salta) BOL (Chuquisaca, Cochabamba, La Paz, Potosí, Santa Cruz, Tarija)	1200–3950	2*X* (2) 3*X*	January–March
	88	*S. brevicaule* Bitter	ARG BOL	ARG (Catamarca, Jujuy, La Rioja, Salta, San Juan) BOL (Chuquisaca, Cochabamba, La Paz, Oruro, Potosí, Santa Cruz, Tarija)	1500–4180	2*X* (2) 4*X* (4) 6*X* (4)	January–April
	89	*S. candolleanum* Berthault	BOL PER	BOL (La Paz, Santa Cruz) PER (Ancash, Huánuco)	1600–4400	2*X* (2) 6*X* (4)	January–March
	90	*S. chacoense* Bitter	ARG BOL BRA PAR PER URU	ARG (Buenos Aires, Catamarca, Chaco, Córdoba, Corrientes, Entre Ríos, Formosa, Jujuy, La Pampa, La Rioja, Misiones, Salta, San Luis, Santa Fé, Santiago del Estero, Tucumán) BOL (Chuquisaca, Cochabamba, La Paz, Potosí, Santa Cruz, Tarija) BRA (Minas Gerais, Paraná, Rio Grande do Sul, Santa Catarina, São Paulo) PAR (Alto Paraguay, Alto Paraná, Amambay, Boquerón, Caaguazú, Caazapá, Central, Concepción, Cordillera, Gran Chaco, Guairá, Itapuá, Paraguarí, Presidente Hayes, San Pedro) PER (Puno) URU (Montevideo, Canelones, Florida)	0–3700	2*X* (2) 3*X* 4*X*	All year
	91	*S. gandarillasii* Cárdenas	BOL	Chuquisaca, Cochabamba, Santa Cruz	1450–3000	2*X* (2)	February–March
	92	*S. gracilifrons* Bitter	PER	Huancavelica	1200–2700	2*X*	January–March
	93	*S. incasicum* Ochoa	PER	Cuzco	[2000] 3700–3800	2*X* (2)	February–March
	94	*S. infundibuliforme* Phil.	ARG BOL	ARG (Jujuy, Salta) BOL (Cochabamba, La Paz, Potosí, Tarija)	2350–4300	2*X* (2) 3*X*	January–April
	95	*S. kurtzianum* Bitter & Wittm.	ARG	Catamarca, La Rioja, Mendoza, San Juan	750–3000	2*X* (2) 3*X*	January–May
	96	*S. lignicaule* Vargas	PER	Cuzco	2510–3460	2*X* (1)	January–April
	97	*S. maglia* Schltdl.	ARG CHI	ARG (Mendoza) CHI (IV, V, VIII Regions)	1630–1820	2*X* 3*X*	

(continued)

Table 4.1 (continued)

Section		Wild Species	Distribution (Countries)	Distribution (Regions)	Altitude (m)	Ploidy (EBN)	Phenology (Flowering & Fruiting)
							CHI lowlands: all year ARG uplands: January–March
	98	*S. medians* Bitter	CHI PER	CHI (I, II Regions) PER (Ancash, Arequipa, Lima, Tacna)	200–3800	2*X* (2) 3*X*	Coast: May–October Andes: November–April
PETOTA	99	*S. microdontum* Bitter	ARG BOL	ARG (Catamarca, Jujuy, La Rioja, Tucumán, Salta) BOL (Chuquisaca, Cochabamba, La Paz, Tarija)	1400–3850	2*X* (2)	December–June
	100	*S. neorossii* Hawkes & Hjert.	ARG	Jujuy, Salta	2530–3800	2*X* (2) 3*X*	January–March
	101	*S. neoweberbaueri* Wittm.	PER	Lima	200–750	3*X*	October
	102	*S. okadae* Hawkes & Hjert.	BOL	Chuquisaca, Cochabamba, La Paz	2450–3200	2*X* (2)	February–March
	103	*S. velardei* Ochoa	PER	Apurimac, Cuzco	[1800] 2450–3400	2*X*	February–May
	104	*S. venturii* Hawkes & Hjert.	ARG	Catamarca, Jujuy, La Rioja, Salta, Tucumán	1900–3000	2*X* (2)	December–March
	105	*S. vernei* Bitter & Wittm.	ARG	Catamarca, Jujuy, Salta, Tucumán	2270–3600	2*X* (2)	December–April
	106	*S.* × *doddsii* Correll	BOL	Chuquisaca, Cochabamba, Santa Cruz	2050–2600	2*X* (2)	January–March
	107	*S.* × *rechei* Hawkes & Hjert.	ARG	La Rioja	1200–3950	2*X* (2) 3*X*	January–March
	VERRUCOSA						
	108	*S. verrucosum* Schltdl.	MEX	Coahuila, Distrito Federal, Gunajuato, Guerrero, Hidalgo, Jalisco, México, Michoacán, Morelos, Nuevo Leon, Oaxaca, Puebla, Querétaro, San Luis Potosí, Tamaulipas, Tlaxcala, Veracruz	[1870] 2100–3500 [4000]	2*X* (2)	June–November

of sympatric species at the pollen-pistil level. Polyploid species are self-compatible due to a "competition interaction" that either reduce or suppress the incompatibility reaction that occurs in pollen grains carrying different S-alleles (Frankel and Galun 1977). Tetraploid and hexaploid species (with the exception of tetraploid forms of *S. andreanum*) are capable of self-fertilization (Hawkes 1990). When diploid self-incompatible potato species are induced to chromosome doubling produces self-compatible tetraploids (Stout and Chandler 1941; Ross 1986). Interestingly, in these tetraploids, pollen tube growth is inhibited when pollen is homozygous for S alleles, but not when it is heterozygous (Lewis 1943, 1947).

Male sterility of hybrid plants is an important post-zygotic isolating mechanism in natural potato species populations. Cytoplasmic-genetic male sterility occurs when dominant nuclear genes from the male parent interact with sensitive cytoplasm from the female parent (Hermundstad and Peloquin 1985; Tucci et al. 1996). Male sterility has been reported in several F1 hybrids derived from crosses involving various wild and cultivated species (Lamm 1941, 1953; Brown 1984; Hermundstand and Peloquin 1985; Tucci et al. 1996, Santini et al. 2000; Carputo et al. 2003a) and between cultivated potatoes (Grun 1973; Hanneman and Peloquin 1981). Genetic and environmental conditions can influence the expression of cytoplasmic-genetic male sterility (Hanneman and Peloquin 1981). Hybrids lacking sensitive cytoplasm or nuclear male sterility genes are male-fertile (Iwanaga et al. 1991; Tucci et al. 1996).

In Angiosperms the development of viable seed depends on double fertilization that generates a diploid embryo and triploid endosperm. The endosperm contains two genomes of the maternal parent and one genome of the paternal parent. Intraspecific, intraploidy crosses in potato typically produce viable seeds containing well-developed endosperm, on the contrary, in most interploidy crosses, seeds are inviable due to endosperm failure (Friedman 1998). Normal endosperm development in potato requires a 2:1 maternal: paternal ratio of a set of genes called endosperm balance factors (Johnston and Hanneman 1980, 1982, 1996). Viable seeds could be generated from crosses between plants that produce gametes with the same endosperm balance number (EBN), resulting in a 2:1 maternal:paternal ratio of endosperm balance factors after male gamete fusion with two nuclei of the central cell to produce triploid endosperm, and consequently allowed further development of a normal embryo. The EBN is an arbitrary value, which is not necessarily a direct indication of species ploidy, assigned to each *Solanum* species based on its behavior in crosses with EBN standards and on the assumption that the 2:1 ratio is essential for normal endosperm development (Hanneman 1994). The ploidy and endosperm balance number combinations in potato are 6x (4EBN), 4x(4EBN), 4x(2EBN), 2x(2EBN) and 2x(1EBN) (Table 4.1). However, endosperm development may also fail in some intraploidy, interspecific crosses, while some interploidy crosses could succeed. The nature of these endosperm balance factors is not yet known, but nuclear genetic models have been proposed (Ehlenfeldt and Hanneman 1988; Camadro and Masuelli 1995).

Wild and cultivated potatoes have meiotic mutants that result in the production of unreduced (2n) gametes (Carputo et al. 2003b). Some meiotic mutations produce 2n eggs (Stelly and Peloquin 1986; Werner and Peloquin 1991), while others produce 2n pollen (Quinn et al. 1974). The fusion of unreduced (2n) gametes during fertilization explain the occurrence of spontaneous polyploidization in wild plant populations (Harlan and de Wet 1975; Veilleux 1985; Bretagnolle and Thompson 1995). Unreduced gametes can be detected microscopically, since diploid pollen grains are larger than monoploid pollen grains (Quinn et al. 1974) and 2n eggs can be identified with stain-clearing techniques (Stelly et al. 1984). Unreduced gametes facilitate the evolution of polyploids by allowing interspecific hybridization across ploidy levels (Mason and Pires 2015). Nevertheless, triploid seeds resulting from the union of an n and a 2n gamete are generally inviable due to endosperm failure (Kohler et al. 2009). Camadro et al. (2004) pointed out the complementary role of EBN and unreduced gametes, not only because it facilitates interspecific gene introgression but also because it maintains the ploidy integrity of the two parental species. Den Nijs and Peloquin (1977) and more recently by Carputo et al. (2003b) proposed an evolutionary scenario for potatoes where n and 2n gametes link together all ploidy and EBN levels, thereby providing an opportunity for gene flow throughout sympatric species with different EBN and chromosome numbers (Camadro et al. 2004). Hawkes (1962) considered that introgression and interspecific hybridization that not led to speciation seems to be a common phenomenon throughout the range of section *Petota*. The lack of strong biological isolating mechanisms, morphologically intermediate characteristics in natural populations, and sympatry of many species suggest that much of the taxonomic confusion in section *Petota* is due most probably to frequent gene flow among the species (Spooner et al. 2019). Interestingly, Rabinowitz et al. (1990) documented high levels of gene flow between wild and cultivated species in Peru, supporting Ugent's hypotheses who proposed that cultivated species were formed and genetically enriched by gene flow from the wild species (Ugent 1970). Traditional Andean farming systems incorporate natural hybrids between cultivated potatoes and the wild potato relatives growing in their surrounding fields (Brush et al. 1981). The unique reproductive characteristics of tuber-bearing potatoes allow to incorporate new genetic combinations by sexual reproduction, while asexual reproduction maintains adapted gene complexes (Spooner et al. 2019).

In natural potato populations, several biological internal barriers such us pollen-style interactions, male sterility, and endosperm failure, prevent the production of interspecific hybrids and maintain the integrity of sympatric species (Camadro et al. 2004). However, on the other side of an evolutionary perspective, the wide natural occurrence of unreduced gametes, self-incompatibility, and little genome differentiation among potato species favor hybridization between wild *Solanum* species (Erazzú et al. 2009; Masuelli et al. 2009; Ispizúa et al. 2015).

Potato wild relatives are critical natural resources that serve as a model system for genebank conservation (Jansky et al. 2013; Bethke et al. 2019). The understanding of several biological mechanisms allows the use of wild species in potato breeding. The introduction of diploid wild potato genes to the tetrahaploid cultivated species has been successfully carried out using various breeding methods and strategies such as haploid production (Peloquin et al. 1989a), the use of unreduced gametes (Mendiburu and Peloquin 1977, Peloquin et al. 1989b, 1996, 2008), the application of the balance of endosperm value to produce hybrids (Johnston and Hanneman 1980, 1982; Hanneman 1994), and the use of embryo rescue techniques and somatic fusion of protoplasts. Carputo and Frusciante (2011) highlighted the classical genetics and traditional approaches applied in potato crop improvement. New insights from genomic research provided promissory methods to explore a wide pool of genetic resources that include not only wild species but also landraces, and increase the efficiency of identifying and introgressing alleles rather than traits (Iorizzo et al. 2014; Mann et al. 2011; Potato Genome Sequencing Consortium 2011; Bethke et al. 2019; Ghislain and Douches 2020).

In the context of present knowledge, wild species have shown their enormous value as a source of traits of agronomic importance and resistances to biotic and abiotic factors for crop improvement (Ross 1986; Hanneman, 1999; Kuhl 2011; Watanabe 2015; Bonierbale et al. 2020; Ortiz 2020). Nevertheless, few wild species have been used in breeding. A better comprehension of wild potato diversity and ecological adaptation, taxonomy, and relationships, and the application of new promissory methods will also contribute to their utilization as models to understand genetic and genomic evolution as well as in cultivated potato improvement (Hardigan et al. 2017) (see other chapters).

4.3 Taxonomy of Potato Species (Section *Petota*) and Close Related Non Tuberous Species (Section *Etuberosum*)

Classical treatments of Bukasov (1978), Correll (1962), Gorbatenko (1989, 2006), Hawkes and Hjerting (1969, 1989), Ochoa (1990, 1999, 2001) and Hawkes (1990), proposed taxa delimitation mainly based on morphological species concept (Spooner and Van der Berg 1992). Hawkes (1990) also considered species intercrossability, and his taxonomic treatment of section *Petota* has been the most comprehensive and traditionally used, where he recognized 228 wild species and seven cultivated species, grouped into 21 Series. Taxonomic treatments differ in author's concepts and interpretation of taxonomic rank used to establish species, botanical varieties or subspecies, hypotheses about species hybrid origin and introgression with other species, as well as the criteria to define the arrangement and number of taxonomical series, the number of species in each series, and the different affiliation of species to these series (Spooner and Van den Berg 1992; Spooner and Hetterscheid 2005; Spooner and Salas 2006).

It is interesting to compare taxonomic interpretations of two sister phylogenetic lineages of genus *Solanum*, potatoes and tomatoes, that separated earlier about eight Ma and later section *Petota* started diversifying around seven Ma (Särkinen et al. 2013). Prevalent taxonomical interpretation (Hawkes 1990) considers more than 200 wild potato species in contrast with the 13 species of tomatoes (Sect. *Lycopersicon*) and four species in the most related groups (Sect. *Juglandifolia* and *Lycopersicoides*) (Peralta et al. 2008). This enormous differences in the interpretation of diversity in sister groups can be explained by the unique reproductive characteristics, genetic structure, and ecological adaptations of potatoes in a wide geographic area. However, another non biological explanation are the concepts to circumscribe species and a complex nomenclature system initially used by potato taxonomists that cause an overestimation of natural diversity (Peralta et al. 2008).

A different philosophy and consistent application of a comprehensive criteria have driven to continuously reduce the number of potato species relative to early taxonomic treatments (Hijmans and Spooner 2001; Spooner and Salas 2006; Ames and Spooner 2010; Fajardo and Spooner 2011; Spooner et al. 2004, 2014, 2016, 2019). More recently comprehensive treatments of section *Petota* and section *Etuberosum* (Spooner et al. 2004, 2016, 2019) not only revised early taxonomical contribution, but also applied a taxonomical integrative approach using different sources of evidences mainly based on phylogeny to propose new group classification. These 3 monographs treated the complete diversity of tuber bearing and stoloniferous wild species in America, based on the analysis of numerous herbarium specimens, including types, and field assessment of cultivated representatives of all recognized species. Other relevant evidences of recent morphological, reproductive and cladistics molecular studies were integrated into these comprehensive treatments. A similar approach has been used to describe wild potato diversity for regional or country Floras (Spooner et al. 2009; Clausen et al. 2013).

The classification of potatoes based on a phylogenetic approach clarified evolutionary relationships, and the recognition of close related cluster of organisms by a parental pattern of ancestry and descent, and diagnosable distinct from other clusters (Cracraft 1989). Most methods for studying cladistics have been based on models of strictly branching cladogeny, however, in complex groups with possible reticulate evolution at chromosomal, genomic and species levels inference of relationships could fail when modeled by a bifurcating tree. Classification of wild potatoes is a difficult goal, since interpretation of relationships is complicated by introgression, interspecific hybridization, auto- and allopolyploidy, sexual incompatibility among many species, a mixture of sexual and asexual reproduction, possible recent species divergence, phenotypic plasticity, and consequent great morphological similarity among species (Spooner and van den Berg 1992; Spooner et al. 2014; Camadro et al. 2004; Camadro 2012). Further phylogenetic analysis using models of reticulated evolution in potatoes, a complex group with hybridization and introgression phenomena, could elucidated their relationships.

The relationships between Sect. *Petota* and Sect. *Etuberosum* have been a subjet of debate. Initially Juzepczuk and Bukasov (1929) included non-tuber-bearing species in ser. Etuberosa within Sect. *Petota* [then referred to as Sect. Tuberarium (Dunal) Bitter]. Morphological similarities of *Solanum* species in Sect. *Petota* and Sect. *Etuberosum* led to considered them as closest relatives. Nevertheless, several concordant molecular studies have clarified relationships among these *Solanum* sections, supporting tomatoes (Sect. *Lycopersicon*) and close related species (sects. *Juglandifolia* and *Lycopersicoides*) as a monophyletic sister clade to Sect. *Petota*, with Sect. *Etuberosum* sister to all the above (Spooner et al. 2016, 2019). These relationships have been corroborated by plastid phylogenies (Spooner et al. 1993; Olmstead and Palmer 1992, 1997; Bohs and Olmstead 1997, 1999; Olmstead et al. 1999), nuclear genes and conserved orthologous sequences phylogenies (Peralta and Spooner 2001; Rodríguez and Spooner 2009; Rodríguez et al. 2009) and recently by seven regions (five plastids and two nuclear) used to generate the Solanaceae megaphylogeny (Särkinen et al. 2013). Further phylogenetic analysis of large data set could test the hypothesis of relationships between Sect. *Petota* and Sect. *Etuberosum*. Within genus *Solanum*, sects. *Etuberosum*, *Petota*, *Juglandifolia*, *Lycopersicoides*, and *Lycopersicon* are all members of a New World broader group of species informally named the Potato Clade (Tepe et al. 2016; Bohs 2005; Särkinen et al. 2013).

We summarized our actual comprehensive taxonomy of Sections *Petota* and *Etuberosum*, including species distributions and habitat, ploidy, and EBN numbers, and phenology (Table 4.1; Fig. 4.1), as well as a complete list of accepted names, synonyms (450), and the inclusion of species in non-formal groups (Tables 4.1 and 4.2). These provisional taxonomic groups are based on hypothesis of phylogenetic relationships among species that reflect the evolutionary history of potatoes, and pending from more data to elucidate interrelationships. Similar non-formal group systems of classification have been widely applied to *Solanum* by Whalen (1984), Bohs (1994, 2005), Knapp (1991, 2000, 2002, 2008, 2013), Peralta et al. (2008), and Spooner et al. (2004, 2016, 2019). These non-formal groups should not be confused with "Groups", which are category taxonomic names for groups of cultivated plants (ICNCP 2016). Non-formal groups in section *Petota* are detailed in Tables 4.1 and 4.2, and the most widespread ones are represented in Fig. 4.1. Detailed morphological description of the accepted wild species, synonyms, taxonomic keys, illustrations, locality distributions, habitat, phenology, uses, and taxonomic characteristics, as well as a detailed explanation of non-formal groups, are found in the three monographic treatments of Spooner and collaborators (2004, 2016, 2019), information that has been also incorporated in the Solanaceae Source website (http://www.solanaceaesource.org).

SECT. *ETUBEROSUM* (Bukasov & Kameraz) A. Child: comprises three species confined to southern South America, erect to ascending herbs that possess thickened rhizomes, from which arise thin stolons but lacking tubers (Tables 4.1 and 4.2; Fig. 4.1).

SECT. *PETOTA* Dumortier: comprises 108 wild herbaceous species (Tables 4.1 and 4.2), erect to ascending, sometimes forming a rosette or semi-rosette, bearing tubers at the ends of stolons. A large geographical distribution of wild potato species from the southwestern United States (latitude 38 °N), Central and South America to Argentina, and adjacent mainland in Chile (39 °S) (Fig. 4.1) indicates a varied range of ecological diversity as well as adaptations to extreme climatic conditions; these species can be found from sea level at both Atlantic and Pacific oceans, and from high altitude deserts to rainforests (Hawkes 1990; Spooner and Hijmans 2001; Hijmans et al. 2002; Spooner et al. 2004,

2016, 2019). Some are widespread such as *S. acaule*, *S. brevicaule* and *S. chacoense* while others, with a restricted range and an endemic nature, are found in areas with specific ecological conditions. *Solanum chacoense*, *S. palustre* and *S. commersonii* are found at very low altitudes, frequently at sea level, while *S. acaule*, *S.* × *aemulans*, *S. brevicaule* and *S. boliviense* and *S. candolleanum*, reach more than 4000 m in the Andes. *Solanum morelliforme* is mainly restricted to Central Mexico, Guatemala, and Honduras but a single population has been identified 4000 km south in Bolivia. This is the only species in sect. *Petota*, growing in both Central and South America (Simon et al. 2011). Cultivated species present a more restricted geographical distribution, from northern South America and down to southern Chile (Fig. 4.2).

Wild potatoes are classified into 16 non-formal species groups (Spooner et al. 2004, 2016, 2019).

ACAULIA GROUP: three species and three nothospecies species with rosette to semi-rosette habit, in some cases erect and taller, typical flower pedicel with articulation appearing toward the distal end or no articulated (*S. acaule*), all polysomic polyploids, distributed in Mexico and Central America, and in South America from Ecuador to Argentina reaching the high puna plateau and possessing high frost tolerance (Vega and Bamberg 1995).

BULBOCASTANA GROUP: *S. bulbocastanum* and *S. cardiophyllum* are characterized by cream to light yellow corollas, diploids and triploids with and EBN = 1, distributed in Guatemala, Honduras, and Mexico (Spooner et al. 2004).

COMMERSONIANA GROUP: *S. commersonii* and *S. malmeanum* possess characteristic stellate corollas, diploids and triploids (*S. commersonii*) with EBN = 1, distributed in Argentina, Brasil, Paraguay, and Uruguay. Both are partially sympatric and are likely sister taxa (Spooner et al. 2016).

CONICIBACCATA GROUP: 19 species characterized by non-glossy parallel shaped leaves, with the distal-most lateral leaflet pairs diminishing toward the leaf base, and typical conical fruits (Spooner et al. 2019; Fajardo and Spooner 2011), diploids, tetraploids, and hexaploids (S. *colombianum*) with EBN = 2 or 4, distributed in Central America and South America from Venezuela to Bolivia.

IOPETALA GROUP: four species distributed in Mexico, polysomic polyploids with 6x(4EBN) crossability, and with no clear morphological characters uniting them, probably because they could have multiple origins (Spooner et al. 2004).

LONGIPEDICELLATA GROUP: two species and one nothospecies, with 4x(2EBN) crossability, but there are no clear specific morphological characters defining them, distributed from the Southwestern U.S.A to South Mexico.

MEGISTACROLOBA GROUP: four species, herbaceous low-growing rosette plants, with terminal leaflets much larger than the lateral leaflets and with the proximal leaflets reduced in size and often broadly decurrent on the rachis, diploids, and also triploids with EBN = 2, distributed in Peru, Bolivia, and northern Argentina.

MORELLIFORME GROUP: *S. morelliforme* and *S. clarum*, small plants with stellate corollas. *Solanum clarum* is present in Guatemala and Mexico, and *S. morelliforme* in Mexico to Guatemala but with a disjunct population in Bolivia (Spooner and Sytsma 1992).

PINNATISECTA GROUP: two species and two nothospecies, characterized by the presence of pinnatifid pseudostipules, diploids, distributed from southern U.S.A. to Mexico (Lara Cabrera and Spooner 2004).

PIURANA GROUP: 29 species, the majority with moniliform tubers and coriaceous, glabrous to subglabrous shiny or glossy leaves (Ames and Spooner 2010), diploids, also triploids, and tetraploids with EBN = 1, 2, or rare 4, distributed in Colombia, Ecuador, and Peru.

POLYADENIA GROUP: two species identified by their glandular leaves (type A trichrome) and strong odor, diploids, distributed in México (Lara Cabrera and Spooner 2004, Spooner et al. 2004).

STENOPYLLIDIA GROUP: three species with typical lunate pseudostipules and with triangular to conical fruits, diploids with EBN = 1, restricted to Mexico (Spooner et al. 2004).

STIPULOIDEA GROUP: *S. stipuloideum* and *S. neocardenasii*, diploids endemic to Bolivia, with typical white corollas (Spooner et al. 2016).

TRIFIDA GROUP: two endemic, diploid, Mexican species supported by cpDNA data (Spooner and Sytsma 1992) and AFLP data (Lara Cabrera and Spooner 2004).

TUBEROSUM GROUP: 21 species and two nothospecies, with dissected leaves, round fruits, and rotate-pentagonal corollas, diploids, triploids, tetraploids, and hexaploids with EBN = 1, 2 or 4, widely distributed in South America (Spooner et al. 2007). Also includes the cultivated potatoes (Table 4.3).

VERRUCOSA GROUP: *S. verrucosum,* distinguished by the corollas with the edges inrolled dorsally and often with berries with raised white dots, diploid species with 2EBN, widely distributed throughout Mexico.

Although provisional taxonomic groups attempt to reflect phylogenetic relationships in sect P*etota*, they are hypothesis that needs further investigations. In South America some *taxa* have been difficult to understand, like *S. boliviense* Dunal initially considered in Acaulia group (Spooner et al. 2016) but recently included in Megastricoloba group (Spooner et al. 2019), provisionally accepted here. Further research including more accessions representative of larger areas could help to solve this taxonomic puzzle. Also, with in Tuberosa group, additional studies are needed in complex taxa, fundamentally in two morphologically very similar species, *S. candolleanum* and *S. brevicaule*, to elucidate relationships and current hypothesis of potato domestication and cultivation origin (Spooner et al. 2005; Rodríguez et al. 2017). The taxonomy of *S. brevicaule* and related species has long been controversial (Correll 1962; Ugent 1970; Grun 1990), and several studies using morphological phenetics (Van den Berg et al. 1998; Alvarez et al. 2008), molecular marker data of RFLPs and RAPDs (Miller and Spooner 1999), and with AFLPs (Spooner et al. 2005) have been focusing on this group. All datasets distinguished *S. candolleanum*, distributed from Central Peru to Northernmost Bolivia, and *S. brevicaule*, distributed from northern Bolivia to northern Argentina, but with little to no support for the many names that were placed in synonymy (Table 4.2). Although Conicibaccata group have been recently revised, further studies are needed in *S. colombianum* and related taxa. Likewise, within Piurana a revision will be necessary to better define the members included in this group. Regarding endemism, new methodologies such environmental niche modeling, could be helpful to define collecting strategies as well as to identify threatened environment and antropic factors that may affect natural populations, like in the case of *S. rhomboideilanceolatum* (Castañeda Álvarez et al. 2015). These approaches can be also applied to explore and define new areas of distributions, as in the case of *S. maglia*, a wild species found in Chile but only one population in the mountains of Mendoza, Argentina. In this case, it will be interesting to test the hypothesis of ploidy in relation of habitat adaptation.

4.4 Cultivated Potatoes Origin, Domestication, Diversity and Taxonomy

South American indigenous farmers domesticated potatoes in the high plateaus of the Andean Punas and also in the lowlands of south-central Chile, where landrace cultivars are still highly diverse with an enormous variety of tuber shapes, sizes, and different colors of skin and internal tissues (INIA 2012; Fonseca et al. 2014; INIAF VDRA and MDRyT 2014; SPDA CCTA and INIA 2015; CIP 2015; MINAGRI 2017; PRODERN 2018; MINAM 2019).

The scarcity of direct botanical evidence has made difficult archeological research focused on potato species used by early American inhabitants and evidences of domestication. Ancient remains of potato have been found in archeological sites in Southern Chile, revealing that potato species have been consumed for at least 13,000 years (Ugent et al. 1987). Similarly, archeological rest indicated that about 10,000 years ago potato species have been used as food supply by native communities of Perú

Table 4.2 Recognized valid wild species names of sect. *Etuberosum* and sect. *Petota* are included in non-formal groups, and a list of synomyms in each species is provided. Numbers indicated species quantity

SECTION	SPECIES		SYNONYMS
ETUBEROSUM	1	*Solanum etuberosum* Lindl.	*S. bustillosii* Phil., *S. etuberosum* var. *antucense* Bitter, *S. etuberosum* var. *bustillosii* (Phil.) Witasek, *S. etuberosum* var. *chillanense* Bitter, *S. kunzei* Phil., *S. looseri* Juz. ex Bukasov, *S. subandinum* F. Meigen, *S. subandinum* Phil., *S. tuberosum* var. *polemoniifolium* Hook.f.
	2	*Solanum palustre* Schltdl.	*S. brevidens* Phil., *S. brevidens* var. *glabrescens* (Dunal) Hawkes, *S. brevidens* var. *glabrescens* (Walp.) Hawkes, *S. bridgesii* A.DC., *S. caldasii* var. *glabrescens* (Walp.) Dunal, *S. palustre* Poepp., *S. palustre* Poepp. ex Walp., *S. palustre* var. *glabrescens* Poepp. ex Walp., *S. pearcei* Phil., *S. tuberosum* subsp. *brevidens* (Phil.) Reiche, *S. tuberosum* subsp. *pearcei* (Phil.) Reiche, *S. tuberosum* var. *brevidens* (Phil.) Reiche, S. *tuberosum* var. *pearcei* (Phil.) Reiche
	3	*Solanum fernandezianum* Phil.	*S. brevistylum* Wittm., *S. tuberosum* subsp. *fernandezianum* (Phil.) Reiche, S. *tuberosum* var. *fernandezianum* (Phil.) Reiche
PETOTA	*ACAULIA*		
	1	*Solanum acaule* Bitter	*S. acaule* subsp. *punae* (Juz.) Hawkes & Hjert., *S. acaule* var. *caulescens* Bitter, *S. acaule* var. *checcae* Hawkes, *S. acaule* var. *punae* (Juz.) Hawkes, *S. acaule* Bitter forma *incuyo* Ochoa, *S. acaule* var. *subexinterruptum* Bitter, *S. depexum* Juz., *S. depexum* var. *chorruense* Hawkes, *S. punae* Juz., *S. schreiteri* Bukasov, *S. uyunense* Cárdenas
	3	*Solanum albicans* (Ochoa) Ochoa	*S. acaule* subsp. *albicans* (Ochoa) Hawkes, *S. acaule* var. *albicans* Ochoa, *S. acaule* subsp. *palmirense* Kardolus
	5	*Solanum demissum* Lindl.	*S. alpicum* Standl. & Steyerm., *S. demissum* forma *adpressoacuminatum* Bukasov ex Rybin, *S. demissum* forma *atrocyaneum* Lechn., *S. demissum* forma *calycotrichum* Hawkes, *S. demissum* forma *microcalyx* Lechn. ex Bukasov, *S. demissum* forma *tlaxpehualcoense* Bukasov ex Rybin, *S. demissum* forma *xitlense* Bukasov ex Rybin, *S. demissum* var. *demissum* forma *calycotrichum* Hawkes, *S. demissum* var. *demissum* forma *longifilamentosum* Hawkes, *S. demissum* var. *demissum* forma *perotanum* Hawkes, *S. demissum* var. *demissum* forma *tolucense* Hawkes, *S. demissum* var. *mastoidostigma* Hawkes, *S. demissum* var. *orientale* Hawkes, *S. semidemissum* Juz. ex Bucasov, *S. stoloniferum* var. *pumilum* M.Martens & Galeotti, *S. utile* Klotzsch
	2	*Solanum* × *aemulans* Bitter & Wittm.	*S. acaule* subsp. *aemulans* (Bitter & Wittm.) Hawkes & Hjert., *S. acaule* var. *aemulans* (Bitter & L. Wittm.) Correll, *S.* × *indunii* K.A.Okada & A.M.Clausen
	4	*Solanum* × *brucheri* Correll	*S.* × *viirsooi* K.A. Okada & A. M. Clausen
PETOTA	6		

(continued)

Table 4.2 (continued)

SECTION	SPECIES		SYNONYMS
		Solanum × *edinense* Berthault	*S. edinense* subsp. *salamanii* (Hawkes) Hawkes, *S. salamanii* Hawkes
	BULBOCASTANA		
	7	*Solanum bulbocastanum* Dunal	*S. bulbocastanum* subsp. *dolichophyllum* (Bitter) Hawkes, *S. bulbocastanum* subsp. *partitum* (Correll) Hawkes, *S. bulbocastanum* var. *dolichophyllum* Bitter, *S. bulbocastanum* var. *glabrum* Correll, *S. bulbocastanum* var. *latifrons* Bitter, *S. bulbocastanum var. partitum* Correll, *S. longistylum* Correll, *S. mexicanum* Sessé & Moc., *S. symphysicaulis* Pav. ex Dunal
	8	*Solanum cardiophyllum* Lindl.	*S. cardiophyllum* subsp. *lanceolatum* (Berthold) Bitter, *S. cardiophyllum* var. *amphixanthandrum* Bitter, *S. cardiophyllum* var. *endoiodandrum* Bitter, *S. cardiophyllum* var. *oligozygum* Bitter, S. *cardiophyllum* var. *pliozygum* Bitter, *S. coyoacanum* Bukasov ex Rybin, *S. lanceolatum* Berthault, *S. lanciforme* Rydb.
	COMMERSONIANA		
	9	*Solanum commersonii* Dunal	*S. acroleucum* Bitter, *S. commersonii* forma *mechonguense* (Bukasov) Correll, *S. commersonii* var. *depauperatum* Bitter, *S. commersonii* var. *ellipticans* Bitter, *S. commersonii* var. *glabratum* Hook.f., *S. commersonii* var. *indigoticascens* Bitter, *S. commersonii* var. *pubescens* Sendtn., *S. commersonii* var. *raphanistrum* Bitter, *S. commersonii* var. *rosulans* Bitter, *S. commersonii* var. *violaceum* Herter, *S. debile* Dunal, *S. henryi* Bukasov & Lechn., *S. henryi* forma *laticalix* Lechn., S. *henryi* forma *pubescens* Lechn., *S. mechonguense* Bukasov, *S. mercedense* Bukasov, *S. nicaraguense* Rydb., *S. ohrondii* Carrière, *S. rionegrinum* Lechn., *S. sorianum* Bukasov, *S. tenue* Sendtn., *S. tenue* var. *pubescens* Sendtn. ex Dunal, *S. tenue* var. *raphanifolium* Dunal
	10	*Solanum malmeanum* Bitter	*S. chacoense* forma *pilosulum* (Hassl.) Hassl., *S. commersonii* forma *malmeanum* (Bitter) Correll, *S. commersonii* subsp. *malmeanum* (Bitter) Firbas & Ross, *S. commersonii* subsp. *malmeanum* (Bitter) Hawkes & Hjert., *S. commersonii* subsp. *pseudostipulatum* Hassl., *S. commersonii* var. *pseudostipulatum* Hassl., *S. commersonii* var. *pubescens* Chodat, *S. guaraniticum* forma *pilosulum* Hassl. *S. millanii* Bukasov & Lechn., *S. pseudostipulatum* (Hassl.) Bukasov
	CONICIBACCATA		
	11	*Solanum agrimonifolium* Rydb.	
	12	*Solanum ayacuchense* Ochoa	
	13	*Solanum bombycinum* Ochoa	

(continued)

Table 4.2 (continued)

SECTION	SPECIES		SYNONYMS
PETOTA	14	*Solanum buesii* Vargas	
	15	*Solanum burkartii* Ochoa	*S. irosinum* Ochoa, *S. irosinum* forma *tarrosum* Ochoa
	16	*Solanum colombianum* Dunal	*S. cacetanum* Ochoa, *S. calacalinum* Ochoa, *S. colombianum* var. *meridionale* Hawkes, *S. colombianum* var. *trianae* Bitter, *S. colombianum* forma *quindiuense* Bukasov, *S. colombianum* var. *zipaquiranum* Hawkes, *S. cuatrecasasii* Ochoa, *S. dolichocarpum* Bitter, *S. filamentum* Correll, *S. jaenense* Ochoa, *S. moscopanum* Hawkes, *S. nemorosum* Ochoa, *S. orocense* Ochoa, *S. otites* Dunal, *S. otites* forma *dizygum* Bitter, *S. otites* forma *trizygum* Bitter, *S. pamplonense* L.E.López, *S. papa* Valenz. ex Palacio, *S. subpanduratum* Ochoa, *S. sucubunense* Ochoa, *S. tundalomense* Ochoa, *S. valenzuelae* Palacio, *S. venezuelicum* Bukasov
	17	*Solanum flahaultii* Bitter	*S. neovalenzuelae* L.E.López
	18	*Solanum garcia-barrigae* Ochoa	*S. donachui* (Ochoa) Ochoa, *S. garcia-barrigae* var. *donachui* Ochoa
	19	*Solanum laxissimum* Bitter	*S. claviforme* Correll, *S. claviformum* Correll, *S. laxissimum* forma *rockefelleri* (Vargas) Correll, *S. neovargasii* Ochoa, *S. rockefelleri* Vargas, *S. santolallae* Vargas, *S. santolallae* var. *acutifolium* Vargas
	20	*Solanum limbaniense* Ochoa	
	21	*Solanum lobbianum* Bitter	
	22	*Solanum longiconicum* Bitter	*S. longiconicum* var. *quadrijugum* Bitter, *S. nanoteranthum* Bitter
	23	*Solanum nubicola* Ochoa	
	24	*Solanum oxycarpum* Schiede	*S. confusum* Correll, *S. nelsonii* Correll, *S. reconditum* Correll
	25	*Solanum pillahuatense* Vargas	
	26	*Solanum rhomboideilanceolatum* Ochoa	
	27	*Solanum salasianum* Ochoa	
	28	*Solanum violaceimarmoratum* Bitter	*S. multiflorum* Vargas, *S. neovavilovii* Ochoa, *S. santolallae* forma *velutinum* Correll, *S. urubambae* Juz., *S. urubambae* forma *chakchabambense* Ochoa, *S. urubambae* forma *velutinum* (Correll) Ochoa, *S. villuspetalum* Vargas, *S. violaceimarmoratum* var. *papillosum* Hawkes
	29	*Solanum woodsonii* Correll	
	IOPETALA		
	30	*Solanum guerreroense* Correll	
	31	*Solanum hougasii* Correll	*S. spectabile* (Correll) Hawkes, *S. verrucosum* var. *spectabile* Correll

(continued)

Table 4.2 (continued)

SECTION	SPECIES		SYNONYMS
PETOTA	32	*Solanum iopetalum* (Bitter) Hawkes	*S. brachycarpum* (Correll) Correll, *S. demissum* forma *longibaccatum* Bukasov ex Rybin, *S. demissum* forma *recurvoacuminatum* Bukasov ex Rybin, *S. demissum* forma *stenantherum* Lechn. ex Bukasov, *S. oxycarpum* var. *brachycarpum* Correll, *S. verrucosum* var. *iopetalum* Bitter
	33	*Solanum schenckii* Bitter	*S. demissum* var. *megalocalyx* Hawkes
	LONGIPEDICELLATA		
	34	*Solanum hjertingii* Hawkes	*S. fendleri* var. *physaloides* Correll, *S. hjertingii* var. *physaloides* (Correll) Hawkes, *S. leptosepalum* Correll, *S. matehualae* Hjert. & Tarn
	35	*Solanum stoloniferum* Schltdl.	*S. ajuscoense* Bukasov ex Rybin, *S. antipovichii* Bukasov ex Rybin, *S. antipovichi* var. *neoantipoviczii* (Bukasov) Hawkes, *S. antipoviczii* Bukasov ex Rybin, *S. boreale* (A.Gray) Bitter, *S. candelarianum* Bukasov, *S. fendleri* A. Gray, *S. fendleri* subsp. *arizonicum* Hawkes, *S. fendleri* var. *texense* Correll, *S. leptosepalum* Correll, *S. longipedicellatum* Bitter, *S. longipedicellatum* var. *longimucronatum* Hawkes, *S. longipedicellatum* var. *pseudoprophyllum* Bitter, *S. malinchense* Hawkes, *S. nannodes* Correll, *S. neoantipoviczii* Bukasov, *S. noctiflorum* Hort. Dunal, *S. orbiculatibaccatum* Lechn., *S. papita* Rydb., *S. polytrichon* Rydb., *S. schizostigma* Bitter, *S. stoloniferum* subsp. *moreliae* Hawkes, *S. tlaxcalense* Hawkes, *S. tuberosum* var. *boreale* A.Gray, *S. wightianum* Rydb.
	36	*Solanum* × *vallis-mexici* Juz.	*S. vallis-mexici* Juz. ex Bukasov
	MEGISTACROLOBA		
	37	*Solanum boliviense* Dunal	*S. alticola* Bitter, *S. alticola* var. *xanthotrichum* Hawkes, *S. astleyi* Hawkes & Hjert., *S. boliviense* subsp. *astleyi* (Hawkes & Hjert.) D.M.Spooner, M. Ugarte & P.W.Scroch, *S. catamarcae* Bitter ex Brücher, *S. decurrentilobum* Cárdenas & Hawkes, *S. ellipsifolium* Cárdenas & Hawkes, *S. megistacrolobum* Bitter, *S. megistacrolobum* forma *purpureum* Ochoa, *S. sanctae-rosae* Hawkes, *S. tilcarense* Hawkes, *S. toralapanum* Cárdenas & Hawkes, *S. toralapanum* var. *subintegrifolium* Cárdenas & Hawkes, *S. ureyi* Cárdenas
	38	*Solanum hastiforme* Correll	
	39	*Solanum raphanifolium* Cárdenas & Hawkes	*S. hawkesii* Cárdenas
	40	*Solanum sogarandinum* Ochoa	*S. sogarandae* Firbas & Ross

(continued)

Table 4.2 (continued)

SECTION		SPECIES	SYNONYMS
PETOTA	MORELLIFORME		
	41	*Solanum clarum* Correll	
	42	*Solanum morelliforme* Bitter & Münch	
	PINNATISECTA		
	43	*Solanum jamesii* Torr.	*S. jamesii* var. *heterotrichium* Bitter, *S. jamesii* var. *sinclairii* Bitter & Correvon, *S. jamesii* subsp. *septentrionale* Bitter
	44	*Solanum pinnatisectum* Dunal	*S. pinnatisectum* var. *heptazygum* Bitter, *S. pinnatisectum* var. *pentazygum* Bitter
	45	*Solanum* × *michoacanum* (Bitter) Rydb.	*Solanum jamesii* var. *michoacanum* Bitter
	46	*Solanum* × *sambucinum* Rydb.	
	PIURANA		
	47	*Solanum acroglossum* Juz.	
	48	*Solanum acroscopicum* Ochoa	*S. lopez-camarenae* Ochoa
	49	*Solanum albornozii* Correll	
	50	*Solanum anamatophilum* Ochoa	*S. peloquinianum* Ochoa
	51	*Solanum andreanum* Baker	*S. baezense* Ochoa, *S. burtonii* Ochoa, *S. cyanophyllum* Correll, *S. correllii* Ochoa, *S. paucijugum* Bitter, *S. pichinchense* Bitter & Sodiro, *S. regularifolium* Correll, *S. serratoris* Ochoa, *S. solisii* Hawkes, *S. suffrutescens* Correll, *S. tuquerrense* Hawkes
	52	*Solanum augustii* Ochoa	
	53	*Solanum cajamarquense* Ochoa	
	54	*Solanum cantense* Ochoa	
	55	*Solanum chilliasense* Ochoa	
	56	*Solanum chiquidenum* Ochoa	*S. ariduphilum* Ochoa, *S. chiquidenum* forma *amazonense* Ochoa, *S. chiquidenum* var. *cachicadense* Ochoa, *S. chiquidenum* var. *gracile* Ochoa, *S. chiquidenum* var. *porconense* Ochoa, *S. chiquidenum* var. *robustum* Ochoa
	57	*Solanum chomatophilum* Bitter	*S. chomatophilum* forma *angustifoliolum* Correll, *S. chomatophilum* forma *pilosum* Correll, *S. chomatophilum* forma *sausianense* Ochoa, *S. chomatophilum* var. *subnivale* Ochoa, *S. huarochiriense* Ochoa, *S. jalcae* Ochoa, *S. jalcae* var. *pubescens* Correll, *S. pascoense* Ochoa, *S. sinclairii* Hort. ex Bitter, *S. taulisense* Ochoa
	58	*Solanum contumazaense* Ochoa	
	59	*Solanum dolichocremastrum* Bitter	*S. chavinense* Correll, *S. huanucense* Ochoa

(continued)

Table 4.2 (continued)

SECTION	SPECIES		SYNONYMS
PETOTA	60	*Solanum huancabambense* Ochoa	
	61	*Solanum humectophilum* Ochoa	
	62	*Solanum hypacrarthrum* Bitter	*S. guzmanguense* Whalen & Sagást., *S. tuberosum* var. *puberulum* Hook.f.
	63	*Solanum immite* Dunal	*S. immite* Dunal var. *vernale* Correll, *S. mathewsii* Bitter, *S. tuberosum* L. var. *multijugum* Hook.f., *S. wittmackii* var. *glauciviride* Bitter, *S. yamobambense* Ochoa
	64	*Solanum minutifoliolum* Correll	
	65	*Solanum mochiquense* Ochoa	*S. chancayense* Ochoa, *S. earl-smithii* Correll, *S. incahuasinum* Ochoa, *S. mochicense* Ochoa
	66	*Solanum multiinterruptum* Bitter	*S. chrysoflorum* Ochoa, *S. moniliforme* Correll, *S. multiinterruptum* Bitter forma *albiflorum* Ochoa, *S. multiinterruptum* forma *longipilosum* Correll, *S. multiinterruptum* var. *machaytambinum* Ochoa
	67	*Solanum olmosense* Ochoa	
	68	*Solanum paucissectum* Ochoa	
	69	*Solanum piurae* Bitter	
	70	*Solanum raquialatum* Ochoa	*S. ingaefolium* Ochoa, *S. rachialatum* Ochoa,
	71	*Solanum scabrifolium* Ochoa	
	72	*Solanum simplicissimum* Ochoa	
	73	*Solanum trinitense* Ochoa	
	74	*Solanum wittmackii Bitter*	*S. vavilovii* Juz. & Bukasov, *S. tuberosum* var. *macranthum* Hook.f., *S. wittmackii* var. *glauciviride* Bitter
	75	*Solanum xblanco-galdosii* Ochoa	
	POLYADENIA		
	76	*Solanum lesteri* Hawkes & Hjert.	
	77	*Solanum polyadenium* Greenmam	*S. polyadenium* subsp. *orizabae* Bitter
	STENOPHYLLIDIA		
	78	*Solanum hintonii* Correll	
	79	*Solanum stenophyllidium* Bitter	*S. brachistotrichium* (Bitter) Rydb., *S. brachistotrichium* var. *ripicolum* (Bitter) Correll, *S. jamesii* var. *brachistotrichium* Bitter, *S. jamesii* subsp. *nayaritense* Bitter, *S. jamesii* var. *ripicolum* Bitter, *S. jamesii* subsp. *septentrionale* var. *ripicola* Bitter *S. nayaritense* (Bitter) Rydb.

(continued)

Table 4.2 (continued)

SECTION	SPECIES		SYNONYMS
PETOTA	80	*Solanum ehrenbergii* (Bitter) Rydb.	*S. cardiophyllum* subsp. *ehrenbergii* (Bitter) Correll
	STIPULOIDEA		
	81	*Solanum stipuloideum* Rusby	*S. capsicibaccatum* Cárdenas, *S. capsicibaccatum* var. *latifoliolatum* Ochoa, *S. circaeifolium* Bitter, *S. circaeifolium* forma *lobatum* Correll, *S. circaeifolium* subsp. *quimense* Hawkes & Hjert., *S. soestii* Hawkes & Hjert.
	82	*Solanum neocardenasii* Hawkes & Hjert.	
	TRIFIDA		
	83	*Solanum tarnii* Hawkes & Hjert.	
	84	*Solanum trifidum* Correll	
	TUBEROSA		
	85	*Solanum amayanum* Ochoa	
	86	*Solanum ancophilum* (Correll) Ochoa	*S. rhomboideilanceolatum* Ochoa var. *ancophilum* Correll
	87	*Solanum berthaultii* Hawkes	*S.* x*flavoviridens* Ochoa, *S. berthaultii* forma *zudanense* (Cárdenas) Correll, *S. litusinum* Ochoa, *S. tarijense* Hawkes, *S. tarijense* var. *pojoense* (Cárdenas) Correll, *S. trigalense* Cárdenas, *S. vallegrandense* Cárdenas, *S. vallegrandense* var. *pojoense* Cárdenas, *S. zudanense* Cárdenas
	88	*Solanum brevicaule* Bitter	*S. abancayense* Ochoa, *S. alandiae* Cárdenas, *S. anomalocalyx* Hawkes, *S. anomalocalyx* var. *brachystylum* Cárdenas & Hawkes, *S. anomalocalyx* var. *llallaguanianum* Cárdenas & Hawkes, *S.anomalocalyx* var. *murale* Cárdenas & Hawkes, *S. avilesii* Hawkes & Hjert., *S. aymaraesense* Ochoa, *S. bill-hookeri* Ochoa, *S. boliviense* subsp. *virgultorum* Bitter, *S. brevimucronatum* Hawkes, *S. calcense* Hawkes, *S. calcense* Hawkes var. *urubambense* Vargas, *S. candelarianum* Cárdenas, *S. coelestispetalum* Vargas, *S. colominense* Cárdenas, *S. famatinae* Bitter & Wittm., *S. gourlayi* Hawkes, *S. gourlayi* subsp. *saltense* A.M. Clausen & K.A.Okada, *S. huancavelicae* Ochoa, *S. hondelmannii* Hawkes & Hjert., *S. hoopesii* Hawkes & K.A.Okada, *S. incamayoense* K.A.Okada & A.M. Clausen, *S. lapazense* Hawkes, *S. leptophyes* Bitter, *S. liriunianum* Cárdenas & Hawkes, *S. membranaceum* Vargas, *S. mollepujroense* Cárdenas & Hawkes, *S. ochoae* Vargas, *S. oplocense* Hawkes, *S. pachytrichum* Hawkes, *S. pampasense* Hawkes, *S. puberulofructum* Correll, *S. ruiz-zeballosii* Cárdenas, *S. sawyeri* Ochoa, *S. setulosistylum* Bitter, *S. sleumeri* Correll, *S. spegazzinii* Bitter, *S. subandigenum* var. *camarguense* Cárdenas, *S. sucrense* Hawkes, *S. torrecillasense* Cárdenas, *S. sparsipilum* (Bitter) Juz. & Bukazov, *S. tuberosum*

(continued)

Table 4.2 (continued)

SECTION	SPECIES		SYNONYMS
			subsp. *sparsipilum* Bitter, *S. ugentii* Hawkes & K.A. Okada, *S. vidaurrei* Cárdenas, *S. virgultorum* (Bitter) Cárdenas & Hawkes
PETOTA	89	*Solanum candolleanum* Berthault	*S. abbottianum Juz.*, *S. achacachense* Cárdenas, *S. amabile* Vargas, *S. ambosinum* Ochoa, *S. ancoripae* Ochoa, *S. antacochense* Ochoa, *S. bukasovii* Juz., *S. canasense* Hawkes, *S. canasense* Hawkes var. *album* Vargas, *S. canasense* Hawkes var. *calcense* Vargas, *S. canasense* Hawkes var. *intihuatanense* Vargas, *S. catarthrum* Juz., *S. chillonanum* Ochoa, *S. cuzcoense* Ochoa, *S. espinarense* Vargas, *S. fragariifructum* Hawkes, *S. hapalosum* Ochoa, *S. lechnoviczii* Hawkes, *S. lechnoviczii* Hawkes var. *latifolium* Vargas, *S. lechnoviczii* var. *xerophyllum* Vargas, *S. longimucronatum* Vargas, *S. longiusculus* Ochoa, *S. marinasense* Vargas, *S. marinasense* Vargas var. *dentifolium* Vargas, *S. multidissectum* Hawkes, *S. multiinterruptum* Bitter forma *longipilosum* Correll, *S. multiinterruptum* Bitter var. *machaytambinum* Ochoa, *S. neohawkesii* Ochoa, *S. orophilum* Correll, *S. pampasense* Hawkes, *S. puchupuchense* Ochoa, *S. punoense* Hawkes, *S. sarasarae* Ochoa, *S. saxatile* Ochoa, *S. sicuanum* Hawkes, *S. soukupii* Hawkes, *S. tapojense* Ochoa, *S. tarapatanum* Ochoa
	90	*Solanum chacoense* Bitter	*S. arnezii* Cárdenas, *S. boergeri* Bukasov, *S. caipipendense* Cárdenas, *S. calvescens* Bitter, *S. commersonii* var. *glabriusculum* Hook.f., *S. cuevoanum* Cárdenas, *S. chacoense* var. *latisectum* forma *plurijugum* Hassler, *S. emmeae* Juz. & Bukasov, *S. garciae* Juz. & Bukasov, *S. gibberulosum* Juz. & Bukasov, *S. guaraniticum* var. *latisectum* forma *glabrescens* Hassler, *S. horovitzii* Bukasov, *S. horovitzii* var. *multijugum* Hawkes, *S. jamesii* var. *grandifrons* Bitter, *S. jujuyense* Hawkes, *S. knappei* Juz. & Bukasov, *S. laplaticum* Bukasov, *S. limense* Correll, *S. muelleri* Bitter, *S. muelleri* forma *densipilosum* Correll, *S. parodii* Juz. & Bukasov, *S. saltense* Hawkes, *S. schickii* Juz. & Bukasov, *S. subtilius* Bitter, *S. tuberosum* var. *glabriusculum* Dunal, *S. tuberosum* subsp. *yanacochense* Ochoa, *S. yungasense* Hawkes
	91	*Solanum gandarillasii* Cárdenas	
	92	*Solanum gracilifrons* Bitter	
	93	*Solanum incasicum* Ochoa	
	94	*Solanum infundibuliforme* Phil.	*S. glanduliferum* Hawkes, *S. infundibuliforme* var. *albiflorum* Ochoa, *S. infundibuliforme* var. *angustepinnatum* Bitter, *S. microphyllum* Hawkes, *S. pinnatifidum* Cárdenas, *S. platypterum* Hawkes
PETOTA	95	*Solanum kurtzianum* Bitter & Wittm.	*S. commersonii* var. *glanduloso-pubescens* Hook.f., *S. commersonii* var. *pilosiusculum* Hook.f., *S. improvidum*

(continued)

Table 4.2 (continued)

SECTION	SPECIES		SYNONYMS
			Brücher, *S. macolae* Bukasov, *S. ruiz-lealii* Brücher, *S. velascanum* Bitter & Wittm.
	96	*Solanum lignicaule* Vargas	*Solanum lignicaule* var. *longistylum* Vargas
	97	*Solanum maglia* Schltdl.	*S. collinum* Dunal, *S. maglia* var. *witasekianum* Bitter, *S. tuberosum* var. *sabinii* A.DC
	98	*Solanum medians* Bitter	*S. arahuayum* Ochoa, *S. medians* var. *angustifoliolum* Ochoa, *S. medians* var. *autumnale* Correll, *S. medians* var. *majorifrons* Bitter, *S. medians* subvar. *protohypoleucum* Bitter, *S. sandemanii* Hawkes, *S. tacnaense* Ochoa, *S. weberbaueri* Bitter, *S. weberbaueri* var. *decurrentialatum* Ochoa, *S. weberbaueri* var. *poscoanum* Cárdenas & Hawkes
	99	*Solanum microdontum* Bitter	*S. bijugum* Bitter, *S. cevallos-tovari* Cárdenas, *S. gigantophyllum* Bitter, *S. higueranum* Cárdenas, *S. microdontum* var. *montepuncoense* Ochoa, *S. simplicifolium* Bitter, *S. simplicifolium* subsp. *gigantophyllum* var. *metriophyllum* Bitter, *S. simplicifolium* subsp. *gigantophyllum* var. *mollifrons* Bitter, *S. simplicifolium* subsp. *gigantophyllum* var. *trimerophyllum* Bitter, *S. simplicifolium* var. *variabile* Brücher & Ross
	100	*Solanum neorossii* Hawkes & Hjert.	
	101	*Solanum neoweberbaueri* Wittm.	
	102	*Solanum okadae* Hawkes & Hjert.	*S. venatoris* Ochoa
	103	*Solanum velardei* Ochoa	
	104	*Solanum venturii* Hawkes & Hjert.	
	105	*Solanum vernei* Bitter & Wittm.	*S. ballsii* Hawkes
	106	*Solanum* × *doddsii* Correll	
	107	*Solanum* × *rechei* Hawkes & Hjert.	
	VERRUCOSA		
	108	*Solanum verrucosum* Schltdl.	*S. macropilosum* Correll, *S. squamulosum* M.Martens & Galeotti

Table 4.3 Cultivated potato species of Section *Petota*, cultivar Group (ICNCP), distribution, ploidy, EBN number, and synonyms of the recognized valid species according to the current taxonomic treatment (Spooner et al. 2007). Complete synonyms, epithets and names not validly published in Ovchinnikova et al. 2011 and Solanaceae Source website (http://www.solanaceaesource.org)

Species	Cultivar Group	Distribution	Ploidy (EBN)	Synonyms
Solanum tuberosum L.	Chilotanum Group	Southern Chile Chonos and Guaitecas Arquipelagos	*4X* (4)	*S. tuberosum* subsp. *tuberosum*
	Andigenum Group	Western Venezuela, Colombia, Ecuador, Perú, Northern Argentina, mid to high Andean elevations (3000–3500 m)	*2X* (2) *3X 4X* (4)	*S. tuberosum* subsp. *andigenum* (Juz. & Bukasov) Hawkes, *S. chaucha* Juz. & Bukasov, *S. phureja* Juz. & Bukasov, *S. phureja* subsp. *estradae* (L. López) Hawkes, *S. phureja* subsp. *hygrothermicum* (Ochoa) Hawkes, *S. stenotomum* Juz. & Bukasov, *S. stenotomum* Juz. & Bukasov subsp. *goniocalyx* (Juz. & Bukasov) Hawkes
Solanum ajanhuiri Juz. & Bukasov		Bolivia and Perú Andean highlands (above 3500 m)	*2X* (2)	*S. ajanhuiri* Juz. & Bukasov forma *janckoajanhuiri* Ochoa, *S. ajanhuiri* Juz. & Bukasov var. *yari* Ochoa
Solanum curtilobum Juz. & Bukasov		Bolivia and Perú Andean highlands (above 3500 m)	*5X* (4)	
Solanum juzepczukii Bukasov		Bolivia, Perú, Northern Argentina Andean highlands (above 3500 m)	*3X* (2)	*S. juzepczukii* Bukasov var. *parco* Hawkes, *S. juzepczukii* Bukasov var. *roseum* Vargas, *S. juzepczukii* Bukasov forma *ckoyuckaisalla* Ochoa, *S. juzepczukii* Bukasov forma *janckock-aisalla* Ochoa, *S. juzepczukii* Bukasov forma *luckipechuma* Ochoa, *S. juzepczukii* Bukasov forma *luckipinkula* Ochoa, *S. juzepczukii* Bukasov forma *wilackaisalla* Ochoa, *S. juzepczukii* Bukasov var. *lucki* Ochoa

(Engel 1970). Fossilized tubers found in Casma Valley of Perú have been directly dated to 7800 cal B.P. (C^{14} calibrated date), and even though starch microremains resembled those of the domesticated potato may still represented a wild species (Ugent et al. 1982). Recently, an archeological study based on the microscopic analysis of starch granules found on ground stone tools in deposits dating between 10,900 and 10,100 cal B.P. at North Creek Shelter (Utah) documented the earliest use of wild potatoes in North America as important food source (Louderback and Pavlik 2017). These archeological findings evidence the early consumption of potato tubers at a time that precedes agriculture (Ugent et al. 1982; Hawkes 1990; Louderback and Pavlik 2017). The analysis of starch microremains recovered from groundstone tools found at Jiskairumoko, an ancient village in Perú, revealed an intensive exploitation of potatoes that took place between 3400 and 1600 B.P. during Late Archaic to Early Formative in the western

Titicaca Basin (Rumold and Aldenderfer 2016). These archeological evidences, based on the consistency of ancient starch remains with those of cultivated potato, documented a time of transition from nomadism to sedentism and food production, and may be related to potato domestication and early cultivation in southern Perú (Rumold and Aldenderfer 2016).

Prevalent hypothesis for cultivated potato's origins advocated multiple, independent domestications from a group of about 20 morphologically similar wild potato species, the "*Solanum brevicaule* complex" (Alvarez et al. 2008), distributed from southern Peru, northwestern Bolivia, and northern Argentina (Brücher 1964; Ugent 1970; Bukasov 1978; Hawkes 1990; Grun 1990; Ochoa 1990, 1999; Van den Berg et al. 1998; Huamán and Spooner 2002; Spooner et al. 2014). New insights from phylogenetic analysis that include a wide sampling of 362 representatives of landraces, putative progenitors, and outgroups, supported a reduction in the number of species in the *Solanum brevicaule* complex and a monophyletic origin of landrace cultivars from a single species in a broad area of southern Peru (Spooner et al. 2005). Landraces developed by early Andean farmers were dispersed from Peru both north and south. Nowadays, potato landraces reveal great morphological, physiological, and genetic diversity, and are distributed throughout the Andes, from western Venezuela to northern Argentina, and in southern Chile (Spooner et al. 2010). Landrace potato populations in Mexico and Central America are recent, post-Columbian introductions (Ugent 1968).

Cultivated potatoes were first introduced from America in the Canary Islands in 1567 (Hawkes and Francisco Ortega 1993; Spooner et al. 2005; Ríos et al. 2007), and soon arrived in Spain in 1573 (Hawkes 1990; Hawkes and Francisco Ortega 1993; Romans 2005). The first botanical description of potato in Europe was made by Caspar Bauhin in 1596, but the origin of the plant described was unknown (Hawkes 1990). In 1597 Gerard made the first description of the potato in English and a detailed illustration in The Herbals, although he mistakenly believed it came from Virginia in North America rather than South America. Later Carolus Clusius (1601) described potatoes and mentioned he received the tubers in 1588 from Phillippe de Sirvry who cultivated potatoes in Belgium and made the first drawing of potato in Europe indicating with his handwriting its common name and origin "*Taratoufli Vienae*, 26 *januarii* 1588, *Papas Peruam Petri Cieca*" (Parodi 1966). Evidences from early herbalist indicated that potatoes were cultivated since mid-XVI century in different European countries and rapidly disseminated worldwide. Regarding the origin of the first potatoes introduced in Europe, two hypotheses have been proposed: from lowland Chile (Juzepczuk and Bukasov 1929) or from the Andes (Salaman 1937; Salaman, and Hawkes 1949), being the Andean origin the most accepted. Ames and Spooner (2008) investigated these two competing hypothesis using historical herbarium potato specimens for a screening with a plastid DNA deletion marker. Interestingly, the first direct evidences from early preserved plants showed not only that potatoes of Andean origin predominated in Europe in the 1700s, but also that potatoes from Chile were introduced as early as 1811 in Europe, and became predominant long before the late blight epidemics begun in 1845 in potato crops causing high mortality and famine (Ames and Spooner 2008). Consequently, after the late blight epidemics, resistances from Chilean landraces were introduced into European potato cultivars.

Alphonse de Candolle, French-Swiss botanist, was a pioneer to investigate the origins of cultivated plants and crop geographic distribution. In his influential contribution, Origin of Cultivated Plants, De Candolle (1882) used evidences from different disciplines, presence of wild relatives, historical sources, linguistics (local names), archeology, and variation patterns, to determine the origin of cultivated plants. De Candolle (1882) was the first to name as distinct the Chilean populations of *S. tuberosum* as *var. chiloense* A.DC. Vavilov (1920, 1940), Russian geneticist and plant geographer, participated in over 100 collecting missions to explore the major agricultural centers worldwide, and built crop origin hypothesis. Vavilov and his Russian colleagues made several expeditions to Central and South America between 1925 and 1930, and generated a large potato collection that initiated the basic germplasm of the N. I. Vavilov Institute of Plant Industry in Saint Petersburg, Russia (Loskutov 1999). Some of the potato accessions are still maintained, as well as the herbarium specimens of the initial collections of high value to elucidate the taxonomy and nomenclature of cultivated potato (Ovchinnikova et al. 2011). Later, between 1938 and 1939 other important potato germplasm collections in South America were made by British Botanists Balls, Gourley and Hawkes (Hawkes 1944; 2004; Hawkes and Hjerting 1969, 1989). Germplasm derived from these initial collections is maintained at the Scottish Crop Research Institute (SCRI) in Dundee, United Kingdom, and specimens are mainly deposited at the Herbarium of Kew Botanical Gardens (KEW), and in many other collections worldwide (Ovchinnikova et al. 2011). Peruvian botanists C. Ochoa and A. Salas collected potatoes throughout South America that initiate the base of germplasm collections at the Universidad Nacional Agraria La Molina, Peru, and later the International Potato Center (Centro Internacional de la Papa, CIP). M. Cárdenas made early collections and descriptions of new potatoes from Bolivia (Cárdenas and Hawkes 1946), where National Agricultural and Forestry Institute (INIAF) maintains an important and diverse collection of cultivated potatoes (Cadima–Fuentes et al. 2013). In Argentina, several potato germplasm collections have been made by K. A. Okada, A. Clausen, and collaborators, which are maintained at INTA Potato Germplasm Bank in Balcarce (Clausen et al. 2010). In Chile, A. Contreras also collected and led germplasm collections that are maintained at the Chilean Germplasm Bank (Contreras 1987). David Spooner and collaborators from different countries also made important potato collections in North, Central and South America, which are maintained in CIP, US Potato Genebank, and other genebanks.

In order to organize large collections of cultivated potatoes, early Russian taxonomists applied a complex method to describe, named, and classified them, based on ploidy, ecogeography, and analysis of morphological and physiological characters. This system of nomenclature considered the homologous series of variation (Vavilov 1922), where geographical distribution and ecological types are major characters to define and name taxa (Juzepczuk and Bukasov 1929; Bukasov 1930; Juzepczuk 1937). Initially, Hawkes also applied this system to describe and name his potato collections, but later he simplified his classifications (Hawkes 2004). The application of this complex nomenclature system in initial cultivated potato collections, created numerous names, sometimes polynomials as well as many invalid names and a complicated classification by ranks. Recently, Ovchinnikova et al. (2011) clarified the nomenclature and taxonomy of cultivated potatoes, recognized 626 epithets associated with all taxa of cultivated potato and placed them in synonymy, and also made lectotype designations for names validly published (Ovchinnikova et al. 2011, Solanaceae Source website http://www.solanaceaesource.org). Four cultivated potato species are recognized: *S. tuberosum* L., *S. ajanhuiri* Juz. & Bukasov, *S. curtilobum* Juz. & Bukasov and *S. juzepczukii* Bukasov (Spooner et al. 2007), the three later species were formed by hybridization of *S. tuberosum* with more distantly related wild species of groups Acaulia and Megistacroloba (Rodríguez et al. 2010; Ovchinnikova et al. 2011) (Table 4.3).

The current taxonomical interpretation considers two main groups of landraces within *Solanum tuberosum* L, named according to the International Code of Nomenclature of Cultivated Plants (2016). The 'Andigenum Group' comprising diploid, triploid, or tetraploid, adapted to short-day flowering and tuberization. The 'Chilotanum Group' includes landraces from Southern Chile, mainly concentrated in the Chonos and Guaitecas Archipelagos, adapted to long-day flowering and tuberization (Huamán and Spooner 2002; Spooner et al. 2007; Ovchinnikova et al. 2011) (Table 4.3).

4.5 Methods and Issues for the Conservation of Potato Genetic Resources

The conservation of potato genetic resources implies two different strategies: the first one, ex situ conservation, is focused on the maintenance of potato genetic diversity outside its natural environment. The second, in situ strategy, comprise the conservation of ecosystems and natural habitats as well as the maintenance and recovery of viable populations of species in their natural surroundings, and in the case of domesticated or cultivated plant species conservation incorporate the surroundings where they have developed their distinctive properties (UNCED 1992; FAO 2009). These strategies have advantages and disadvantages, but the most remarkable characteristic is that both are complementary rather than exclusive. Nowadays it is accepted that a holistic approach is more effective in conservation programs of genetic resources.

The ex situ conservation of potato genetic resources is performed mainly in genebanks, and in a few cases in botanical gardens and museums. A fundamental goal of genebanks is to preserve germplasm and made it available for different purposes including research, breeding, agriculture production, industry, etc. (Ellis et al. 2020). The Second Report of the State of the Genetic Resources for Food and Agriculture (FAO 2010) registered 174 potato genebanks around the world and a total number 98,285 accessions, with possible material duplication, but it was estimated that 24,500–29,500 unique potato accessions are conserved worldwide. The same report showed that six genebanks hold 41% of the global potato accessions: The French National Institute for Agricultural Research (INRA) in France (11%), Vavilov Institute in Russia (9%), The International Potato Center (CIP) in Peru (8%), The Leibniz Institute of Plant Genetics and Crop Plant Research (IPK) in Germany (5%), USDA-ARS in the USA (5%), and The National Institute of Agrobiological Sciences (NIAS) in Japan (3%) and other 20 genebanks hold over 1000 potato accessions each. These genebanks collectively held collections of 15% wild relatives of potato, 20% cultivated potato accessions, 16% research, and breeding materials, 14% advanced breeding lines, and 35% uncategorized accessions (Ellis et al. 2020). According to the last report of the Global Strategy for Ex Situ Conservation of Potato, Latin American genebanks contain principally native cultivars while those in Europe and North America contain modern cultivars, breeding materials, and wild relatives (Ellis et al. 2020). Castañeda-Álvarez et al. (2015) determined that 43.8% of wild potato species are under-represented in genebanks, some of them with no accessions available such as *S. ayacuchense*, *S. neovavilovii*, *S. olmosense* and *S. salasianum*. To improve the worldwide representation of wild potatoes conserved ex situ in genebanks, Instituto Nacional de Innovación Agraria (INIA, Peru) and CIP joined to collect more than 300 new accessions in Peru; two of them belong to *S. ayacuchense* (Zorrilla et al. 2019a). The type of biological sample preserved in genebanks varies depending on the biological status of the accessions, if they come from wild or cultivated species. In genebanks seeds are the most common biological sample for conservation of potato wild relatives, to assure the preservation of genetic diversity of the original population instead of individual characteristics. The most important variables, taken into account for adequate seed conservation, are population size, storage temperature, seed humidity, and seed quality as these variables can affect the expected viability of the seed accession lot. The number of regenerations, sexual multiplication, should be the minimum in order to preserve the accessions identity. Genebanks follow standard conservation methods recommended by Biodiversity International and other institutions (FAO-IPGRI, Engels, and Visser 2003; FAO 2014; CGIAR 2020). Seeds of cultivated and wild potatoes are orthodox (Holle 1988), meaning that seeds can be dried to low moisture contents and stored at cold temperature without damage during different periods of time (Roberts and Ellis 1984). For instance, seeds are stored at −18 °C for the maintenance as base collections or safety copy in the genebank Svalbard Global Seed Vault which preserves potato duplicated seeds as a "black box" for long term conservation (Ellis et al. 2020). In this case, the depositors are responsible for processing, packing, and shipping the seeds before storage; subsequent seed multiplication when needed, and distributing these seeds from their stocks under conditions similar to the International Treaty for Plant Genetic Resources for Food and Agriculture (ITPGRFA). The International Potato Center (CIP) in Perú stores potato seeds in the active collection at 4 °C, and in the base collection at −20 °C. The active collection is used for distributing germplasm and the base collection for the regeneration of new seed stocks. CIP has defined standard procedures for seed processing, packing, and storage of potato wild relatives, considering the extraction, cleaning of impurities, drying to 5% water seed content before entering the corresponding storage room (Salas et al. 2008). Similarly, the US potato genebank stores dried seeds with 5% of water content at −20 °C, while a subset of the most requested accessions is maintained at 4 °C, and another copy is sent to genebank Fort Collins Seed Vault for safety duplication (del Río A. pers. com.).

Seed regenerations are necessary to maintain seed stocks, and are performed when viability and/or the amount of seeds drop below the established minimum value. Indeed, regeneration is performed at CIP when viability drops below 85% and/or when the amount of seeds drops below 5000 units. The same criteria as in CIP, is taking into account for the regeneration of wild potato seeds at the Potato Genebank at Balcarce Agricultural Experimental Station of the National Institute of Agricultural Research (INTA) (Digilio A. pers. com.). At the US Potato Genebank, accessions are regenerated every 20 years (del Rio, pers.com.). Seed regeneration procedures define population size and type of crosses (open pollination, reciprocal sibling crosses, or mass pollination), depending on the species reproductive biology (cleistogamous, autogamous, exogamous, or alogamous). CIP has defined 25 plants as the minimum population size for regeneration; meanwhile, the US potato genebank uses 20 plants. Camadro (2012) recommended using 15–25 plants in controlled crosses during the multiplication process. The results of studies that assessed the effects of population size on genetic diversity during regeneration of the accessions, found that a sample of 25 to 30 plants is the optimal number to capture and maintain most of the alleles (Bamberg and del Río 2004), and similar criteria is used by the Potato Genebank at Balcarce. A concern in genebanks are the effects of accession multiplication, since populations derived from different regeneration events can differ significantly (Cadima-Fuentes 2014; Zorrilla et al. 2019b). Currently, the genetic diversity of Solanaceous crops from the principal genebanks is being studied within a similar objectives and approaches outlined in the G2PSOL initiative. As part of this project, the effects of sample size on the genetic diversity of conserved accessions in genebanks is an important issue expected to be solved.

A different strategy is used to preserve accessions of cultivated potatoes, botanical tubers named tuber-seeds are the main type of biological sample in genebanks for maintaining potato characteristics as a clone (asexually) where unique allelic combinations in the individual are preserved. Andean farmers traditionally have been used tuber-seeds as the principal method for conservation of their cultivated potatoes, since is the type of preservation that requires fewer resources in terms of equipment and supplies. Even though tubers conservation in genebanks is less demanding, only allows tuber conservation between 3 and 9 months at temperatures between 0 and 10 °C, depending of the dormancy and quality of the tuber before entering the storage room (JICA 2016). In the same way, the lowland tetraploid landraces, *S. tuberosum* L. Chilotanum Group, is conserved using botanical tubers at Institute of Agricultural Research of Chile and, Agricultural and Livestock Service of Chile (Muñoz et al. 2016).

Other methods, like in vitro slow growth have been used for the conservation of landraces, modern cultivars, and/or specific genotypes, and it eliminates the challenges of external biotic and abiotic factors. This procedure is based on tissue-culture conservation methodology, minimizing tissue growth so subcultures are reduced. Explants, under aseptic conditions, are introduced into glass tubes containing slow growth media and transferred into cultivated chamber where environmental variables, temperature, light intensity, and photoperiod, are regulated. In vitro conservation allows potato samples survive up to two years, depending on the composition of the media and environmental conditions. The culture media frequently used for potato in vitro conservation is Murashige and Skoog supplemented with sorbitol and sucrose, and samples are maintained at temperatures between 6 to 10 °C and low light intensity (Clausen et al. 2010; Bamberg et al. 2016; Muñoz et al. 2019). CIP's potato method (2019) is currently the most efficient in vitro method for conserving potato accessions, which prolonged transference periods from once every 6–8 weeks up to two years. CIP currently maintains the largest in vitro potato collection with 8354 potato in vitro accessions, the majority of which (89.8%) are landraces or "papas nativas" originating mainly from the Andean region, with the remaining accessions being improved varieties and breeding lines (Ellis et al. 2020). In vitro slow growth conservation has been used by farmers of highland communities at the Potato Park in Cusco, Peru, to maintain their native potatoes tuber-seed clean of pest and diseases.

Cryopreservation is another methodology for long-term germplasm preservation where different types of explants are maintained at ultra-low temperatures in liquid nitrogen or liquid nitrogen vapor (Wang et al. 2020), and using osmoprotectant solutions in order to dehydrate the tissues and decrease the formation of ice crystals that would be lethal for cells. Potato seeds and meristems can be preserved and maintain its genetic stability by cryopreservation for longer periods of time compared to any other method (Digilio et al. 2018). This conservation strategy allows the storage of large quantities of samples in reduced spaces and at low costs, and is being employed in the largest genebanks with adequate infrastructure, equipment, trained personnel, and a continuous supply of liquid nitrogen. At CIP, the droplet PVS2 vitrification method preceded by a pre-culture treatment at 6 °C is used for application of cryopreservation for the long-term conservation of a wide diversity of potato genotypes (Panta et al. 2014, 2015) and has a recovery rate of 55–61% (Vollmer et al. 2019). A high-quality management of cryopreservation systems includes periodical viability reassessment, clear recovery criteria and the monitoring of success and contamination rates has been implemented (Vollmer et al. 2016, 2017). Currently, CGIAR in vitro genebanks unite the in vitro slow growth conservation and cryopreservation as part of the clonal crop conservation strategy, where the first is implemented to maintain for a medium-term the active collection, and the latter for the long term conservation of the base collection (Benson et al. 2011).

Another fundamental objective of genebanks is the characterization of genetic resources, in potato different molecular markers have been used to assess genetic variability, population structure, and taxa relationships, but single nucleotide polymorphisms (SNPs) are most frequently used due to their affordable cost and data quality, SNP arrays allow to assess about one million genome-wide SNP's simultaneously in an individual assay, and because provide fundamental information not only for conservation but also for genetic analysis and breeding (Ellis et al. 2018).

DNA storage is another type of sample used for the conservation of genetic resources. However, only fragments of nucleic acids that can be later amplified, cloned, and inserted into another plant. This type of conservation is employed by a limited number of genebanks.

Herbarium specimens are another type of samples representative of potato genetic resources diversity. These collections kept voucher samples of the originally collected material and later regenerations. Herbarium specimens are important to document potato diversity and also permit the correct taxonomic characterization and identification of genebanks collections. Interestingly, DNA samples have been obtained from historical specimens using isolation techniques to determine potato species identity with molecular markers (Ames and Spooner 2008).

The maintenance of genetic resources in their natural environment where the species naturally evolved, is the strategy for in situ conservation. The main goal is to establish genetic reserves that guarantee the dynamic species evolutionary process, favoring the appearance of new allelic variants in natural populations that allow species adaptation in front of changing environments. The most effective method, in economic and political terms, is to implement Genetic Reserves inside existing Protected Areas (Maxted et al. 1997). In order to establish protected areas for in situ conservation, it is fundamental to know species taxonomy, distribution, phenology, genetic characteristics, plant demography, ecology, etc. as well as species diversity in the community of the Protected Areas (Dulloo et al. 2008). Clausen et al. (2018) identified 12 wild potato species growing in different Protected Areas of Argentina, and all these species were included in at least one protected area. More recently, studies of wild potato population species are in progress for the future establishment of a genetic reserves within Los Cardones National Park in Salta province (Kozub et al. 2019), in the Natural Reserve Villavicencio in Mendoza province (Marfil et al. 2015), and in the Natural Reserve Paititi in the southeast of Buenos Aires province (Garavano 2018). In the last area, the wild species *S. commersonii* could lose a high percentage of its distribution range due to agricultural activities. In Bolivia, Cadima-Fuentes et al. (2013), found that only 7 of the 21 Bolivian species were detected in parks and protected areas and recommended an increase in the inventories in these areas. In Peru, Sotomayor and Zorrilla (2019) are developing GIS-based studies for the identification of a network of areas with high diversity of wild potato species that need to be protected from threats such as urbanization, land-use change, mining, etc. One way to protect them is to become "Agrobiodiversity Zones", community-owned territories recognized by the government as prioritized for in situ conservation activities. The establishment of on farm conservation zones not only focused on the conservation of cultivated landraces but also in the fundamental role of local communities in the processes of crop evolution (De Haan et al. 2016a, b, c). One of these initiatives is "The Potato Park", recognized as Agrobiodiversity Zones by the Peruvian Minister of Agriculture and Irrigation (MINAGRI), based on their native, cultural and ecological wealth where indigenous peoples preserve their cultural traditions to manage and maintain genetic resources of their fields and ecosystems. The Potato Park located in the Cusco Inca Sacred valley in Calca province (Peru) comprise and extension of more than 7000 ha, where around 1330 potato landraces and other native Andean crops are cultivated and maintained in a local genebank, managed by four Quechua communities or "ayllus". This example of in situ conservation is characterized by strong interactions between the crops, their wild relatives, and the farmers, promoting different strategies for conservation and sustainable use of these genetic resources (MINAM 2019). Currently, different initiatives and efforts, such as the Papa Andina experience are focused on sustainable agriculture-food system development considering the needs for nutrition security while promoting better crop management and productivity, and the optimization of the potato value chain (Devaux et al. 2020). At present different projects in Andean countries are focused on the conservation of their potato genetic resources.

The in situ strategy could be essential for the conservation of genetic resources, and considering different scenarios of climatic change the distribution of wild potato species is likely to diminish at low altitudes and wild species population could be affected or became extinct (Jarvis et al. 2008). Different approaches, such as environmental niche modeling methods, have been used to evaluate the conservation status 73 wild potatoes maintained in genebanks, revealing high priority for collecting 32 species (43.8%), while 20 species have medium priority for collection and only three species have good representation in ex situ collections (Castañeda-Álvarez et al. 2015). These studies not only support collecting strategies but also help to define zones for in situ conservation through the detection of areas where species are potentially threatened by several factors (urbanization, agriculture expansion, overgrazing, etc.). In front of uncertain scenarios, could be possible to securely preserve diversity in potato, and predict what diversity has been lost or is in imminent danger of being lost, and also estimated the actual economic value, as well as, potential future value of the potato diversity that is not securely conserved (Ellis et al. 2020). The understanding of the diversity and distribution of wild and cultivated potato also contributes to the development of policies on biosafety as it has occurred in Peru (MINAM 2019).

4.6 Conclusions

At present, there is a better understanding of the diversity, distribution, and genetic diversity of potatoes. A recent comprehensive taxonomy of all wild and cultivated potatoes, based on the integration of multiple evidences and phylogenetic relationships between taxa, proposed a framework for further investigation of complex groups as well as endemic species in restricted areas, which have been poorly collected.

Integrative taxonomy also provides hypothesis to study patterns of species diversity and distributions, physical and ecological isolation factors, biological barriers and physiological plasticity, and in particular the role of ploidy, gene flow, and interspecific hybridization in the adaptation of species to different and extreme environment.

Genetic and genomic studies have contributed to understand potato evolution. Recent investigations provide promissory approaches to compare and differentiate potato species genomes that help not only to elucidate the process of evolution, but also to detect alleles related to important agronomic traits. Few wild species have been used to improve cultivated potatoes, and these valuable resources can be incorporate into breeding plans.

Results from different disciplines demonstrated the value of complementary evidences from archaeological remains and phylogeny to support the hypothesis of a single origin of domestication and culture of potatoes occurred in southern Peru, and encouraged to continue research to elucidate the sources and areas of initial cultivation.

Genebanks worldwide guarantee the conservation of valuable potato genetic resources for food security. In addition, advances in ex situ conservation techniques like in vitro and cryoconservation made more efficient sample preservation reducing costs and maintenance space. A stable taxonomic framework could be useful for harmonizing potato nomenclature used in global genebanks and to identify unique and redundant material to promote an efficient conservation through collection homologation.

A better understanding of potato diversity is fundamental for the establishment of protected areas, where populations could be monitored to ensure their persistence in natural habitat or in local communities. Agrobiodiversity Zones are also fundamentals for preservation and sustainable use of potato genetic resources and for local communities' alimentary subsistence.

Acknowledgements This work is dedicated to David Spooner, botanist, indefatigable explorer and passionate taxonomist who led research using new methods and analytical approaches to elucidate the phylogeny and diversity of crop species and wild relatives, fundamentally potatoes, and inspire people and students with his insights.The authors acknowledge Dr. Claudio Galmarini, for his critical revision of the present chapter. Iris Edith Peralta research supported by CONICET and SIIP UNCuyo projects.

References

Alefeld FGC (1866) Landwirthschaftliche flora. Weigant & Hempel, Berlin

Álvarez NMB, Peralta IE, Salas A, Spooner DM (2008) A morphological study of species boundaries of the wild potato *Solanum brevicaule* complex: replicated field trials in Peru. Plant Syst Evol 274:37–45

Ames M, Spooner DM (2008) DNA from herbarium specimens settles a controversy about origins of the European potato. Am J Botany 95:252–257

Ames M, Spooner DM (2010) Phylogeny of *Solanum* series *Piurana* and related species in *Solanum* section *Petota* based on five conserved ortholog sequences. Taxon 59:1091–1104

Aversano R, Contaldi F, Ercolano MR, Grosso V, Iorizzo M, Tatino F, Xumerle L, Dal Molin A, Avanzato C, Ferrarini A, Delledonne M,

Sanseverino W, Cigliano RA, Capella-Gutierrez S, Gabaldón T, Frusciante L, Bradeen JM, Carputo D (2015) The *Solanum commersonii* genome sequence provides insights into adaptation to stress conditions and genome evolution of wild potato relatives. Plant Cell 27:954–968

Bauhin C (1596) Pinax Theatri Botanici, sive Index in Theophrasti, Dioscoridis, Plinii et botanicorum qui à seculo scripserunt opera. Basilea

Bamberg JB, Del Rio AH (2004) Genetic heterogeneity estimated by RAPD polymorphism of four tuber-bearing potato species differing by breeding system. Am J Potato Res 81(6):377–383

Bamberg JB, Martin MW, Abad J, Jenderek MM, Tanner J, Donnelly DJ, Novy RG (2016) In vitro technology at the US Potato Genebank. In Vitro Cell Dev Biol Plant 52(3):213–225

Benson EE, Harding K, Debouck D, Dumet D, Escobar R, Mafla G, Panis B, Panta A, Tay D, Van den Houwe I, Roux N (2011) Refinement and standardization of storage procedures for clonal crops Global Public Goods Phase 2: Part I. Project landscape and general status of clonal crop in vitro conservation technologies. System-wide Genetic Resources Programme, Rome, Italy

Brown CR (1984) Tetrad sterility: a cytoplasmic-genic male sterility attractive to bumble bees. In: Winiger FA, Stockli A (eds) Abstracts of the conference papers of the 9th Trienn Conf Europ Assoc Potato Res, Interlaken, Switzerland, 1–6 July 1984, pp 101–102

Brücher H (1964) El origen de la papa (*Solanum tuberosum*) Physis 24:439–452

Burgos G, Felde TZ, Andre C, Kubow S (2020) The Potato and Its Contribution to the Human Diet and Health. In: Campos H, Ortiz O (eds) The potato crop, Chap. 2, pp 37–74. https://doi.org/10.1007/978-3-030-28683-5_2

Bethke PC, Halterman DA, Jansky SH (2019) Potato Germplasm enhancement enters the genomics era. Agronomy 9: 575. https://doi.org/10.3390/agronomy9100575

Bohs L (1994) *Cyphomandra* (Solanaceae). Fl. Neotrop Monogr 63:1–175

Bohs L, Olmstead RG (1997) Phylogenetic relationships in *Solanum* (Solanaceae) based on ndhF sequences. Syst Bot 22:5–17

Bohs L, Olmstead RG (1999) *Solanum* phylogeny inferred from chloroplast DNA sequence phylogeny. In: Nee M, Symon DE, Lester RN, Jessop JP (eds) Solanaceae IV: advances in biology and utilization. Royal Botanic Gardens, Kew, pp 97–110

Bohs L (2005) Major clades in *Solanum* based on ndhF sequence data. In: Keating RC, Hollowell VC, Croat TB (eds) A Festschrift for William G. D'Arcy: A Legacy of a Taxonomist. Missouri Botanical Garden, St. Louis, pp 27–49

Bonierbale MW, Amoros WR, Salas E, de Jong W (2020) Potato breeding. In: Campos H, Ortiz O (eds) The potato crop, Chap. 6, pp 163–217. https://doi.org/10.1007/978-3-030-28683-5_6

Bretagnolle F, Thompson JD (1995) Gametes with the somatic chromosome number: mechanisms of their formation and role in the evolution of autopolyploid plants. New Phytol 129:1–22

Brush SB, Carney HJ, Huamán Z (1981) Dynamics of Andean potato agriculture. Econ Bot 35:70–88

Bukasov SM (1930) The cultivated plants of Mexico, Guatemala and Colombia. Trudy Po Prikladnoj Botanike Genetike i Selekcii, Supplement 47(191–226):513–525

Bukasov SM (1978) Systematics of the potato. Trudy Po Prikladnoj Botanike Genetike i Selekcii 62:3–35

Cadima-Fuentes X, Van Zonneveld M, Scheldeman MX, Castañeda N, Patiño F, Beltrán M, Van Damme P (2013) Endemic wild potato (*Solanum* spp.) biodiversity status in Bolivia: reasons for conservation concerns. J Nat Conserv 22:113–131

Cadima-Fuentes X (2014) Conserving the genetic diversity of Bolivian wild potatoes Doctoral dissertation, PhD thesis, Wageningen University, Wageningen. http://library.wur.nl/WebQuery/clc/2075456

Cai D, Rodríguez F, Teng Y, Ané C, Bonierbale M, Mueller LA, Spooner DM (2012) Single copy nuclear gene analysis of polyploidy in wild potatoes (*Solanum* section *Petota*). BMC Evol Biol 12:70

Camadro EL, Peloquin SJ (1981) Cross-incompatibility between two sympatric polyploid *Solanum* species. Theor Appl Genet 60:65–70

Camadro EL, Masuelli RW (1995) A genetic model for the endosperm balance number (EBN) in the wild potato *Solanum acaule* Bitt. and two related diploid species. Sex Plant Reprod 8:283–288

Camadro EL, Carputo D, Peloquin SJ (2004) Substitutes for genome differentiation in tuber-bearing *Solanum*: interspecific pollen-pistil incompatibility, nuclear-cytoplasmic male sterility, and endosperm. Theor Appl Genet 109:1369–1376

Camadro EL (2012) Relevance of the genetic structure of natural populations, and sampling and classification approaches for conservation and use of wild crop relatives: potato as an example. Botany 90:1065–1072

Cárdenas M, Hawkes JG (1946) New or little-known wild potato species from Bolivia and Peru. J Linn Soc Bot 53:91–108

Carputo D, Frusciante L (2011) Classical genetics and traditional breeding. In: Bradeen J, Kole C (eds) Genetics, genomics and breeding of potato. Science Publishers, pp 20–40

Carputo D, Parisi M, Consiglio F, Iovene M, Caruso G, Monti L, Frusciante L (2003a) Aneuploid hybrids from 5x–4x crosses in potato: chromosome number, fertility, morphology and yield. Am J Potato Res 80:93–101

Carputo D, Frusciante L, Peloquin SJ (2003b) The role of 2n gametes and endosperm balance number in the origin and evolution of polyploids in the tuber-bearing Solanums. Genetics 163:287–294

Castañeda-Álvarez NP, de Haan S, Juárez H, Khoury CK, Achicanoy HA, Sosa CC, Bernau V, Salas A, Heider B, Simon R, Maxted N, Spooner DM (2015) *Ex situ* conservation priorities for the wild relatives of potato (*Solanum* L. Section *Petota*). PLos One 10(14): e0122599

CGIAR (2020) Crop genebank knowledge base. Strengthening capacity to manage genebanks. https://cropgenebank.sgrp.cgiar.org/index.php/procedures-mainmenu-242

CIP (2015) Catálogo de variedades de la papa andina de Chugay, La Libertad, Perú Centro Internacional de la papa, Asociación Pataz, Instituto Nacional de Innovación Agraria

CIP (2019) https://cipotato.org/genebankcip/. Accessed 15 Mar 2019

Cipar MS, Peloquin SJ, Hougas RW (1964) Variability in the expression of self-incompatibility in tuber-bearing diploid *Solanum* species. Am Potato J 41:155–162

Clausen AM, Ispizúa VN, Digilio A (2010) Native andean potato varieties in Argentina: conservation and evaluation of an endangered genetic resource. Am J Plant Sci Biotechnol. Plant Sci Biotechnol South America: Focus Argentina I. Vol. 3 Special Issue 1:72–82

Clausen AM, Peralta IE, Spooner DM (2013) Grupo VIII. Potato. In: Anton AM, Zuloaga FO (eds) Flora Argentina (Flora Vascular de la República Argentina) Vol 13, pp 264–289

Clausen AM, Ispizua VN, Atencio HM, Calandroni M, Digilio A (2018) Especies silvestres de papa (*Solanum* sect. *Petota* y sect. *Etuberosum*) identificadas en áreas protegidas de la Argentina. Bol Soc Argent Bot 53:67–75

Clusius C (1601) Rariorum plantarum historia: quae accesserint, proxima pagina docebit. Ioannem Moretum, Antwerp

Contreras A (1987) Germoplasma chileno de papas (*Solanum* spp.). Anales Simposio Recursos Fitogenéticos, Valdivia, 1984. Universidad Austral de Chile, Valdivia: International Board of Plant Genetic Resources, pp 43–75

Contreras A, Spooner DM (1999) Revision of *Solanum* section *Etuberosum*. In: Nee M, Symon DE, Lester RN, Jessop JP (eds) Solanaceae IV advances in biology and utilization. Royal Botanic Gardens, Kew, pp 227–245

Correll DS (1962) The potato and its wild relatives contributions from the Texas Research Foundation. Bot Stud 4:1–606

Cracraft J (1989) Speciation and its ontology: the empirical consequences of alternative species concepts for understanding patterns and processes of differentiation. In: Otte D, Endler JA (eds) Speciation and its consequences, a view for evolutionary biology and ecology, biology and philosophy, vol 2, pp 415–434

De Candolle AP (1882) Origine des plantes cultivées

De Haan S, Rodríguez F (2016a) Potato origin and production. In: Singh J, Kaur L (eds) Advances in potato chemistry and technology. Academic Press, pp 1–32

De Haan S, Polreich S, Rodríguez F, Juarez H, Plasencia F, Ccanto R, Alvarez C, Otondo A, Sainz H, Venegas C, Kalazich J (2016b) A long-term systematic monitoring framework for on-farm conserved potato landrace diversity. In: Maxted N, Ehsan Dulloo M, Ford-Lloyd BV (eds) Enhancing crop genepool use: Capturing wild relative and landrace diversity for crop improvement. CAB International, Oxfordshire, pp 289–296

De Haan S, Rodríguez F, Becerra LA, Polreich S, Scurrah M, Nuñez J, Juarez H, Plasencia F, Bernardo L, Meza K (2016c) Conservation dynamics of roots and tuber crops under on-farm management. Indian J Plant Genet Resour 29:289–291

De Haan S, Burgos G, Rodríguez F, Creed H, Liria M, Bonierbale M (2019) The nutritional role of potato varietal diversity in Andean food systems: a case study. Am J Potato Res. https://doi.org/10.1007/s12230-018-09707-2

Devaux A, Ordinola M, Horton D (eds) (2011) Innovation for development: the Papa Andina experience. International Potato Center (CIP)

Devaux A, Goffart JP, Petsakos A, Kromann P, Gatto M, Okello J, Suarez V, Hareau G (2020) Global food security, contributions from sustainable potato agri-food systems. In: Campos H, Ortiz O (eds) The potato crop, Chap. 1, pp 3–35. https://doi.org/10.1007/978-3-030-28683-5_1

Digilio A, Molina-García AD, Deladino L, Schneider Teixeira A (2018) Effective cryopreservation approach for the Andean potato shoot tip *in vitro* culture. Abstracts/cryobiology 85:172

Dodds PN, Clarke AE, Newbigin E (1996) A molecular perspective on pollination in flowering plants. Cell 85:141–144

Dulloo ME, Labokas J, Iriondo JM, Maxted N, Lane A, Laguna E, Jarvis A, Kell SP (2008) Genetic reserve location and design. In: Iriondo JM, Dulloo E, Maxted N (eds) Conserving plant genetic diversity in protected areas. CAB International Publishing, Wallingford, pp 23–64

Ehlenfeldt MK, Hanneman RE Jr (1988) Genetic control of endosperm balance number (EBN): three additive loci in a threshold-like system. Theor Appl Genet 75:825–832

Ellis D, Chavez O, Coombs J, Soto J, Gomez R, Douches D, Panta A, Silvestre R, Anglin NL (2018) Genetic identity in genebanks: application of the SolCap 12K SNP array in fingerprinting the global in trust potato collection. Genome 61:523–537

Ellis D, Salas A, Chavez O, Gomez R, Anglin N (2020) *Ex Situ* conservation of potato [*Solanum* Section *Petota* (Solanaceae)] genetic resources in genebanks. In: Campos H, Ortiz O (eds) The potato crop, Chap. 4, pp 109–138. https://doi.org/10.1007/978-3-030-28683-5_4

Engel F (1970) Exploration of the Chilca Canyon, Peru. Curr Anthropol 11:55–58

Engels JMM, Visser L (eds) (2003) A guide to effective management of germplasm collections. IPGRI. Handbooks for Genebanks No 6. IPGRI, Rome, Italy

Erazzú LE, Camadro EL, Clausen AM (2009) Persistence over time, overlapping distribution and molecular indications of interspecific hybridization in wild potato populations of Northwest Argentina. Euphytica 168:249–262

Fajardo D, Spooner DM (2011) Phylogenetic relationships of *Solanum* series Conicibaccata and related species in *Solanum* section *Petota* inferred from five conserved ortholog sequences. Syst Bot 36:163–170

FAO (2009) International treaty on plant genetic resources for food and agriculture. http://www.fao.org/3/a-i0510e.pdf

FAO (2010) The second report on the state of the world's plant genetic resources for food and agriculture. FAO, Rome

FAO (2014) Genebank standards for plant genetic resources for food and agriculture, rev. FAO, Rome

FAO-IPGRI Technical Guidelines for the Safe Movement of Germplasm. https://cropgenebank.sgrp.cgiar.org/images/file/learning_space/potato_tech_guid_safe_move_germplasm.pdf

Fonseca C, Burgos G, Rodríguez F, Muñoa L, Ordinola M (2014) Catálogo de variedades de papa nativa con potencial para la seguridad alimentaria y nutricional de Apurímac y Huancavelica. Lima: Centro Internacional de la Papa

Frankel R, Galun E (1977) Pollination mechanisms, reproduction and plant breeding. Springer, Berlin Heidelberg New York

Friedman WE (1998) The evolution of double fertilization and endosperm: an "historical" perspective. Sex Plant Reprod 11:6–16

Garavano ME (2018) Estudio de *Solanum commersonii Dunal* en un ecosistema serrano del Sistema de Tandilia (Buenos Aires) para implementar su conservación *in situ.* MSc thesis, Universidad Nacional University de Mar del Plata, Argentina, library. http://intrabalc.inta.gob.ar/dbtw-wpd/advanced.htm

Gavrilenko T (2007) Potato cytogenetics. In: Vruegdenhil D (ed) Potato Biology and biotechnology: advances and perspectives. Elsevier, Amsterdam, pp 203–206

Gavrilenko T (2011) Application of molecular cytogenetics in fundamental and applied research of potato. In: Bradeen J, Kole C (eds) Genetics, Genomics and breeding of potato. Science Publishers, Enfield, pp 184–206

Gebhardt C, Ritter E, Barone A, Debener T, Walkemeier MN, Ganal MW, Tanksley SD, Salamini F (1991) RFLP maps of potato and their alignment with the homoeologous tomato genome. Theor Appl Genet 83:49–57

Gerard J (1597) The herball or generall historie of plantes (1st ed.). London

Ghislain M, Andrade D, Rodríguez F, Hijmans RJ, Spooner DM (2006) Genetic analysis of the cultivated potato *Solanum tuberosum* L. *Phureja* Group using RAPDs and nuclear SSRs. Theor Appl Genet 113 (8):1515–1527

Ghislain M, Douches DS (2020) The genes and genomes of the potato. In: Campos H, Ortiz O (eds) The potato crop, Chap. 5, pp 139–162. https://doi.org/10.1007/978-3-030-28683-5

Goldberg EE, Kohn JR, Lande R, Robertson KA, Smith SA, Igic B (2010) Species selection maintains self-Incompatibility. Science 330:493–495

Gorbatenko LE (1989) Systematic conspectus of section *Petota* Dumort of the genus *Solanum* L. in South America (in Russian). Trudy Prik Bot 126:92–108

Gorbatenko LE (2006) Potato species of South America: ecology, geography, introduction, taxonomy, and breeding value. Russian Academy of Agricultural Sciences, State Scientific Centre of the Russian Federation, St. Petersburg

G2PSOL: http://www.g2psol.eu/image/users/432653/ftp/my_files/GPSOL_Leaflet.pdf?id=30388884

Grun P (1990) The evolution of cultivated potatoes. Econ Bot 44(3):39–55

Grun P (1973) Cytoplasmic sterilities that separate the Group Tuberosum cultivated potato from its putative tetraploid ancestor. Evolution 27:633–643

Hanneman RE Jr, Peloquin SJ (1981) Genetic-cytoplasmic male sterility in progeny of 4x–2x crosses in cultivated potatoes. Theor Appl Genet 59:53–55

Hanneman RE Jr (1994) Assignment of endosperm balance numbers to the tuber-bearing Solanums and their close non-tuber-bearing relatives. Euphytica 74:19–25

Hanneman RE Jr (1999) The reproductive biology of the potato and its implications for breeding. Potato Res 42:283–312

Hardigan MA, Parker F, Laimbeer E, Newton L, Crisovan E, Hamilton JP, Vaillancourt B, Wiegert-Rininger K, Wood JC, Douches DS, Farré EM, Veilleux R, Buella CR (2017) Genome diversity of tuber-bearing Solanum uncovers complex evolutionary history and targets of domestication in the cultivated potato. Proc Nat Acad Sci USA 114(46):E9999–E10008. https://doi.org/10.1073/pnas.1714380114

Harlan J, De Wet J (1975) On Ö Winge and a prayer: the origins of polyploidy. Bot Rev 41:361–390

Hawkes JG (1944) Potato collecting expeditions in Mexico and South America. II. Systematic classification of the collections. Imperial Bureau of Plant Breeding and Genetics, Aberystwyth, pp 1–142

Hawkes JG (1962) Introgression in certain wild potato species. Euphytica 11:752–757

Hawkes JG (1990) The potato: evolution, biodiversity and genetic resources. Belhaven Press, London

Hawkes JG (2004) Hunting the wild potato in the South American Andes. Memories of the British Empire potato collecting expedition to South America 1938–1939. Nijmegen: Botanical and Experimental Garden, University of Nijmegen

Hawkes JG, Hjerting JP (1969) The potatoes of Argentina, Brazil, Paraguay and Uruguay. A biosystematic study. Oxford University Press, Oxford

Hawkes JG, Hjerting JP (1989) The potatoes of Bolivia, their breeding value and evolutionary relationships. Oxford University Press, Oxford, UK

Hawkes JG, Francisco-Ortega J (1993) The early history of the potato in Europe. Euphytica 70:1–7

Hermundstad SA, Peloquin SJ (1985) Germplasm enhancement with potato haploids. J Hered 76:463–467

Hijmans RJ, Spooner DM (2001) Geography of wild potato species. Am J Bot 88:2101–2112

Hijmans RJ, Spooner DM, Salas AR, Guarino L, De la Cruz J (2002) Atlas of wild potatoes. Systematic and ecogeographic studies on crop genepools. International Plant Genetic Resources Institute, Rome

Hijmans RJ, Gavrilenko T, Stephenson S, Bamberg J, Salas A, Spooner, (2007) Geographic and environmental range expansion through polyploidy in wild potatoes (*Solanum* section *Petota*). Global Ecol Biogeogr 16:485–495

Hogenboom NG (1973) A model for incongruity in intimate partner relationships. Euphytica 22:219–233

Hogenboom NG (1979) Incompatibility and incongruity in *Lycopersicon*. In: Hawkes JG, Lester RN, Skelding AD (eds) The biology and taxonomy of the Solanaceae. Academic Press, London, pp 435–444

Holle M (1988) Seed conservation of potato genetic resources – IBPGR standards: theoretical ideals and practical reality. Report of the 29th planning conference on strategies for the conservation of potato genetic resources, CIP, Lima, pp 115–128

Hosaka K, Hanneman RE Jr (1998) Genetics of self-compatibility in a self-incompatible wild diploid potato species *Solanum chacoense*.2. Localization of an S locus inhibitor (Sli) gene on the potato genome using DNA markers. Euphytica 103:265–271

Huamán Z, Spooner DM (2002) Reclassification of landrace populations of cultivated potatoes (*Solanum* sect. *Petota*). Amer J Bot 89:947–965

INIA (2012) Catálogo de nuevas variedades de papa: sabores y colores para el gusto peruano. INIA, CIP, Red LatinPapa

INIAF, VDRA and MDRyT (2014) Catálogo de accesiones de papa *Solanum tuberosum* subsp. *Andigenum* (Juz. & Bukasov) Hawkes del Banco de Germoplasma de tubérculos y raíces de Bolivia

ICNCP, International Code of Nomenclature for Cultivated Plants Ninth Edition (2016) Scripta Horticulturae 18

Iorizzo M, Gao L, Mann H, Traini A, Chiusano ML, Kilian A, Aversano R, Carputo D, Bradeen JM (2014) A DArT marker-based linkage map for wild potato *Solanum bulbocastanum* facilitates structural comparisons between *Solanum* A and B genomes. BMC Genet 15:123

Iovene M, Savarese S, Cardi T, Frusciante L, Scotti N, Simon P, Carputo D (2007) Nuclear and cytoplasmic genome composition of *Solanum bulbocastanum* (+) *S. tuberosum* somatic hybrids. Genome 50:443–450

Ispizúa VN, Camadro EL, Clausen AM (2015) Variation patterns among natural populations of wild potatoes at Inca Cueva (Jujuy, Argentina). Genet Resour Crop Evol 62:235–253. https://doi.org/10.1007/s10722-014-0149-7

Iwanaga M, Ortiz R, Cipar MS, Peloquin SJ (1991) A restorer gene for genetic-cytoplasmic male sterility in cultivated potatoes. Am Potato J 68:19–28

Jansky SH, Dempewolf H, Camadro EL, Simon R, Zimnoch-Guzowska E, Bisognin DA, Bonierbale M (2013) A case for crop wild relative preservation and use in potato. Crop Sci 53:746–754

Jarvis A, Lane A, Hijsmans RJ (2008) The effect of climate change on crop wild relatives. Agric Eco Syst Environm 126:13–23

JICA (2016) Potato seed tuber production techniques manual. https://www.jica.go.jp/nepal/english/office/others/c8h0vm0000bjww96-att/tm_4.pdf

Johnston SA, Hanneman RE Jr (1980) Support of the endosperm balance number hypothesis utilizing some tuber-bearing *Solanum* species. Am Potato J 57:7–14

Johnston SA, Hanneman RE Jr (1982) Manipulations of endosperm balance number overcome crossing barriers between diploid *Solanum* species. Science 217:446–448

Johnston SA, Hanneman RE Jr (1996) Genetic control of endosperm balance number (EBN) in the Solanaceae based on trisomic and mutation analysis. Genome 39:314–321

Johnston SA, den Nijs TM, Peloquin SJ, Hanneman RE Jr (1980) The significance of genic balance to endosperm development in interspecific crosses. Theor Appl Genet 57:5–9

Juzepczuk SW (1937) New species of the genus *Solanum* L. in the group *Tuberarium* Dun. Izvestiya Akademii Nauk SSSR 2:295–331

Juzepczuk SW, Bukasov SM (1929) A contribution to the question of the origin of the potato. Trudy Vsesoyuznogo S"zeda po Genetike i Selektsii Semenovodstvu i Plemennomu Zhivotnovodstvu 3: 593–611

Knapp S (1991) A revision of *Solanum* sessile species group (section *Geminata* pro parte: Solanaceae). Bot J Linn Soc 105:179–210

Knapp S (2000) A revision of *Solanum thelopodium* species group (section *Anthoresis* sensu Sheite, pro parte): Solanaceae. Bull Nat Hist Mus. London (bot) 30:13–30

Knapp S (2002) *Solanum* section *Geminata* (G.Don) Walpers (Solanaceae). Flora Neotropical Monograph 84:1–45

Knapp S (2008) A revision of *Solanum havanense* species group (section *Geminata* (G.Don)Walp pp) and new taxonomic additions to the *Geminata* clade (*Solanum*: Solanaceae). Ann Mo Bot Gard 95:405–458

Knapp S (2013) A revision of the Dulcamaroid clade of *Solanum* L. (Solanaceae). PhytoKeys 22:1–432

Kohler C, Scheid OM, Erilova A (2009) The impact of the triploid block on the origin and evolution of polyploid plants. Trends Genet 26:142–148

Kozub PC, Ibañez VN, Digilio A, Atencio HM, Garavano ME, Sánchez ME, Marfil CF (2019) Wild potato genetic reserves in protected areas: prospection notes

from Los Cardones National Park, Salta, Argentina. Rev FCA UNCuyo 51:461–474
Kuhl JC (2011) Mapping and tagging of simply inherited traits. In: Bradeen J, Kole C (eds) Genetics, genomics and breeding of potato. Science Publishers, Enfield, pp 90–112
Lamm R (1941) Varying cytological behavior in reciprocal *Solanum* crosses. Hereditas 27:202–208
Lamm R (1953) Investigations on some tuber-bearing *Solanum* hybrids. Hereditas 39:97–112
Lara Cabrera SI, Spooner DM (2004) Taxonomy of North and Central American diploid wild potato (*Solanum* sect. *Petota*) species: AFLP data. Plant Syst Evol 248:129–142
Linnaeus C (1753) Species plantarum, 2 vol. Stockholm
Lewis D (1943) Physiology of incompatibility in plants. III. Autopolyploids. J Genet 45:171–185
Lewis D (1947) Competition and dominance of incompatibility alleles in diploid pollen. Heredity 1:85–108
Louderback LA, Pavlik BM (2017) Starch granule evidence for the earliest potato use in North America. Proc Natl Acad Sci USA 114(29):7606–7610
Loskutov IG (1999) Vavilov and his institute. A history of the world collection of plant genetic resources in Russia. International Plant Genetic Resources Institute, Rome
Luu DT, Qin X, Morse D, Cappadocia M (2000) S-RNase uptake by compatible pollen tubes in gametophytic self-incompatibility. Nature 407:649–651
Mann H, Iorizzo M, Gao L, D'Agostino N, Carputo D, Chiusano ML, Bradeen JM (2011) Molecular linkage maps: strategies, resources and achievements. In: Bradeen J, Kole C (eds) Genetics, genomics and breeding of potato. Science Publishers, Enfield, pp 68–86
Mason AS, Pires JC (2015) Unreduced gametes: Meiotic mishap or evolutionary mechanism? Trends Genet 31:5–10
Marfil CF, Hidalgo V, Masuelli RW (2015) *In situ* conservation of wild potato germplasm in Argentina: example and possibilities. Global Ecol Conserv 3:461–476
Masuelli RW, Camadro EL, Erazzu LE, Bedogni MC, Marfil CF (2009) Homoploid hybridization in the origin and evolution of wild diploid potato species. Plant Syst Evol 277:143–151
Matsubayashi M (1991) Phylogenetic relationships in the potato and its related species. In: Tsuchiya, Gupta PK (eds) Chromosome engineering in plants: genetics, breeding, evolution, part B. Amsterdam, Elsevier, pp 93–118
Maxted N, Hawkes JG, Ford-Lloyd BV, Williams JT (1997) A practical model for *in situ* genetic conservation. In: Maxted N, Hawkes JG, Ford-Lloyd BV, Williams JT (eds) Plant genetic conservation: the *in situ* approach. Chapman & Hall, London, pp 339–367
Mendiburu A, Peloquin SJ (1977) Bilateral sexual polyploidization in potatoes. Euphytica 26:573–583
Miller JT, Spooner DM (1999) Collapse of species boundaries in the *Solanum brevicaule* complex: molecular data. Plant Syst Evol 214:103–130
MINAM (2019) Línea de base de la diversidad genética de la papa peruana con fines de bioseguridad. Ministerio del Ambiente. Perú
Ministerio de Agricultura y Riego (MINAGRI), Grupo Yanapai, Instituto Nacional de Innovación Agrarias (INIA), Centro Internacional de la papa (2017) Catálogo de variedades de la papa andina del sureste del Departamento de Junín, Perú. Lima, Perú CIP
Muñoz M, Folch C, Rodríguez F, Kalazich J, Orena S, Santos J, Vargas R, Fahrenkrog A, Puga A (2016) Genotype number and allelic diversity overview in the national collection of Chilean potatoes. Potato Res 59:227. https://doi.org/10.1007/s11540-016-9329-5
Muñoz M, Díaz O, Reinún W, Winkler A, Quevedo R (2019) Slow growth *in vitro* culture for conservation of Chilotanum potato germplasm. Chil J Agric Res 79 (1):26–35
den Nijs TPM, Peloquin SJ (1977) 2n gametes in potato species and their function in sexual polyploidization. Euphytica 26:585–600
Ochoa CM (1990) The potatoes of South America: Bolivia. Cambridge Univ. Press, Cambridge
Ochoa CM (1999) Las Papas de Sudamerica: Peru (parte I). Lima, Peru: International Potato Center
Ochoa CM (2001) Las Papas de Sudamérica: Bolivia. Volumen 127 de Travaux de l'Institut français d'études andines, Institut Français d'Études Andines, CIP Lima
Olmstead RG, Palmer JD (1992) A chloroplast DNA phylogeny of the Solanaceae: subfamilial relationships and character evolution. Ann Mo Bot Gard 79:346–360
Olmstead RG, Palmer JD (1997) Implications for the phylogeny, classification, and biogeography of *Solanum* from cpDNA restriction site variation. Syst Bot 22:19–29
Olmstead RG, Sweere JA, Spangler RE, Bohs L, Palmer JD (1999) Phylogeny and provisional classification of the Solanaceae based on chloroplast DNA. In: Nee M, Symon DE, Lester RN, Jessop JP (eds) Solanaceae IV: advances in biology and utilization. Royal Botanic Gardens, Kew, pp 111–137
Ortiz R (2020) Genomic-led potato breeding for increasing genetic gains: achievements and outlook. Crop Breed Genet Genom. 2:e200010. https://doi.org/10.20900/cbgg20200010
Ovchinnikova A, Krylova E, Gavrilenko T, Smekalova T, Zhuk M, Knapp S, Spooner DM (2011) Taxonomy of cultivated potatoes (*Solanum* section *Petota*: Solanaceae). Bot J Linn Soc 165:107–155
Panta A, Panis B, Ynouye C, Swennen R, Roca W (2014) Development of a PVS2 droplet vitrification method for potato cryopreservation. Cryo Letters 35:255–266
Panta A, Panis B, Ynouye C, Swennen R, Roca W, Tay D, Ellis D (2015) Improved cryopreservation method for the long-term conservation of the world potato germplasm collection. Plant Cell Tis Org Cult 120:117–125
Parodi LR (1966) La Agricultura aborigen argentina. Eudeba, Buenos Aires

Parque de la Papa, Cusco, Perú. https://parquedelapapa.org/

Peloquin SJ, Yerk GL, Werner JE (1989) Ploidy manipulations in potato. In: Adolph KW (ed) Chromosomes: Eukaryotic, prokaryotic and viral. CRC Press, Bosa Raton, FL, pp 167–178

Peloquin SJ, Yerk GL, Werner JE, Darmo E (1989) Potato breeding with haploids and 2n gametes. Genome 32:1000–1004

Peloquin SJ, Gabert AC, Ortiz R (1996) Nature of 'pollinator' effect in potato (*Solanum tuberosum* L.) haploid production. Ann Bot 77:539–542

Peloquin SJ, Boiteux LS, Simon PW, Jansky SH (2008) A chromosome-specific estimate of transmission of heterozygosity by 2n gametes in potato. J Hered 99:177–181

Pendinen G, Gavrilenko T, Jiang J, Spooner DM (2008) Allopolyploid speciation of the tetraploid Mexican potato species *S. stoloniferum* and *S. hjertingii* revealed by genomic *in situ* hybridization. Genome 51:714–720

Pendinen G, Spooner DM, Jiang J, Gavrilenko T (2012) Genomic *in situ* hybridization (GISH) reveals both autopolyploid and allopolyploid origins of different North and Central American hexaploid potato (*Solanum* section *Petota*) species. Genome 55:407–415

Peralta IE, Spooner DM (2001) Granule-bound starch synthase (GBSSI) gene phylogeny of wild tomatoes (*Solanum* L. section *Lycopersicon* [Mill.] Wettst. subsection *Lycopersicon*). Amer J Bot 88:1888–1902

Peralta IE, Spooner DM, Knapp S (2008) The taxonomy of tomatoes: a revision of wild tomatoes (*Solanum* section *Lycopersicon*) and their outgroup relatives in sections *Juglandifolia* and *Lycopersicoides*. Syst Bot Monogr 84:1–186 + 3 plates

Popenoe H, King SR, León J, Kalinowski LS (1990) Lost crops of the Incas. Little known plants of the Andes with promise for worldwide cultivation. In: Vietmeyer ND (ed) The National Academies Press, Washington, DC

Potato Genome Sequencing Consortium (2011) Genome sequence and analysis of the tuber crop potato. Nature 475:189–195

PRODERN (2018) La papa nativa en Apurímac, identificación participativa de variedades en los distritos de Huayana y Pomococha, Perú

Quinn A, Mok D, Peloquin SJ (1974) Distribution and significance of diplandroids among the diploid Solanums. Am Pot J 51:16–21

Rabinowitz D, Linder CR, Ortega R, Begazo D, Murguia H, Douches DS, Quiros CF (1990) High levels of interspecific hybridization between *Solanum sparsipilum* and *S. stenotomum* in experimental plots in the Andes. Am Pot J 67:73–81

Ríos D, Ghislain M, Rodriguez F, Spooner DM (2007) What is the origin of the European potato? Evidence from Canary Island landraces. Crop Sci 47:127–1280

Rivard SR, Cappadocia M, Landry BS (1996) A comparison of RFLP maps based on anther culture derived, selfed, and hybrid progenies of *Solanum chacoense*. Genome 39:611–621

Roberts EH, Ellis RH (1984) The implication of the deterioration of orthodox seeds during storage for genetic resources conservation. In: Holden JHW, Williams JT (eds) Crop genetic resources. George Allen and Unwin, London, pp 18–37

Rodríguez A, Spooner DM (1997) Chloroplast DNA analysis of *Solanum bulbocastanum* and *S. cardiophyllum*, and evidence for the distinctiveness of *S. cardiophyllum* subsp. *ehrenbergii* (sect. *Petota*). Syst Bot 22:31–43

Rodríguez F, Spooner DM (2009) Nitrate reductase phylogeny of potato (*Solanum* sect. *Petota*) genomes with emphasis on the origins of the polyploid species. Syst Bot 34:207–219

Rodríguez F, Wu F, Ané C, Tanksley S, Spooner DM (2009) Do potatoes and tomatoes have a single evolutionary history, and what proportion of the genome supports this history? BMC Evol Biol 9:191. https://doi.org/10.1186/1471-2148-9-191

Rodríguez F, Ghislain M, Clausen AM, Jansky SH, Spooner DM (2010) Hybrid origins of cultivated potatoes. Theor Appl Genet 121:1187–1198

Rodríguez F, Núñez J, Vowinkel A, Sanseverino W, Simon R, Spooner DM, Bonierbale M (2017) Insights on the origin of cultivated potatoes. Poster presented at The Plant and Animal Genome XXV Conference (PAG), San Diego. 14–18 January 2017

Romans A (2005) The potato book. Frances Lincoln, London

Ross H (1986) Potato breeding, problems and perspectives. Advances in Plant Breeding, Suppl. 13. Berlin and Hamberg: Paul Parey

Rumold CU, Aldenderfer MS (2016) Late archaic–early formative period microbotanical evidence for potato at Jiskairumoko in the Titicaca Basin of southern Peru. Proc Natl Acad Sci USA 13(48):13672–13677

Rybin VA (1929) Karyological investigation on some wild growing and indigenous cultivated potatoes of America. Trudy Po Prikladnoi Botanike, Genetike i Selektsii 20:655–720

Rybin VA (1933) Cytological investigation of the South American cultivated and wild potatoes, and its significance for plant breeding. Trudy po Prikladnoi Botanike, Genetike i Selektsii Ser. 2. Genet. Rast 2:3–100

Salaman RN (1937) The potato in its early home and its introduction into Europe. J Roy Hort Soc 62: p 61–67, 112–113, 156–162, 253–266

Salaman RN, Hawkes JG (1949) The character of the early European potato. Proc Linn Soc London 161:71–84

Salas A, Gaspar O, Rodríguez W, Vargas M, Centeno R, Tay D (2008). Guías para la regeneración de germoplasma: especies de papa silvestre. In: Dulloo ME, Thormann I, Jorge MA, Hanson J (eds) Crop specific regeneration guidelines [CD-ROM]. CGIAR Systemwide Genetic Resource Programme (SGRP), Rome,

Italy. 8 p. https://www.genebanks.org/resources/publications/potato-sp/UNCED, 1992. Convention on Biological Diversity, United Nations Conference on Enviromental and Development, UNCED, Genova

Santini M, Camadro EL, Marcellán ON, Erazzú LE (2000) Agronomic characterization of diploid hybrid families derived from crosses between haploids of the common potato and three wild Argentinean tuber-bearing species. Am J Potato Res 77:211–218

Särkinen T, Bohs L, Olmstead RG, Knapp (2013) A phylogenetic framework for evolutionary study of the Nightshades (Solanaceae): A Dated 1000-tip Tree BMC Evol Biol 13:214. https://doi.org/10.1186/1471-2148-13-214.

Simon R, Fuentes AF, Spooner DM (2011) Biogeographic implications of the striking discovery of a 4000 kilometer disjunct population of the wild potato *Solanum morelliforme* in South America. Syst Bot 36:1062–1067

Solanaceae Source website http://www.solanaceaesource.org

Schmiediche PE, Hawkes JG, Ochoa CM (1980) Breeding of the cultivated potato species *Solanum* × *juzepczukii* Buk. and *S.* × *curtilobum* Juz. et Buk. I. A study of the natural variation of *S.* × *juzepczukii*, *S.* × *curtilobum* and their wild progenitor *S. acaule* Bitt. Euphytica 29:685–704

Schmiediche PE, Hawkes JG, Ochoa CM (1982) The breeding of the cultivated potato species *Solanum* × *juzepczukii* and *S.* × *curtilobum*. II. The resynthesis of *S.* × *juzepczukii* and *S.* × *curtilobum*. Euphytica 31:395–707

Sotomayor D, Zorrilla C (2019) Identifying high priority areas for *in situ* conservation of wild relatives of potato. https://doi.org/10.6084/m9.figshare.7549976.v1

SPDA, CCTA, INIA (2015) Los cultivos de la sierra y el cambio climático, siete casos de la sierra centro y sur del Perú. Sociedad Peruana de Derecho Ambiental (SPDA)

Spooner DM, Sytsma KJ, Conti E (1991) Chloroplast DNA evidence for genome differentiation in wild potatoes (*Solanum* sect. *Petota*: Solanaceae). Amer J Bot 78:1354–1366

Spooner DM, Sytsma KJ (1992) Reexamination of series relationships of Mexican and Central American wild potatoes (*Solanum* sect. *Petota*): evidence from chloroplast DNA restriction site variation. Syst Bot 17:432–448

Spooner DM, Van den Berg RG (1992) An analysis of recent taxonomic concepts in wild potatoes (*Solanum* sect. *Petota*). Gen. Res. Crop Evol 39:23–37

Spooner DM, J. Anderson GJ, Jansen RJ, (1993) Chloroplast DNA evidence for the interrelationships of tomatoes, potatoes, and pepinos (Solanaceae). Am J Bot 80(6):676–688

Spooner DM, Castillo R (1997) Reexamination of series relationships of South American wild potatoes (Solanaceae: *Solanum* sect. *Petota*): evidence from chloroplast DNA restriction site variation. Am J Bot 84:671–685

Spooner DM, Hijmans RJ (2001) Potato systematics and germplasm collecting, 1989–2000. Am J Potato Res 78(237–268):395

Spooner DM, Van den Berg RG, Rodríguez A, Bamberg J, Hijmans RJ, Lara-Cabrera SI (2004) Wild potatoes (*Solanum* section *Petota*) of North and Central America. Syst Bot Monogr 68:1–209 + 9 plates

Spooner DM, Hetterscheid WLA (2005) Origins, evolution, and group classification of cultivated potatoes. In: Motley TJ, Zerega N, Cross H (eds) Darwin's harvest: new approaches to the origins, evolution, and conservation of crops. Colombia University Press, New York, pp 285–307

Spooner DM, Nuñez J, Rodríguez F, Naik PS, Ghislain M (2005) Nuclear and chloroplast DNA reassessment of the origin of Indian potato varieties and its implications for the origin of the early European potato. Theor Appl Genet 110:1020–1026

Spooner DM, Salas A (2006) Structure, biosystematics, and genetic resources. In: Gopal J, Khurana SMP (eds) Handbook of potato production, improvement, and post-harvest management. Haworth's Press, Inc., Binghampton, New York

Spooner DM, Clausen AM, Peralta IE (2009). Taxonomic treatment of *Solanum* section *Petota* (wild potatoes). In: Zuloaga FO, Morrone O, Belgrano MJ (eds) Catálogo de plantas vasculares del Cono Sur (Argentina, Chile, Paraguay, Uruguay, y sur del Brasil). Monogr Syst Bot, Mo Bot Gard 107:3011–3053

Spooner DM, Van den Berg RG, Rodríguez A, Bamberg J, Hijmans RJ, Lara-Cabrera SI. (2004) Wild potatoes (*Solanum* section *Petota*) of North and Central America. Syst Bot Monogr 68, 209p + 9 plates

Spooner DM, McLean K, Ramsay G, Waugh R, Bryan GJ (2005) A single domestication for potato based on multilocus amplified fragment length polymorphism genotyping. Proc Natl Acad Sci USA 102:14694–14699

Spooner DM, Núñez J, Trujillo G, del Rosario Herrera M, Guzmán F, Ghislain M (2007) Extensive simple sequence repeat genotyping of potato landraces supports a major reevaluation of their gene pool structure and classification. Proc Natl Acad Sci USA 104:19398–19403

Spooner DM, Rodríguez F, Polgár Z, Ballard HE Jr, Jansky SH (2008) Genomic origins of potato polyploids: GBSSI gene sequencing data. The Plant Genome, a suppl. To Crop Sci 48:S27–S36

Spooner DM, Gavrilenko T, Jansky SH, Ovchinnikova A, Krylova E, Knapp S, Simon R (2010) Ecogeography of ploidy variation in cultivated potato (*Solanum* sect. *Petota*). Am J Bot 97:2049–2060

Spooner DM, Ghislain M, Simon R, Jansky SH, Gavrilenko T (2014) Systematics, diversity, genetics, and evolution of wild and cultivated potatoes. Bot Rev 80:283–383

Spooner DM, Alvarez N, Peralta IE, Clausen AM (2016) Taxonomy of wild potatoes and their relatives in southern South America (*Solanum* sects. *Petota* and *Etuberosum*). Syst Bot Monogr 100:1–240 + 10 plates

Spooner DM, Ruess H, Arbizu C, Rodríguez F, Solis-Lemus C (2018) Greatly reduced phylogenetic structure in the cultivated potato clade of *Solanum* section *Petota*. Am J Bot 105:60–70

Spooner DM, Shelley J, Rodríguez F, Simon R, Ames M, Fajardo D, Castillo R (2019) Taxonomy of Wild Potatoes in Northern South America (*Solanum* section *Petota*). Syst Bot Monogr 108:305p + 5 plates

Stout AB, Chandler C (1941) Change from self-incompatibility to self-compatibility accompanying change from diploidy to tetraploidy. Science 94:118

Stelly D, Peloquin SJ (1986) Formation of 2n megagametophytes in diploid tuber-bearing Solanums. Am J Bot 73:1351–1363

Stelly D, Peloquin SJ, Palmer R, Crane C (1984) Mayer's hemalum-methy salicylate: a stain-clearing technique for observations within whole ovules. Stain Technol 59:155–161

Tepe EJ, Anderson GJ, Spooner DM, Bohs L (2016) Relationships among wild relatives of tomato, potato, and pepino. Taxon 65:262–276

Tucci M, Carputo D, Bile G, Frusciante L (1996) Male fertility and freezing tolerance of hybrids involving *Solanum tuberosum* haploids and diploid *Solanum* species. Potato Res 39:345–353

Ugent D (1968) The potato in Mexico: geography and primitive culture. Econ Bot 22:109–123

Ugent D (1970) The potato: what is the origin of this important crop plant, and how did it first become domesticated? Science 170:1161–1166

Ugent D, Pozorski S, Pozorski T (1982) Archaeological potato tuber remains from the Casma Valley of Peru. Econ Bot 36:182–192

Ugent D, Dillehay T, Ramirez C (1987) Potato remains from a late Pleistocene settlement in southcentral Chile. Econ Bot 41:17–27

UNCED (1992) Convention on Biological Diversity, United Nations Conference on Enviromental and Development, UNCED, Genova

Van den Berg RG, Miller JT, Ugarte ML, Kardolus JP, Villand J, Nienhuis J, Spooner DM (1998) Collapse of morphological species in the wild potato *Solanum brevicaule* complex (sect. *Petota*). Am J Bot 85:92–109

Vavilov NI (1922) The law of homologous series in variation. J Genet 12:47–90

Vavilov NI (1940) The new systematics of cultivated plants. In: Huxley J (ed) The new systematics. Clarendon Press, Oxford, pp 549–566

Vega SE, Bamberg JB (1995) Screening the US potato collection for frost hardiness. Amer. Potato J. 72:13–21

Veilleux R (1985) Diploid and polyploid gametes in crop plants: mechanisms of formation and utilization in plant breeding. Plant Breed Rev 3:253–288

Vollmer R, Villagaray R, Egusquiza V, Espirilla J, Garcia M, Torres A, Ellis D (2016) The potato cryobank at the International Potato Center (CIP): a model for long term conservation of clonal plant genetic resources collections of the future. Cryo Letters 37(5):318–329

Vollmer R, Villagaray R, Cárdenas J, Castro M, Chávez O, Anglin N L, Ellis D (2017). A large-scale viability assessment of the potato cryobank at the International Potato Center (CIP). *In vitro* Cell Dev Biol-Plant 53 (4):309-317

Vollmer R, Villagaray R, Castro M, Anglin NL, Ellis D (2019) Cryopreserved potato shoot tips showed genotype-specific response to sucrose concentration in rewarming solution (RS). Plant Cell Tis Org Cult (PCTOC), pp 1–11

Wang MR, Lambardi M, Engelmann F, Pathirana R, Panis B, Volk GM, Wang QC (2020) Advances in cryopreservation of *in vitro*-derived propagules: technologies and explant sources. Plant Cell Tiss Org Cult, pp 1–14. https://doi.org/10.1007/s11240-020-01770-0

Watanabe K (2015) Potato genetics, genomics, and application. Breed Sci 65:53–68. https://doi.org/10.1270/jsbbs.65.53

Werner J, Peloquin SJ (1991) Occurrence and mechanisms of 2n egg formation in 2x potato. Genome 4:975–982

Whalen MD (1984) Conspectus of species groups in *Solanum* subgenus *Leptostemonu*m. Gentes Herb 12:179–282

Zhang L, Rodriguez F (2015) The potato. In: Beaudri MC (ed) Metheny KB. An Encyclopedia. Rowman & Little field Publishers, The Archaeology of Food, pp 415–418

Zorrilla C, Sotomayor D, Gómez R, Salas A, Vergara P, Ellis D (2019a) Ensuring the long-term conservation of wild relatives of potato in Peru. https://doi.org/10.6084/m9.figshare.7549922.v1

Zorrilla C, Salas A, Roca W, Tay D (2019b) Changes in the genetic structure of seed populations in six South American wild potato species as a consequence of seed multiplication at CIP Genebank. https://doi.org/10.6084/m9.figshare.7844714.v1

On the Value of Wild *Solanum* Species for Improved Crop Disease Resistance: Resistances to Nematodes and Viruses

5

James M. Bradeen

Abstract

The genus *Solanum* includes important crop species and numerous wild species. The wild *Solanum* species are a genetic diversity treasure trove useful for the improvement of crops, with disease resistance being of critical importance. In this chapter, I review the status of research and breeding efforts for improved nematode and virus resistance in *Solanum* crops, especially potato and tomato. In total, 33 disease resistance genes are described; virtually all originate from wild relatives of potato and tomato. This observation underscores the utility of wild *Solanum* species and the need to prioritize their conservation through in situ and ex situ approaches. Trends in research well positioned to impact trait discovery in wild *Solanum* species and introgression into crop species are outlined. Included are the potential to mine genebank collections for novel disease resistance alleles using target DNA sequencing approaches, visualization of deep evolutionary and allele diversification patterns across the Solanaceae, and streamlined gene mapping and cloning methodologies. The potential impacts of Marker Assisted Breeding, genetic transformation, gene-editing, and conversion of cultivated potato to a diploid species are explored. A growing world population and changing global climate that requires crop plants to tolerate increasingly chaotic production environments present the need for urgent investment in the genetic improvement of crops, with crop wild relatives being critical donors of useful traits.

5.1 Introduction

A changing global climate and growing world population present an urgent need to genetically improve crop plants, making them more resilient to chaotic growing conditions and more resistant to pests and diseases. The genetic diversity found in wild relatives of crop plants, whether maintained as native populations or as genebank collections, represents an invaluable resource for crop improvement. Discovery of useful traits in crop wild relatives through 'mining' efforts and their introgression into crop plants is an immediate and critical scientific goal. This chapter outlines progress towards the discovery of useful genetic disease resistance in wild *Solanum* species and the improvement of crop species.

Consistent with global production statistics, research and breeding aimed at improving genetic disease resistance among *Solanum* crops have focused heavily on improvement of potato and

J. M. Bradeen (✉)
Department of Plant Pathology, University of Minnesota, 495 Borlaug Hall/1991 Upper Buford Circle, St. Paul, MN 55108, USA
e-mail: jbradeen@umn.edu

© Springer Nature Switzerland AG 2021

D. Carputo et al. (eds.), *The Wild Solanums Genomes*, Compendium of Plant Genomes,
https://doi.org/10.1007/978-3-030-30343-3_5

tomato, with some emphasis on eggplant and very little effort on Solanums of small scale or regional significance. Wild *Solanum* species have played an unusually important role as donors of genetic disease resistance for crop improvement, underscoring the value of in situ and ex situ conservation of *Solanum* species. Researchers have employed a variety of strategies to speed the identification of useful disease resistance in cultivated or wild *Solanum* germplasm and to improve crops (see also Chap. 12). Identification of genetic resistance, especially in wild *Solanum* species has depended almost solely upon phenotypic screening and currently represents a significant bottleneck in crop improvement. Contemporary approaches leveraging the evolutionary history of disease resistance genes might provide insights to streamline discovery of useful disease resistance and is a key research emphasis in our laboratory. Other researchers are leveraging targeted sequencing of resistance loci and association mapping.

Once useful traits are identified, the corresponding genes are frequently mapped and/or cloned. Molecular mapping entails creation of populations or the establishment of diversity panels segregating for disease resistance followed by the application of various forms of molecular markers to identify marker-trait associations. Markers associated with disease resistance genes may then be converted to forms that are readily applicable to screening large numbers of breeding lines, with the presence/absence or allele state of the marker serving as an initial proxy for enhanced disease resistance upon which a breeder might make selections. Termed "Marker Associated Selection" (MAS) or "Marker Associated Breeding" (MAB), this approach can significantly reduce the length of time required for breeding crops with improved disease resistance and can reduce overall costs by limiting phenotypic testing to smaller advanced breeding populations. Cloning of disease resistance genes offers the potential for crop improvement via genetic transformation, although regulatory hurdles and consumer acceptance continue to limit these approaches despite the significant potential advantages of reduced pesticide usage. Gene-editing approaches may be used to modify existing loci in crop plants to enhance disease resistance. At present, the regulatory and consumer acceptance landscape associated with gene-edited crops is not entirely clear on a global level, but researchers continue to generate needed background knowledge and protocols to make gene-editing accessible.

Here I provide an overview of the current status of genetic disease resistance research and breeding in *Solanum* species. I present comprehensive reviews of important nematode and virus resistance traits as case studies documenting the value of wild *Solanum* species as donors in crop improvement. I also present an assessment of trends and emerging approaches for the discovery of useful traits in wild *Solanum* species and their integration into *Solanum* crops.

5.2 Genetic Resistance Against Nematodes

Fuller et al. (2008) summarize genetic resistance to nematodes in a wide variety of plant species, including *Solanum* crops. Ramakrishnan et al. (2015) provide an overview of molecular markers, including markers linked to nematode resistance, that are utilized in modern potato breeding. Gebhardt and Valkonen (2001) described map locations for numerous nematode resistance genes in *Solanum* crops and crop wild relatives. Williamson (1998) provides an overview of genetic resistance to nematodes in tomato and related wild species. Research on nematode resistance in eggplant has been minimal, but the recent cloning of the NB-LRR gene *SacMi* from the wild *S. aculeatissimum* and its introgression into cultivated eggplant (*S. melongena*) shows potential for the management of *Meloidogyne incognita* (Zhou et al. 2018). Here, genetic resistance of significance or potential significance to improvement of tomato and potato and genetic resistance that has been extensively studied are reviewed.

5.2.1 Tomato Root-Knot Nematodes

At least 34 species of nematodes parasitize tomato (Jones et al. 2014), with root-knot nematodes (*Meloidogyne* spp.) being the most economically important. The most widely deployed source of genetic resistance to tomato root-knot nematodes is the *Mi-1* gene, introgressed into cultivated tomato from the wild *S. peruvianum* (Smith 1944). The gene has proven durable for decades (Jones et al. 2014). The *Mi-1* gene enhances resistance to *M. incognita, M. arenaria,* and *M. javanica*, three of the most important root-knot nematode species. *Mi-1* was initially mapped to tomato chromosome 6 (Klein-Lankhorst et al. 1991; Messeguer et al. 1991) using RFLP and isozyme markers. Efforts to fine map and eventually clone the gene were initially hampered by repression of recombination in areas around the *Mi-1* gene, a pattern consistent with the wide interspecific cross that allowed introgression of *Mi-1* into the cultivated tomato background. However, a combination of brute force (screening nearly 23,000 tomato F2 genotypes) and discovery of recombinants in populations of *S. peruvianum* with the aid of molecular markers associated with the *Mi-1* locus (Kaloshian et al. 1998), ultimately confined the resistance phenotype to a region of less than 65 kb. In a follow-up study, the authors created a YAC- and BAC-based physical map of the *Mi-1* region (Milligan et al. 1998). Sequencing of two overlapping BAC clones revealed a cluster of three NB-LRR genes, one of which was an obvious pseudogene. The two intact genes, candidates for *Mi-1,* were subjected to independent complementation analyses. One candidate gene, dubbed *Mi-1.2,* conferred root-knot nematode (*M. javanica*) resistance to the normally susceptible tomato variety Moneymaker (Milligan et al. 1998).

Mi-1 encodes an NB-LRR protein of 1257 amino acids that lacks a TIR domain (Milligan et al. 1998). The Mi-1 protein triggers a hypersensitive response that either deters feeding or kills nematodes directly (Paulson and Webster 1972). Interestingly, the gene was found to also be effective against the potato aphid (*Macrosiphum euphorbiae*) (Rossi et al. 1998) and sweetpotato whitefly (*Bemisia tabaci*) (Nombela et al. 2003). *Mi-1* is temperature-sensitive with resistance function completely absent above 30 °C (Dropkin 1969), limiting its use in warmer climates. Phenotypic screening for nematode resistance is both costly and time-consuming, especially on a large scale. The utilization of *Mi-1* for improvement of tomato has been significantly advanced using MAS. Williamson et al. (1994) developed *Rex-1,* a CAPS-based, co-dominant PCR marker linked to *Mi-1. Rex-1* was the first DNA marker linked to *Mi-1* and replaced use of the isozyme marker *Aps-1* by breeding programs. *Rex-1* maps approximately 0.9 cM from *Mi-1* (Williamson et al. 1994). Subsequent fine mapping and cloning (Milligan et al. 1998) of *Mi-1* yielded significant opportunity for development of improved markers for this gene. Most recently, Devran et al. (2016) reported the development of a co-dominant KASP marker linked to *Mi-1.*

5.2.2 Potato Root-Knot Nematodes

A broad array of nematode species economically impact potato. Like tomato, root-knot nematodes (*Meloidogyne* spp.) can cause significant damage to potato. While southern root-knot nematode (*M. incognita*) is a problem especially in tropical and subtropical regions, the Columbia root-knot nematode (*M. chitwoodi*) and the northern root-knot nematode (*M. hapla*) are of particular concern in temperate regions (Stevenson et al. 2001) where most worldwide potato production takes place. As is true of tomato, wild relatives of potato appear to be a particularly rich source of genetic resistance to root-knot nematodes. Austin et al. (1993) utilized somatic hybridization to transfer *M. chitwoodi* Race 1 resistance from the wild *S. bulbocastanum* to cultivated potato. Later, Brown et al. (1996) generated a low-resolution RFLP map from a BC2 population derived from the initial somatic hybrids, mapping *M. chitwoodi* Race 1 resistance to chromosome 11. This gene was subsequently dubbed $R_{Mc1(blb)}$ (Brown et al. 1999). In a follow-up study, Zhang et al. (2007) employed a bulked segregant analysis approach to identify six molecular markers that cosegregate with resistance in an F1 *S. bulbocastanum* mapping population. These markers included one AFLP fragment and five sequence-tagged sites with known physical association with NB-LRR genes on chromosome 11 of the wild potato *S. demissum*. The authors concluded that the availability of these markers will speed introgression of $R_{Mc1(blb)}$ into commercially available potato varieties (Zhang et al. 2007). Janssen et al. (1997) also reported resistance to *M. chitwoodi* Race 1, this time from the wild *S. fendleri*. Segregation ratios in testcross and F2 populations suggested a single resistance gene was responsible for the observed phenotype. The authors dubbed the gene R_{Mc2}, a name that was later changed to $R_{MC1(fen)}$ (Brown et al. 1999) to ensure a consistent nomenclature. Finally, Brown et al. (1999) identified resistance to both *M. chitwoodi* Race 1 and Race 2 in *S. hougasii,* a wild species that can be directly crossed with cultivated potato (Brown et al. 1991). Subsequent segregation analyses (Brown et al. 1999) suggested that resistance to *M. chitwoodi* Race 1 and Race 2 is conditioned by independent genes. Brown et al. (1999) proposed the gene names $R_{Mc1(hou)}$ and $R_{Mc2(hou)}$, respectively. Research on *M. chitwoodi* resistance from *S. bulbocastanum, S. fendleri,* and *S. hougasii* appears to have not proceeded further. The structural basis for the genes underlying root-knot nematode resistance is unknown, and, to date, no prominent potato cultivar with *M. chitwoodi* resistance derived from these species has been released.

This series of studies aimed at improving root-knot nematode resistance in potato underscores the value of leveraging wild crop relatives for improved disease resistance. In each instance, the resistance phenotype appears to be conferred by a single locus, although that has been demonstrated mechanistically in only a few cases. Nevertheless, genetic resistance to root-knot nematodes in wild relatives of potato is a valuable resource that requires careful management to ensure long-term efficacy (Barbary et al. 2015).

5.2.3 Potato Cyst Nematodes

Potato cyst nematodes, including *Globodera rostochiensis* (golden nematode) and *G. pallida* (pale cyst nematode), are a significant threat to potato production, resulting in regulatory and quarantine policies in most countries (Stevenson et al. 2001). Several other *Solanum* spp., including tomato, eggplant, and weedy species, are hosts or potential hosts of potato cyst nematodes, complicating cultural control. Important genetic resistance to *G. rostochiensis* pathotype Ro1 includes the *Gro1* gene family derived from the wild *S. spegazzinnii*; *GroV1* from the wild *S. vernei*; and *H1* from *S. tuberosum* ssp. *andigena. H1* is also effective against *G. rostochiensis* pathotype Ro4. *Solanum spegazzinii* is also the source of other resistance loci including the *G. rostochiensis* resistance genes *Gro1.2* on chromosome 10 and *Gro1.3* on chromosome 11 (Kreike et al. 1993) and the *G. pallida* resistance gene *Gpa*, located on chromosome 5 (Kreike et al. 1994).

Segregation of resistance to *G. rostochiensis* in an F1 population derived from an interspecific hybrid between cultivated potato and a resistant *S. spegazzinnii* genotype suggested that the trait was controlled by a single locus (Barone et al. 1990). Researchers subsequently mapped the locus to chromosome 7 between RFLP markers CP51 and TG20 (Barone et al. 1990). Next, this same research group employed a bulked segregant analysis approach to find tightly linked markers, ultimately localizing the gene to a 100 kb segment (Ballvora et al. 1995). Recognizing commonalities between the cloned disease resistance genes *N* (from tobacco) and *RPS2* (from Arabidopsis), both of which encode NB-LRR proteins, Leister et al. (1996) developed a degenerative PCR approach to amplifying disease resistance gene fragments from genomic DNA of potato. Recovered fragments were employed as RFLP probes, allowing authors to detect fragments that cosegregated with *Gro1*. The authors also document that *Gro1* exists as a physical cluster of sequence-related gene copies (Leister et al. 1996). Later, Paal et al. (2004) used the PCR fragments generated by Leister et al. (1996) to identify homologous genomic clones from a *G. rostochiensis* resistant potato genotype, which were then sequenced revealing a cluster of NB-LRR encoding genes. Subcloning of individual genes and complementation analysis ultimately revealed that gene *Gro1.4* imparted resistance to *G. rostochiensis* pathotype Ro1 (Paal et al. 2004). *Gro1.4,* and other members of the *Gro1* family, encode a TIR-NB-LRR protein. Molecular markers linked to *Gro1* resistance are used for MAS by public potato breeders.

The resistance gene *H1* imparts resistance to *G. rostochiensis* pathotypes Ro1 and Ro4. The gene originated from *S. tuberosum* ssp. *andigena* clone CPC1673 (Ellenby 1952) and it has been widely deployed in potato breeding programs worldwide. Gebhardt et al. (1993) mapped *H1* in a diploid population segregating for *G. rostochiensis* resistance. Four RFLP markers were utilized to locate the gene on chromosome 5. Even at this relatively early phase of plant molecular biology, the authors noted the potential for employing MAS for breeding potato varieties with enhanced nematode resistance (Gebhardt et al. 1993). Subsequently, leveraging a high-density AFLP map for potato bulked segregant analysis, and an F1 diploid potato population segregating for *H1*, Bakker et al. (2004) identified a series of markers tightly (<1 cM) linked to *H1.* Some of these markers were converted to CAPS markers suitable for MAS. Molecular markers linked to *H1* have been tested and further refined by numerous groups [e.g., (Milczarek 2012; Milczarek et al. 2011; Mori et al. 2011; Park et al. 2018)]. Recently, Meiyalaghan et al. (2018) reported the development of two high-resolution DNA melting markers linked to *H1.* This class of marker is co-dominant, differentiating between resistant and susceptible genotypes while also indicating allele dosage.

GroV1 is a single locus that imparts resistance to *G. rostochiensis* pathotype Ro1. Originating from *S. vernei,* this locus was introgressed into the cultivated potato and a diploid F1 mapping population segregating for *GroV1* was used to map the gene to chromosome 5 using RFLP markers (Jacobs et al. 1996). Subsequent bulked segregant analysis using RAPD and derived SCAR markers identified a series of closely associated markers, which the authors postulate are appropriate for MAS applications (Jacobs et al. 1996). *GroV1* maps in close proximity to *H1* and the authors speculate that *GroV1* and *H1* might be alleles of a common locus. Like *H1, GroV1* has yet to be cloned, but markers linked to *GroV1* are available and widely used for MAS in public potato breeding programs.

Solanum tuberosum ssp. *andigena* clone CPC1673, the source of the *H1* gene, is also the source of *Gpa2*. *Gpa2* imparts resistance to *G. pallida* in a population-specific manner (Arntzen et al. 1992). Using a diploid mapping population segregating for *Gpa2* resistance, Rouppe van der Voort et al. (1997) first demonstrated segregation consistent with a single locus and then positioned the gene relative to AFLP and RFLP markers on chromosome 12. The chromosome location clearly distinguished *Gpa2* (chromosome 12) from *H1* (chromosome 5). Interestingly, however, the authors noted a linkage between *Gpa2* and *Rx1,* a gene for PVX resistance that was also thought to derive from *S. tuberosum* ssp. *andigena* clone CPC1673. Because enhanced PVX resistance had long been a target for potato breeders, Rouppe van der Voort et al. (1997) speculated that selection for *Rx1* might have resulted in introgression of *Gpa2* as well. Later, van der Vossen et al. (2000) utilized markers associated with *Gpa2* to generate a BAC-based physical map of the region. The authors identified four NB-LRR encoding genes in the region. One gene corresponded to *Rx1,* which had been previously cloned (Bendahmane et al. 1999). The other three genes were subjected to complementation analyses using in vitro and in vivo approaches. A single gene was identified as *Gpa2* (Vossen et al. 2000). *Gpa2* encodes a non-TIR NB-LRR and the authors note both tight physical linkage and high sequence homology between *Rx1* and *Gpa2.*

Recently, the *G. pallida* pathotype Pa1 resistance gene *H2* was mapped to potato chromosome 5 (Strachan et al. 2019). *H2* derives from the wild potato species *S. multidissectum*

(Dunnett 1961). An F1 population between a *G. pallida* resistant *S. multidissectum* clone and the *G. pallida* susceptible cultivated potato 'Picasso' showed 1:1 segregation for nematode resistance, suggesting a single genomic region is responsible. Strachan et al. (2019) next employed a bulked segregant analysis approach paired with the genome enrichment strategies RenSeq (Witek et al. 2016) and GenSeq (Chen et al. 2018). Through sequence-capture facilitated by DNA 'baits' developed from known genes, RenSeq specifically enriches NB-LRR gene fragments in an experimental DNA sample (e.g., a resistant bulk) while GenSeq employs a parallel approach to enrich for single or low copy genes. In both strategies, the enriched DNA sample is then subjected to sequencing and SNP discovery. Employing the two approaches in independent experiments, Strachan et al. (2019) localized *H2* to the distal short arm of potato chromosome 5. The authors next converted 11 SNPs associated with *H2* into KASP markers, applying the markers to the segregating F1 mapping population. Anchoring the KASP markers to the potato reference genome sequence allowed the authors to define a 4.7 Mb interval that contains *H2* (Strachan et al. 2019).

Although tomato is a potential host to potato cyst nematodes, economic damage is typically minimal. Nevertheless, studies show that resistance genes from tomato species can be important in the control of potato cyst nematodes. *Hero,* derived from the wild tomato *S. pimpinellifolium* (Ellis and Maxon Smith 1971) is a broad-spectrum potato cyst nematode resistance gene, conferring resistance against a large majority of *G. rostochiensis* (Ganal et al. 1995) and *G. pallida* (Ernst et al. 2002) genotypes. Segregation analyses indicated that *Mi-1* and *Hero* are distinct loci (Ellis and Maxon Smith 1971). Indeed, in a molecular analysis of resistant lines *S. pimpinellifolium* LA121 (from which *Hero* was initially identified) and *S. lycopersicum* LA1792 (the cultivated tomato line generated by introgression of *Hero*), and susceptible line *S. lycopersicum* 'Ailsa Craig', a cultivated tomato closely related to LA1792 but lacking the *Hero* gene, Ganal et al. (1995) mapped *Hero* to tomato chromosome 4, in close association with RFLP markers CT229, TG15, and TG370, confining *Hero* to a 12 cM region. Subsequent marker enrichment in the region and fine mapping in an F2 population between 'Ailsa Criag' and LA1792 reduced the genetic size of the *Hero* region to 8.4 cM (Ganal et al. 1995). Additional molecular mapping by Ernst et al. (2002) indicated that *Hero* is part of a cluster of sequence-related gene copies, with homologs identified in both resistant and susceptible lines. Next, the authors generated a contig comprising 19 cosmid clones and spanning 180 kb. Sequence analysis identified 14 NB-LRR genes, 6 of which were obvious pseudogenes. Additional molecular markers derived from sequence analysis allowed the identification of recombinants within the NB-LRR gene cluster, ultimately allowing the authors to define the *Hero* region to just 60 kb, a region harboring 3 intact NB-LRR genes, each a candidate for *Hero* (Ernst et al. 2002). Complementation analyses were employed to eliminate two of the three candidates and to confirm the third as *Hero. Hero* encodes a 1283 amino acid NB-LRR protein that lacks a TIR domain (Ernst et al. 2002).

5.3 Genetic Resistance Against Viruses

Several viruses cause significant economic loss in tomato (Jones et al. 2014) and potato (Valkonen 2007), with dozens more causing minor or localized losses. In a recent review, Hanssen et al. (2010) highlight emerging viral diseases of tomato. Valknonen (2007) provides similar treatment of emerging viral diseases of potato and Palukaitis (2012) provides a review of genetic resistance to an array of potato viruses. Decades of research aimed at crop improvement have resulted in the identification of genetic resistance, often derived from wild *Solanum* species, to many different viruses. However, elucidation of the genetic basis of resistance, resistance gene mapping, and cloning, and the development of molecular markers for MAS have been reserved only for a few economically significant viruses and model systems. This subset of well-studied virus resistance genes is the primary focus of this section.

5.3.1 Alphaflexiviridae

Potato Virus X (PVX) is member of the viral genus *Potexvirus.* Of global distribution, PVX is among the most widespread potato viruses and can cause significant yield losses in potato, especially in combination with other viruses such as Potato Virus A or Potato Virus Y (Stevenson et al. 2001). Spread primarily through mechanical means, planting virus-free seed and careful cultural practices are effective means of control for PVX. Accordingly, identifying and utilizing genetic resistance to PVX appears to not be a significant goal for potato breeders. Nevertheless, the genetic control of resistance to PVX has provided an important model system to understand the function and evolution of plant disease resistance genes.

The PVX resistance gene *Rx1* originated from *S. tuberosum* spp. *andigena* and maps to potato chromosome 12 (Bendahmane et al. 1997). Cloning of *Rx1* entailed the creation of a BAC-based physical map for the genetically-defined region responsible for the resistance phenotype. A single BAC clone, BAC77, was postulated to contain the *Rx1* gene (Kanyuka et al. 1999). Subsequent transient testing in potato and *Nicotiana benthamiana* confirmed that BAC77 contained the *Rx1* gene (Bendahmane et al. 1999). The *Rx1* gene was further physically localized to an 11 kb fragment derived from BAC77 using the same transient expression approach. Sequencing of the fragment and homologous cDNAs revealed that *Rx1* encodes an NB-LRR protein of 937 amino acids (Bendahmane et al. 1999).

Cloning of the *Rx1* gene provided opportunity to analyze resistance gene function and evolution. At a macro-phenotypic level, *Rx1*-mediated resistance lacks a clear hypersensitive response. Instead, *Rx1* induces "extreme resistance", a resistance phenotype that extends beyond the point of infection and lacks the controlled cell death characteristic of the hypersensitive response. In a set of experiments using potato and *N. benthamiana* plants stably expressing *Rx1*, Bendahmane et al. (1999) demonstrated that transient co-expression of the cognate viral coat protein resulted in a clear hypersensitive response at the site of infiltration, leading the authors to conclude that *Rx1* is capable of inducing hypersensitivity. The authors also note, however, that under natural infection the Rx1 protein encounters the cognate coat protein only in a very localized (single cell) fashion. Subsequent experiments demonstrated that the extreme resistance phenotype induced by *Rx1* under natural infection conditions is epistatic to the hypersensitive phenotype induced by the tobacco *N* gene, which confers resistance to Tobacco Mosaic Virus (Bendahmane et al. 1999). In summary, experiments leveraging the cloned *Rx1* gene allowed analysis of molecular mechanisms inducing resistance phenotypes. These observations later proved critical to development of a comprehensive view of NB-LRR gene function at a molecular level.

The cloning of *Rx1* also provided exciting insights into the evolution of NB-LRR genes. Like many other NB-LRR genes, *Rx1* was discovered to reside in a cluster of sequence-related gene copies. As noted above, the cloning of the nematode-resistance gene *Gpa2* entailed creation of a BAC contig that spanned the *Gpa2* gene cluster on potato chromosome 12 (Vossen et al. 2000). In total, this region comprises four related NB-LRR gene copies. While one gene copy was demonstrated to encode Gpa2 (Vossen et al. 2000), the cluster also encompassed the cloned *Rx1* gene. Gpa2 and Rx1 proteins are 88% identical, with differences mainly concentrated in the LRR domain (Vossen et al. 2000). The LRR domain is directly or indirectly involved in identification of the presence of a pathogen through specific protein:protein interactions (Ellis et al. 1999). Accordingly, minor changes in the LRR domains of Gpa2 and Rx1 correlate with detection of very different pathogen types (a nematode vs. a virus, respectively). This observation demonstrated that NB-LRR gene clusters in plant genomes are important in generating small-scale sequence variants that sometimes result in novel resistance protein function. This observation further demonstrates the difficulty of linking NB-LRR proteins to specific pathogen classes based strictly on sequence analyses. These observations have proven to be of general applicability to the NB-LRR superfamily of resistance genes.

The PVX resistance gene *Rx2* originates from *S. acuale* and maps to potato chromosome 5 (Ritter et al. 1991). It was experimentally demonstrated that *Rx1* and *Rx2* recognize the same region of the PVX coat protein and are functionally redundant (Bendahmane et al. 1995; Querci et al. 1995). This observation led Bendahmane et al. (2000) to speculate that *Rx1* and *Rx2,* despite originating from distinct *Solanum* species and mapping to unique genome locations could be related at a DNA sequence level. Using primers developed from the cloned *Rx1* gene and low stringency PCR conditions, the authors generated a library of 200 PCR fragments from *S. acaule.* These were surveyed using a transient expression vector by Agrobacterium infiltration into *N. benthamiana* leaves stably expressing the cognate PVX coat protein. The authors identified two PCR fragments that induced a hypersensitivity response, candidates for *Rx2* (Bendahmane et al. 2000). Using an F1 population derived from potato 'Bzura' (which carries the *R2* gene) and segregating for *R2* resistance, one of the candidate fragments was mapped to both the genetic region conditioning the *Rx2* phenotype and to an associated RFLP marker (Bendahmane et al. 2000). Finally, *N. benthamiana* genotypes stably transformed with the intact *Rx2* candidates were generated and employed to demonstrate the anticipated strain-specific resistance to PVX. The cloned *Rx2* gene encodes and NB-LRR protein of 938 amino acids (Bendahmane et al. 2000).

The successful cloning of *Rx2* and comparison with *Rx1* and related gene copies revealed a common evolutionary origin, despite their distinct genome distributions. The authors further demonstrated sequence exchange between the *Rx1* and *Rx2* clusters, presumably through a process of gene conversion, speculating that this is an important mechanism for the generation of allelic diversity at NB-LRR loci (Bendahmane et al. 2000). These observations proved important to understanding general evolutionary processes that shape the adaptation of NB-LRR genes in response to changing pathogen pressures.

5.3.2 Bunyaviridae

Tomato spotted wilt virus (TSWV) is a type species of the viral genus *Tospovirus.* TSWV has a wide host range and is among important viruses impacting tomato production worldwide (Jones et al. 2014). Genetic resistance to TSWV is considered to be the most effective measure of disease management in tomato and resistance has been identified in wild relatives of tomato. The *Sw5* resistance gene, derived from the wild *S. peruvianum,* has been widely deployed, but resistance-breaking viral strains have emerged (Jones et al. 2014). Stevens et al. (1992) demonstrated that resistance introgressed from *S. peruvianum* segregated as a single gene in a series of experimental populations. The authors further demonstrated that resistance segregating in these populations is distinct from previously described resistances, proposing the gene name *Sw-5* (Stevens et al. 1992). In a follow-up study, Stevens et al. (1995) mapped *Sw-5* to the long arm of tomato chromosome 9 using RAPD and RFLP markers. The authors also speculated on the potential use of molecular markers linked to *Sw-5* in tomato breeding programs (Stevens et al. 1995). Chague et al. (1996) confirmed the chromosome 9 map location of *Sw-5* using a bulked segregant analysis approach. These authors describe four RAPD and SCAR markers linked to *Sw-5* and suitable for use in MAS. YAC and cosmid physical maps of the *Sw5* region were constructed, allowing further refinement of the region encompassing *Sw-5* to a maximum of 100 kb (Brommonschenkel and Tanksley 1997). Working independently, Folkertsma et al. (1999) constructed a BAC-based physical map of the *Sw-5* region. These authors further demonstrated the presence of numerous NB-LRR genes in the cloned region using the degenerate primer PCR approach pioneered by Leister et al. (1996). Subsequent sequencing of a BAC clone that spanned the genetically defined *Sw-5* region revealed two candidate NB-LRR genes (Spassova et al. 2001). Transformation of a TSWV-susceptible *Nicotiana tabacum* genotype with each of the candidate genes independently revealed that candidate gene *Sw-5b* was necessary and sufficient to impart virus resistance. The cloned *Sw-5* gene encodes a 1246 amino acid protein that contains both NB-ARC and LRR domains but lacks a TIR domain (Spassova et al. 2001). de Oliveirra et al. (2018) provide a comprehensive review of the cloning of *Sw-5,* its utility in tomato breeding, gene function, Avr factors, and paralogs and orthologs.

5.3.3 Geminiviridae

Tomato Yellow Leaf Curl Virus (TYLCV) is in the viral genus *Begomovirus.* Vectored by whitefly (*Bemisia tabaci*), TYLCV has also been recently shown to be vertically transmissible through tomato seeds (Kil et al. 2016). This virus infects a wide array of dicot species and causes economically significant disease issues in tomato, especially in tropical and subtropical environments. TYLCV was recently listed among the most economically significant crop diseases in the world (Rybicki 2015). Verlaan (2013) provides a good overview of the significance of TYLCV to tomato production and challenges in disease management. Ji et al. (2007b) review sources of resistance and specific resistance genes that are effective against Begomoviruses infecting tomato, including TYLCV. Yan et al. (2018) provide a comprehensive overview of genetic resistance to TYLCV using an experimental approach. While cultivated tomato is uniformly TYLCV susceptible, genetic resistance to TYLCV has been identified in wild tomato species and utilized by tomato breeding programs.

The TYLCV resistance gene *Ty-1* was introgressed from the wild tomato species *S. chilense.* Used extensively in tomato breeding, introgression of this gene into cultivated tomato seems to trace back to a single accession, LA1969. Numerous studies confirm that *Ty-1* maps to tomato chromosome 6. However, early mapping studies provided contradictory explanations of exact chromosome locale (Perez de Castro et al. 2007; Zamir et al. 1994). This conundrum was later explained with the help of Fluorescence In Situ Hybridization (FISH), which revealed a large chromosome block introgressed from *S. chilense.* This block contains the *Ty-1* gene and encompasses two chromosomal rearrangements relative to the *S. lycopersicum* genome, resulting in local suppression of recombination in the *S. lycopersicum* background. Suppression of recombination in the *Ty-1* region led different research studies to draw different conclusions based on a small number of observed recombinant genotypes (Verlaan et al. 2011). FISH confirmed that *Ty-1* maps to the long arm of tomato chromosome 6, in close proximity to the centromere. FISH also confirmed that the *Ty-1* region overlapped the map location of another TYLCV resistance gene, *Ty-3* (Verlaan et al. 2011). *Ty-3* was introgressed from *S. chilense* accession LA2779 or LA1932 and had been previously mapped to the long arm of chromosome 6 (Ji et al. 2007a). Based on their results, Verlaan et al. (2011) speculated the *Ty-1* and *Ty-3* might be allelic. Later, these authors fine mapped both loci to a 71 kb region. The corresponding region of the sequenced tomato genome encoded five genes, each a candidate for *Ty-1* and *Ty-3* (Verlaan et al. 2013). To test for resistance function of the candidate genes, a Tobacco Rattle Virus (TRV)-based Viral Induced Genome Silencing (VIGS) strategy was deployed. Two VIGS constructs, corresponding to two closely related RNA Dependent RNA Polymerases (RDRs), effectively silenced *Ty-1.* Importantly the same constructs also silenced the function of *Ty-3*, confirming the proposed allelic nature of *Ty-1* and *Ty-3* (Verlaan et al. 2013). *Ty-1* and *Ty-3* encode an RDRγ protein encompassing an atypical DFDGD motif in the catalytic domain. Thus, Ty-1 and Ty-3 belong to a novel class of plant disease resistance proteins (Verlaan et al. 2013). Butterbach et al. (2014) studied the mode of action of Ty-1, documenting that this protein acts to enhance transcriptional gene silencing through innate RNAi defense mechanisms that target viral coat proteins. The result is low viral titer in infected plants and hypermethylation of specific regions of the TYLCV DNA genome (Butterbach et al. 2014).

Ty-2 is a TYLCV resistance gene introgressed from the wild *S. habrochaites* accession B6013. The gene was first mapped to tomato chromosome 11 using RFLP markers (Hanson et al. 2000). Subsequent fine mapping, accomplished by screening 11,000 plants for recombination in the *Ty-2* region delineated by Hanson et al. (2000), allowed Yang et al. (2014) to localize *Ty-2* to a 300 kb genomic region. The authors document significant suppression of recombination in the *Ty-2* region, consist of the interspecific nature of the genetic materials they studied (Yang et al. 2014). Using populations of cultivated tomato segregating for *Ty-2* resistance, Yamaguchi et al. (2018) further localized the gene to a 200 kb genomic region derived from *S. habrochaites.* The region encompassed two NB-LRR genes with tight linkage to the *Ty-2* resistance phenotype. Genome fragments corresponding to one of the candidate genes conferred TYLCV resistance to transgenic tomato and segregation of the fragment correlated with segregation of resistance in the F1 generation, confirming the identity of *Ty-2* (Yamaguchi et al. 2018).

The TYLCV resistance gene *Ty-4* originated from *S. chilense* accession LA1932. The gene has been mapped to the long arm of tomato chromosome 3 (Ji et al. 2009). Ji et al. (2009) developed PCR-based markers for *Ty-4* and reported their utility for MAS. However, *Ty-4* appears to have a comparatively minor resistance impact on TYLCV relative to other known resistance genes (Ji et al. 2009). *Ty-4* has not been cloned to date.

The TYLCV resistance gene *Ty-5* derives from *S. peruvianum.* In a QTL mapping study using cultivated tomato lines segregating for resistance derived from this wild species, *Ty-5* was mapped to tomato chromosome 4 (Anbinder et al. 2009). The region, one of five QTLs identified, had the largest effect, accounting for 39.7–46.6% of observed phenotypic variation. The QTL was associated with PCR marker *SlNAC1.* Interestingly, resistance associated with *SlNAC1* behaved as a partially dominant or as a recessive trait depending on the TYLCV-susceptible parent employed in creation of segregating populations (Anbinder et al. 2009). Later, Hutton et al. (2012) studied the genetic basis of TYLCV resistance derived from the cultivated tomato 'Tyking'. 'Tyking' resistance to TYLCV is of uncertain provenance, with a tomato landrace from the Canary Islands or *S. peruvianum* being putative sources (Hutton et al. 2012). Working in F3 populations carrying 'Tyking' resistance, Hutton et al. (2012) document both a recessive genetic nature of resistance to TYLCV and a strong correlation with marker *SlNAC1* associated with the *Ty-5* locus. These observations led Hutton et al. (2012) to conclude that TYLCV resistance present in 'Tyking' is a likely allele of the *Ty-5* locus reported by Anbinderr et al. (2009). They dubbed the allele *ty-5.* Cloning of *ty-5* was accomplished by Lapidot et al. (2015). By mapping in 51 recombinant populations segregating for *ty-5,* the authors localized the region responsible to a 425 bp genome fragment. This fragment encompassed part of the promoter and part of the coding region of *Pelo,* a gene encoding the messenger RNA surveillance factor Pelota. A single transversion in the coding region resulting in a valine-to-glycine substitution was demonstrated through a series of knockout and overexpression analyses to be responsible for the recessive resistant phenotype (Lapidot et al. 2015). Pelota is involved in the recycling phase of ribosome-driven protein biosynthesis. During this phase, the assembled ribosomal subunits disassociate and become available for a new round of translation initiation. Lapidot et al. (2015) speculate that *ty-5* mediated resistance interferes with viral protein synthesis, resulting in reduced titer.

The TYLCV resistance gene *Ty-6* was introgressed into cultivated tomato from the wild *S. chilense* accession 1938 (Scott and Hutton 2015). The gene has also been found effective against the begomovirus Tomato Mottle Virus (ToMoV) (Gill et al. 2019; Scott and Hutton 2015). Using a series of F2 mapping populations and SNP markers, *Ty-6* has been localized to tomato chromosome 10 (Gill et al. 2019). While several SNP markers associated with the *Ty-6* locus were identified, their ability to distinguish between resistant and susceptible genotypes appears largely population-dependent, limiting their utility for MAS (Gill et al. 2019). *Ty-6* has not been cloned.

5.3.4 Potyviridae

Potato Virus Y (PVY), the type specimen for the genus *Potyvirus,* is globally distributed and among the most economically important potato viruses (Stevenson et al. 2001). Numerous strains have been identified. PVY is transmitted in a rapid, non-persistent manner by numerous aphid species, making control of the vector largely ineffective. Genetic resistance and planting of virus-free certified seed are considered the most important methods for control (Stevenson et al. 2001; Valkonen 2007). PVY genetic resistance has been reported in several *Solanum* species.

The PVY resistance gene Ry_{adg} originates from *S. tuberosum* ssp. *andigena* and confers "extreme resistance" characterized by non-virus strain specificity and low, often non-detectable viral loads. Using a bulked segregant analysis approach, Ry_{adg} was mapped to potato chromosome 11 near homologs of the *N* gene of tobacco. Four RFLP markers linked to Ry_{adg} were identified (Hämäläinen et al. 1997). Using PCR primers developed by Leister et al. (1996), *N*-like gene products that cosegregated with Ry_{adg} were subsequently isolated from PVY-resistant *S. tuberosum* ssp. *andigena* line 2x(v-2)7 (Hämäläinen et al. 1998). Generation of the SCAR marker RYSC3 from *N*-like fragments associated with Ry_{adg} was accomplished by Kasai et al. (2000). Marker RYSC3 has been tested and validated for MAS in several potato breeding programs [e.g., (del Rosario Herrera et al. 2018; Fulladolsa et al. 2015; Nie et al. 2016; Ortega and Lopez-Vizcon 2012; Ottoman et al. 2009; Sagredo et al. 2009)]. To date, Ry_{adg} has not been cloned and the structural organization of the $\mathrm{Ry}_{\mathrm{adg}}$ protein is unknown.

Another PVY resistance gene widely used in breeding programs is Ry_{sto}. This gene derives from the wild *S. stoloniferum.* Although originally thought to map to potato chromosome 11 near Ry_{adg} (Brigneti et al. 1997), Ry_{sto} was later localized to potato chromosome 12 (Song et al. 2005). Using bulked segregant analysis of dihaploids derived through anther culture of a tetraploid potato carrying the Ry_{sto} gene in simplex, Song et al. (2005) first identified AFLP markers linked to the resistance phenotype. Next, the authors surveyed a collection of previously mapped SSR markers, seeking linkage between SSR markers and the AFLP markers associated with Ry_{sto}. A single SSR marker, an STM0003 allele of 111 bp, was associated in tight linkage with the AFLP markers identified through bulked segregant analysis. Finally, the map position was confirmed relative to chromosome 12 RFLP markers (Song et al. 2005). Importantly, AFLP and SSR markers associated with Ry_{sto} were well correlated with phenotypic resistance observed in an array of potato genotypes, while the SCAR marker RYSC3 (associated with Ry_{adg}) showed poor correlation with resistance phenotypes (Song et al. 2005). Later, Song and Schwarzfischer (2008) converted two AFLP markers associated with Ry_{sto} to the STS markers YES3-3A and YES3-3B, demonstrating their utility for MAS, an observation later confirmed by Fulladolsa et al. (2015). Ry_{sto} has not been cloned to date and encodes a protein of unknown structure.

Other forms of extreme resistance with some utility in potato breeding have been reported. Like Ry_{sto}, $Ry\text{-}f_{sto}$ originated from *S. stoloniferum* and maps to potato chromosome 12 (Flis et al. 2005). However, comparison of map locations between $Ry\text{-}f_{sto}$ and Ry_{sto} suggests these are likely two distinct genes (Song et al. 2005). Ry_{chc}, from the wild *S. chacoense,* has been mapped to potato chromosome 9 (Sato et al. 2006). Conferring resistance phenotypes distinct from extreme resistance, two hypersensitivity genes conferring strain-specific resistance to PVY have also been mapped. *Ny-1* maps to potato chromosome 9 (Szajko et al. 2008) and *Ny-2* maps to potato chromosome 11 (Szajko et al. 2014).

5.3.5 Virgaviridae

Tobacco Mosaic Virus (TMV) is the type member of the viral genus *Tobamovirus* and infects many Solanaceaous species. In tomato, TMV can cause yield losses of 20% and reduce fruit quality. Closely related, Tomato Mosaic Virus (ToMV) also infects a wide range of Solanaceous and non-Solanaceous species, causing symptoms on tomato that mimic those of TMV. TMV and ToMV are readily spread by mechanical means and are remarkably stable in the environment. Resistance breeding has been a major goal for management of these viruses and three major resistance genes, *Tm-1* and the alleles *Tm-2* and *Tm-2*^{2} have been described and cloned.

The *Tm-1* resistance locus was introgressed into tomato from the wild *S. habrochaites* PI126445. The gene, which maps to chromosome 2 near the centromere, has been widely used in tomato breeding, but resistance-breaking isolates are now widespread. Fraser and Loughlin (1980) demonstrated that *Tm-1* resistance is gene dosage-dependent, with homozygous *Tm-1/Tm-1* plants displaying higher levels of TMV resistance than heterozygotes, documenting that the *Tm-1* phenotype is semi-dominant. By comparing genome sequences of wild-type and resistance-breaking TMV isolates, Meshi et al. (1988) demonstrated that single base pair changes leading to amino acid substitutions in the viral RNA replicases were sufficient to impart resistance to *Tm-1*. These authors further speculated that the *Tm-1* mode of action entailed direct interaction between the Tm-1 protein and the viral RNA replicase proteins. Earlier, Motoyoshi and Oshima (1977) had demonstrated that *Tm-1* is expressed in tomato protoplasts, eliminating replication of TMV isolates recognized by Tm-1 but having no effect on replication of other tomato-infecting viruses. Yamafuji et al. (1991) later explored co-inoculation of *Tm-1/Tm-1* tomato protoplasts with wild-type TMV and resistance-breaking TMV isolates. These authors demonstrated that Tm-1 specifically inhibited replication of wild-type but not resistant-breaking TMV isolates, consistent with the semi-dominant phenotype described on a whole plant level. Importantly, this observation suggested that the *Tm-1* resistance mechanism differed from the hypersensitivity response associated with many NB-LRR genes, consistent with the suggestion of Meshi et al. (1988) that Tm-1 is a direct inhibitor of viral replication. Building on this emerging line of research, Ishibashi et al. (2007) first demonstrated that an extract from *Tm-1* tomato protoplasts could inhibit the in vitro replication of wild-type TMV and, using column chromatography, isolated a cell fraction that contained the inhibitory element. Sequencing of proteins from the fraction allowed cloning of the corresponding gene. Subsequent tests demonstrated that the cloned gene is *Tm-1*. Consistent with earlier observations and hypotheses, the Tm-1 protein directly or indirectly binds to TMV RNA replicases, impeding function. *Tm-1* is the first reported resistance gene to encode an inhibitor of RNA replication (Meshi et al. 1988).

The *Tm-2* resistance locus encodes the alleles *Tm-2* and *Tm-2*^{2}. The *Tm-2* allele derives from the wild *S. peruvianum* PI126926, while the *Tm-2*^{2} allele derives from the wild *S. peruvianum* accession PI128650. The *Tm-2*^{2} allele has been widely deployed in tomato breeding lines where it has proven durable. Although viral isolates capable of breaking *Tm-2*^{2} resistance have been identified, these display compromised fitness and have not had significant impact on tomato production (Hall 1980; Lanfermeijer et al. 2003). In contrast, the *Tm-2* allele has not been durable, with resistance-breaking viral isolates being widely reported. The *Tm-2* locus was mapped near the chromosome of chromosome 9 (Tanksley et al. 1992). Suppression of recombination near the centromere complicated fine mapping and map-based cloning for the *Tm-2* locus. Lanfermeijer et al. (2003) employed a clever Ac/Ds transposon-tagging approach to disrupt *Tm-2*^{2}resistance function and to recover the mutated allele. Briefly, these authors generated tomato lines hemizygous for *Ds* (integrated on chromosome 9) and the stabilized Activator (sAc) and homozygous for *Tm-2*^{2}. These genotypes were then crossed with a transgenic tomato line homozygous for the ToMV *MP* movement protein gene, which encodes the cognate *Tm-2*^{2} AVR product. The resulting population is expected to segregate for gene expression on the basis of introgression of Ds into coding regions. Because the combination of a functional Tm-2^2 protein and the MP gene product is lethal (Weber and Pfitzner 1998), all surviving seedlings were expected to encode a *Tm-2*^{2}allele disrupted by the insertion of Ds. From approximately 30,000 seeds screened, the authors identified 5 putative Tm-2^2 mutants, all of which were susceptible to ToMV infection (Lanfermeijer et al. 2003). Plasmid rescue and subsequent DNA sequence analysis identified an NB-LRR gene encoding a protein of 861 amino acids. The functionality of the cloned *Tm-2*^{2}gene was confirmed by complementation analysis of susceptible tomato genotypes transformed with the gene fragment under control of a native promoter or the constitutive CaMV promoter (Lanfermeijer et al. 2003). Subsequent to the cloning of the *Tm-2*^{2}allele, the *Tm-2* allele was cloned using PCR primers developed from the *Tm-2*^{2}allele (Lanfermeijer et al. 2005). Like the *Tm-2*^{2} allele, the *Tm-2* allele encodes an NB-LRR protein of 861 amino acids. At a DNA sequence level, the two alleles differ by 7 nucleotides, which result in 4 amino acid changes. The identity of the *Tm-2* allele was confirmed by complementation analyses using viral isolates that differentiated between the *Tm-2*^{2}allele and the *Tm-2* allele (Lanfermeijer et al. 2005).

5.4 Trends and Emerging Approaches

Rapid advances are being made in both the discovery of useful resistance traits in wild relatives of crop plants and their transfer from wild species into cultivated germplasm. In a recent review, Bethke et al. (2017) describe the potential and challenges of using wild species for potato improvement. Here, I outline some of the most significant technological and research advances that I predict will play an increasingly important role in discovery of resistance traits in *Solanum* species and their application to *Solanum* crop improvement.

5.4.1 Resistance Trait Discovery

It has been approximately one decade since the genome sequences of potato (Consortium 2011) and tomato (Consortium 2012) were made publicly available. Generating these genome sequences was a substantial technological feat. Since then, advances in DNA sequencing technology, DNA sequencing strategy, and informatics have made whole-genome sequencing more widely available. In addition to potato and tomato, the genome sequences of eggplant (Hirakawa et al. 2014), the wild potato species *S. commersonii* (Aversano et al. 2015), *S. chacoense* (Leisner et al., 2018) and the wild tomato species *S. pennellii* (Bolger et al. 2014) and *S. pimpinellifolium* (Razaimi et al. 2018) have been publicly released, with genome sequences of additional *Solanum* species currently in preparation (see also Chaps. 7–10).

The availability of these resources has enabled genome-wide survey of disease resistance genes (Andolfo et al. 2013). My laboratory at the University of Minnesota has developed an approach to analysis of NB-LRR genes, predicted from whole-genome sequences, across plant families, revealing deep evolutionary relationships and more contemporary patterns of resistance gene allelic diversification that can be leveraged for trait discovery in wild species. Our concept was first outlined for wild potato species (Quirin et al. 2012) and subsequently refined for analysis of the Rosaceae (van Eck and Bradeen 2019). We are now pursuing a detailed characterization of the Solanaceae, including the genus *Solanum* and more distant relatives such as *Petunia* spp., *Nicotiana* spp., and *Capsicum* spp.

The sequence-capture method RenSeq has been developed for targeted sequencing of the resistance gene component of plant genomes (Witek et al. 2016). First employed to map a late blight resistance gene in the wild *Solanum americanum,* the method is well positioned to enable mining of resistance alleles from wild *Solanum* genotypes, populations, species, or entire genebank collections. The emerging datasets, in turn, will be utilized for understanding how genetic diversity at resistance loci is structured across phylogenetic, environmental, or disease pressure gradients; in what species, populations, or genotypes novel genetic resistance exists; and what species, populations, or genotypes should be targeted for resource-intensive phenotypic discovery of useful resistance traits.

In what is sure to become a groundbreaking discovery in wheat, Arora et al. (2019) combined a RenSeq approach with association mapping approaches, leveraging phenotypic disease resistance data generated across a diversity panel of diploid *Aegilops tauschii*. Using this combination of approaches, the authors reported the cloning of four NB-LRR resistance genes in a period of approximately six months. Application of similar methodologies to *Solanum* species would appear well-positioned to speed the discovery of useful disease resistance traits, particularly for improvement of potato and tomato.

5.4.2 Resistance Trait Application

Resistance breeding is likely to remain an important goal for *Solanum* crops, with wild *Solanum* species remaining important donors of useful genes. Traditional breeding methodologies, including the application of MAS or MAB, will certainly play a prominent role in the introgression of useful resistance traits from wild *Solanum* species.

Biotechnological methodologies, including Agrobacterium transformation, are well developed in *Solanum* crop species and are positioned to play a role in introgression of resistance traits from wild species. Given significant regulatory and consumer acceptance challenges associated with genetic transformation, however, these methods are likely to be utilized on a very limited basis depending on the genetics of the crop and the economic value of the trait. In general, while genetic transformation is tractable in tomato, the self-pollinating, diploid nature of the species and the wide availability of effective breeding strategies for trait introgression in this crop make it a less attractive target for genetic transformation for cultivar development. In contrast, given the autotetraploid nature of cultivated potato, its intolerance of inbreeding, and commercial markets dependent on well-specified and difficult-to-replicate tuber phenotypes, genetic transformation of commercially prominent potato cultivars with disease resistance genes derived from wild species is a logical application. Indeed, despite the associated challenges, genetic transformation has been utilized for improvement of disease resistance traits in commercially available potato (Hameed et al. 2018), and the approach would appear to have significant potential in developing countries facing food security challenges (Ghislain et al. 2019).

Gene-editing is also likely to provide new strategies for improvement of disease resistance in *Solanum* crops. Scheben et al. (2017) summarize efforts to improve crop plants using CRISPR/Cas editing methods. Nadakuduti et al. (2018) and Hameed et al. (2018) provide reviews of gene-editing in potato while Reem and Van Eck (2019) provide a review of gene-editing methodology in tomato. As understanding of pathogen effector targets in the plant cell expands, gene-editing is likely to be leveraged to disrupt 'Susceptibility' targets, thereby enhancing disease resistance. This approach has already been demonstrated in rice (Jiang et al. 2013), wheat (Wang et al. 2014), grape (Malnoy et al. 2016), citrus (Jia et al. 2017) and tomato (Nekrasov et al. 2017). Analyses of NB-LRR allelic diversity across *Solanum* species in my laboratory (*unpublished*) underscores that most loci are present, albeit in different allelic forms and in differing quantities, in most species. As knowledge of disease resistance gene allelic diversity and phenotypic resistance across *Solanum* species and genebank collections accumulates the potential to utilize gene-editing to modify existing NB-LRR alleles in crop Solanums to mimic alleles found in wild species is an exciting possibility. In some pathosystems, a strong correlation between NB-LRR transcript accumulation and disease resistance exists (Bradeen et al. 2009). In the future, gene-editing may be useful in modifying the promoters of existing resistance alleles to enhance transcript accumulation. While these applications of gene-editing to improvement of disease resistance in *Solanum* crops are exciting and foreseeable, significant technological challenges and knowledge gaps still exist.

An international effort to convert the cultivated potato of widespread dominance to a diploid species (Jansky et al. 2016) will have significant impact on potato breeding and the use of wild *Solanum* species for potato improvement. Among likely impacts for resistance breeding, converting cultivated potato to a diploid species will increase the number of wild *Solanum* species with which it can be directly crossed—effectively expanding the genepool available for potato improvement. Diploid potato will also allow backcross breeding to eliminate introgression of non-target traits from wild species and to minimize linkage drag. Commercial production of potato from true botanical seed, rather than 'seed' tubers, a possibility for diploid potato, may reduce the spread of potato pathogens, especially viral pathogens, potentially providing new cultural approaches to disease management that will reduce the need for resistance breeding.

5.5 Conclusions

This chapter details research on some of the most important genes for the control of nematodes and viruses impacting *Solanum* crop production. This body of research, summarized in Table 5.1, provides opportunity for meta-analysis. In total, I report on 33 resistance genes including 15 nematode resistance genes and 18 virus resistance genes. All but two of these genes (the PVY resistance genes *Ny-1* and *Ny-2*) originate from wild *Solanum* species or *S. tuberosum* ssp. *andigena* (a closely related taxon of the cultivated potato of worldwide dominance, *S. tuberosum* ssp. *tuberosum*). This observation underscores the value of wild *Solanum* species and the importance of genetic conservation through in situ or ex situ means. In total, 29 of the reported resistance genes have been mapped. Disease resistance genes are well distributed across the potato genome, with nematode and virus resistance genes detailed in this chapter mapping to every chromosome except chromosomes 1 and 8. In this data set, chromosomes 5, 9, and 11 are especially well represented, with 6, 5, and 5 resistance genes mapping to these chromosomes, respectively. These observations are consistent with previous research demonstrating genome-wide distribution of disease resistance genes in *Solanum* and the presence of disease resistance gene 'hot spots' on particular chromosomes (Gebhardt and Valkonen 2001; Grube et al. 2000). Among genes reported in this chapter, 5 nematode resistance genes and 10 virus resistance genes have been cloned. Each of the 5 reported nematode resistance genes and 6 of the 10 reported virus resistance genes encode an NB-LRR protein. Across plant species and pathogen types, NB-LRR proteins predominate among cloned disease resistance genes (Hulbert et al. 2001), consistent with the findings reported here. Still, the discovery of other resistance genes that deviate from the canonical NB-LRR form (e.g., *Tm-1*, *Ty-1/3*, and *Ty-5*), highlights the value of multiple approaches to gene mapping and cloning. The increasing availability of whole-genome sequence data for *Solanum* species, the emergence of targeted sequence approaches for discovery of disease resistance gene variants, the development of strategies for visualizing and leveraging deep evolutionary histories of disease resistance genes, and association mapping of disease resistance traits in germplasm core collections are well-positioned to speed the discovery of useful genes in collections of wild *Solanum* species. Improvements in breeding and biotechnological strategies and efforts to convert cultivated potato to a diploid species are all likely to enhance the potential for improved disease resistance in crop Solanums. With a growing world population, increasingly unpredictable climate, and ever-evolving pests and pathogens, these solutions are urgently needed and wild *Solanum* germplasm must be recognized and protected as an indispensable genetic treasure trove.

Table 5.1 Important genes for the control of nematodes and viruses impacting *Solanum* crop production

Pathogen/Pathogen Type	*Solanum* crop	Gene name	Donor species	Chromosome Location	Structure	GenBank Accession	Reference(s)
Nematodes							
Globodera pallida	Potato	*Gpa*	*S. spegazzinii*	Chr 5	Unknown	Not Available	Kreike et al. (1994)
Globodera pallida	Potato	*Gpa2*	S. tuberosum ssp. andigena CPC1673	Chr 12	NB-LRR	AF195939.1	Vossen et al. (2000)
Globodera pallida pathotype Pa1	Potato	*H2*	*S. multidissectum*	Chr 5	Unknown	Not Available	Strachan et al. (2019)
Globodera rostochiensis	Potato	*Gro1.2*	*S. spegazzinii*	Chr 10	Unknown	Not Available	Kreike et al. (1993)
Globodera rostochiensis	Potato	*Gro1.3*	*S. spegazzinii*	Chr 11	Unknown	Not Available	Kreike et al. (1993)
Globodera rostochiensis	Tomato	*Hero*	*S. pimpinellifolium*	Chr 4	NB-LRR	AJ457051.1	Ernst et al. (2002)
Globodera rostochiensis pathotype Ro1	Potato	*Gro1.4*	*S. spegazzinii*	Chr 7	NB-LRR	AY196151.1	Paal et al. (2004)
Globodera rostochiensis pathotype Ro1	Potato	*GroV1*	*S. vernei*	Chr 5	Unknown	Not Available	Jacobs et al. (1996)
Globodera rostochiensis pathotype Ro1, Ro4	Potato	*H1*	*S. tuberosum* ssp. andigena CPC1673	Chr 5	Unknown	Not Available	Gebhardt et al. (1993), Bakker et al. (2004), Meiyalaghan et al. (2018)
Meloidogyne chitwoodi Race 1	Potato	$R_{Mc1(blb)}$	*S. bulbocastanum*	Chr 11	Unknown	Not Available	Austin et al. (1993), Brown et al. (1996), Zhang et al. (2007)
Meloidogyne chitwoodi Race 1	Potato	$R_{MC1(fen)}$	*S. fendleri*	Unknown	Unknown	Not Available	Janssen et al. (1997)
Meloidogyne chitwoodi Race 1	Potato	$R_{Mc1(hou)}$	*S. hougasii*	Unknown	Unknown	Not Available	Brown et al. (1999)
Meloidogyne chitwoodi Race 2	Potato	$R_{Mc2(hou)}$	*S. hougasii*	Unknown	Unknown	Not Available	Brown et al. (1999)

(continued)

Table 5.1 (continued)

Pathogen/Pathogen Type	*Solanum* crop	Gene name	Donor species	Chromosome Location	Structure	GenBank Accession	Reference(s)
Meloidogyne incognita	Eggplant	*SacMi*	*S. aculeatissimum*	Unknown	NB-LRR	Not Available	Zhou et al. (2018)
Meloidogyne javanica	Tomato	*Mi-1 (Mi1.2)*	*S. peruvianum*	Chr 6	NB-LRR	AF039682.1	Milligan et al. (1998)
Viruses							
Potato Virus X (PVX)	Potato	*Rx1*	*S. tuberosum* ssp. andigena	Chr 12	NB-LRR	AJ011801.1	Bendahmane et al. (1999)
Potato Virus X (PVX)	Potato	*Rx2*	*S. acuale*	Chr 5	NB-LRR	AJ249448.1	Bendahmane et al. (2000)
Potato Virus Y (PVY)	Potato	*Ny-1*	*S. tuberosum*	Chr 9	Unknown	Not Available	Szajko et al. (2008)
Potato Virus Y (PVY)	Potato	*Ny-2*	*S. tuberosum*	Chr 11	Unknown	Not Available	Szajko et al. (2014)
Potato Virus Y (PVY)	Potato	$Ry\text{-}f_{sto}$	*S. stoloniferum*	Chr 5	Unknown	Not Available	Flis et al. (2005)
Potato Virus Y (PVY)	Potato	Ry_{adg}	*S. tuberosum* ssp. *andigena*	Chr 11	Unknown	Not Available	Hämäläinen et al. (1997), Kasai et al. (2000)
Potato Virus Y (PVY)	Potato	Ry_{chc}	*S. chacoense*	Chr 9	Unknown	Not Available	Sato et al. (2006)
Potato Virus Y (PVY)	Potato	Ry_{sto}	*S. stoloniferum*	Chr 12	Unknown	Not Available	Song et al. (2005), Song and Schwarzfischer (2008), Fulladolsa et al. (2015)
Tobacco Mosaic Virus (TMV) and Tomato Mosaic Virus (ToMV)	Tomato	*Tm-1*	*S. habrochaites* PI126445	Chr 2	RNA replicase inhibitor	AB287296.1	Ishibashi et al. (2007)
Tobacco Mosaic Virus (TMV) and Tomato Mosaic Virus (ToMV)	Tomato	*Tm-2*	*S. peruvianum* PI126926	Chr 9	NB-LRR	AF536200.1	Lanfermeijer et al. (2005)
Tobacco Mosaic Virus (TMV) and Tomato Mosaic Virus (ToMV)	Tomato	*Tm-2^2* (allelic to *Tm-2)*	*S. peruvianum* PI128650	Chr 9	NB-LRR	AF536201.1	Lanfermeijer et al. (2003)

(continued)

Table 5.1 (continued)

Pathogen/Pathogen Type	*Solanum* crop	Gene name	Donor species	Chromosome Location	Structure	GenBank Accession	Reference(s)
Tomato Spotted Wilt Virus (TSWV)	Tomato	*Sw5*	*S. peruvianum*	Chr 9	NB-LRR	AY007366.1	Spassova et al. (2001)
Tomato Yellow Leaf Curl Virus (TYLCV)	Tomato	*Ty-1*	*S. chilense* LA1969	Chr 6	RDRγ protein	Not Available	Verlaan et al. (2013)
Tomato Yellow Leaf Curl Virus (TYLCV)	Tomato	*Ty-2*	*S. habrochaites* B6013	Chr 11	NB-LRR	LC126693.1	Yamaguchi et al. (2018)
Tomato Yellow Leaf Curl Virus (TYLCV)	Tomato	*Ty-3* (allelic to *Ty-1*)	*S. chilense* LA2779 or LA1932	Chr 6	RDRγ protein	Not Available	Verlaan et al. (2013)
Tomato Yellow Leaf Curl Virus (TYLCV)	Tomato	*Ty-4*	*S. chilense* LA1932	Chr 3	Unknown	Not Available	Ji et al. (2009)
Tomato Yellow Leaf Curl Virus (TYLCV)	Tomato	*Ty-5*	*S. peruvianum*	Chr 4	Ribosomal release factor eRF1	KC447285.1	Lapidot et al. (2015)
Tomato Yellow Leaf Curl Virus (TYLCV)	Tomato	*Ty-6*	*S. chilense* LA1938	Chr 10	Unknown	Not Available	Scott and Hutton (2015), Gill et al. (2019)

References

Anbinder I, Reuveni M, Azari R, Paran I, Nahon S, Shlomo H, Chen L, Lapidot M, Levin I (2009) Molecular dissection of *Tomato leaf curl virus* resistance in tomato line TY172 derived from *Solanum peruvianum*. Theor Appl Genet 119:519–530

Andolfo G, Sanseverino W, Rombauts S, Van der Peer Y, Bradeen JM, Carputo D, Frusciante L, Ercolano MR (2013) Overview of tomato candidate pathogen recognition genes reveals important *Solanum* R locus dynamics. New Phytol 197:223–237

Arntzen F, Vinke J, Hoogendoorn J (1992) Inheritance and level of resistance to potato cyst nematodes (*Globodera pallida*), derived from *Solanum tuberosum* ssp. *andigna* CPC 1673. In: Jacobs T, Parlevliet J (eds) Durability of disease resistance. Springer-Science+Business Media, B.V., Wageningen, The Netherlands, p 304

Arora S, Steuernagel B, Gaurav K, Chandramohan S, Long Y, Matny O, Johnson R, Enk J, Periyannan S, Singh N, Asyraf Md Hatta M, Athiyannan N, Cheema J, Yu G, Kangara N, Ghosh S, Szabo L, Poland J, Bariana H, Jones J, Bentley A, Ayliffe M, Olson E, Xu S, Steffenson B, Lagudah E, Wulff B (2019) Resistance gene cloning from a wild crop relative by sequence capture and association genetics. Nat Biotech 37:139–143

Austin S, Pohlman JD, Brown CR, Mojtahedi H, Santo GS, Douches DS, Helgeson JP (1993) Interspecific somatic hybridization between *Solanum tuberosum* L. and *S. bulbocastanum* Dun. as a means of transferring nematode resistance. Am Potato J 70:485–495

Aversano R, Contaldi F, Ercolano MR, Grosso V, Iorizzo M, Tatino F, Xumerle L, Molin AD, Avanzato C, Ferrarini A, Delledonne M, Sanseverino W, Cigliano RA, Capella-Gutierrez S, Gabaldón T, Frusciante L, Bradeen JM, Carputo D (2015) The *Solanum commersonii* genome sequence provides insights into adaptation to stress conditions and genome evolution of wild potato relatives. Plant Cell 27:954–968

Bakker E, Achenbach U, Bakker J, van Vliet J, Peleman J, Segers B, van der Heijden S, van der Linde P, Graveland R, Hutten R, van Eck H, Coppoolse E, Van der Vossen E, Bakker J, Goverse A (2004) A high-resolution map of the *H1* locus harbouring resistance to the potato cyst nematode *Globodera rostochiensis*. Theor Appl Genet 109:146–152

Ballvora A, Hesselbach J, Niewohner J, Leister D, Salamini F, Gebhardt C (1995) Marker enrichment and high-resolution map of the segment of potato chromosome VII harbouring the nematode resistance gene *Gro1*. Mol Gen Genet 249:82–90

Barbary A, Djian-Caporalino C, Palloixd A, Castagnone-Sereno P (2015) Host genetic resistance to root-knot nematodes, *Meloidogyne* spp., in Solanaceae: from genes to the field. Pest Manag Sci 71:1591–1598

Barone A, Ritter E, Schachtschabel U, Debener T, Salamini F, Gebhardt C (1990) Localization by restriction fragment length polymorphism mapping in potato of a major dominant gene conferring resistance to the potato cyst nematode *Globodera rostochiensis*. Mol Gen Genet 224:177–182

Bendahmane A, Kanyuka K, Baulcombe DC (1997) High-resolution genetical and physical mapping of the Rx gene for extreme resistance to potato virus X in tetraploid potato. Theor Appl Genet 95:153–162

Bendahmane A, Kanyuka K, Baulcombe DC (1999) The Rx gene from potato controls separate virus resistance and cell death responses. Plant Cell 11:781–791

Bendahmane A, Kohm BA, Dedi C, Baulcombe DC (1995) The coat protein of potato virus X is a strain-specific elicitor of *Rx1*-mediated virus resistance in potato. Plant J 8:933–941

Bendahmane A, Querci M, Kanyuka K, Baulcombe DC (2000) Agrobacterium transient expression system as a tool for the isolation of disease resistance genes: application to the *Rx2* locus in potato. Plant J 21:73–81

Bethke P, Halterman D, Jansky S (2017) Are we getting better at using wild potato species in light of new tools? Crop Sci 57:1241–1258

Bolger A, Scossa F, Bolger M, Lanz C, Maumus F, Tohge T, Quesneville H, Alseekh S, Sørensen I, Lichtenstein G, Fich E, Conte M, Keller H, Schneeberger K, Schwacke R, Ofner I, Vrebalov J, Xu Y, Osorio S, Aflitos S, Schijlen E, Jiménez-Goméz J, Ryngajllo M, Kimura S, Kumar R, Koenig D, Headland L, Maloof J, Sinha N, van H, RC, Lankhorst R, Mao L, Vogel A, Arsova B, Panstruga R, Fei Z, Rose J, Zamir D, Carrari F, Giovannoni J, Weigel D, Usadel B, Fernie A (2014) The genome of the stress-tolerant wild tomato species *Solanum pennellii*. Nature Genetics 46:1034–1038

Bradeen JM, Iorizzo M, Mollov DS, Raasch J, Colton Kramer L, Millett BP, Austin-Phillips S, Jiang J, Carputo D (2009) Higher copy numbers of the potato *RB* transgene correspond to enhanced transcript and late blight resistance levels. Mol Plant Microb Interact 22:437–446

Brigneti G, Garcia Mas J, Baulcombe DC (1997) Molecular mapping of the potato virus Y resistance gene Rysto in potato. Theor Appl Genet 94:198–203

Brommonschenkel SH, Tanksley SD (1997) Map-based cloning of the tomato genomic region that spans the *Sw-5* tospovirus resistance gene in tomato. Mol Gen Genet 256:121–126

Brown C, Mojtahedi H, Santo G (1991) Resistance to Columbia root-knot nematode in *Solanum* spp. and in hybrids of *S. hougasii* with tetraploid cultivated potato. Am Pot J 68:445–452

Brown C, Mojtahedi H, Santo G (1999) Genetic analysis of resistance to meloidogyne chitwoodi introgressed from *Solanum hougasii* into cultivated potato. J Nematol 31:264–271

Brown CR, Yang CP, Mojtahedi H, Santo GS, Masuelli R (1996) RFLP analysis of resistance to Columbia root-knot nematode derived from *Solanum bulbocastanum* in a BC2 population. Theor Appl Genet 92:572–576

Butterbach P, Verlaan MG, Dullemans A, Lohuis D, Visser RGF, Bai Y, Kormelink R (2014) Tomato yellow leaf curl virus resistance by *Ty-1* involves increased cytosine methylation of viral genomes and is compromised by cucumber mosaic virus infection. Proc Natl Acad Sci U S A 111:12942–12947

Chague V, Mercier JC, Guenard M, de Courcel A, Vedel F (1996) Identification and mapping on chromosome 9 of RAPD markers linked to *Sw-5* in tomato by bulked segregant analysis. Theor Appl Genet 92:1045–1051

Chen X, Lewandowska D, Armstrong MR, Baker K, Lim TY, Bayer M, Harrower B, McLean K, Jupe F, Witek K, Lees AK, Jones JD, Bryan GJ, Hein I (2018) Identification and rapid mapping of a gene conferring broad-spectrum late blight resistance in the diploid potato species *Solanum verrucosum* through DNA capture technologies. Theor Appl Genet 131:1287–1297

Consortium PGS (2011) Genome sequence and analysis of the tuber crop potato. Nature advance online publication:189–195

Consortium TG (2012) The tomato genome sequence provides insights into fleshy fruit evolution. Nature 485:635–641

de Oliveirra AS, Boiteux LS, Kormelink R, Rresend RO (2018) The *Sw-5* gene cluster: tomato breeding and research toward orthotospovirus disease control. Front Plant Sci 9:1055

del Rosario HM, Vidalon L, Montenegro J, Riccio C, Guzman F, Bartolini I, Ghislain M (2018) Molecular and genetic characterization of the *Ryadg* locus on chromosome XI from Andigena potatoes conferring extreme resistance to potato virus Y. Theor Appl Genet 131:1925–1938

Devran Z, Goknur A, Mesci L (2016) Development of molecular markers for the *Mi-1* gene in tomato using the KASP genotyping assay. Hort Environ Biotechnol 57:156–160

Dropkin VH (1969) The necrotic reaction of tomatoes and other hosts resistant to *Meloidogyne:* reversal by temperature. Phytopath 59:1632–1637

Dunnett J (1961) Inheritance of resistance to potato root eelworm in a breeding line stemming from *Solanum multidissectum* Hawes. Report of the Scottish Plant Breeding Station, pp 39–46

Ellenby C (1952) Resistance to the Potato Root Eelworm, *Heterodera rostochiensis* Wollenweber. Nature 170:1016

Ellis JG, Lawrence GJ, Luck JE, Dodds PN (1999) Identification of regions in alleles of the flax rust resistance gene L that determine differences in gene-for-gene specificity. Plant Cell 11

Ellis PR, Maxon Smith JW (1971) Inheritance of resistance to potato cyst-eelworm (*Heterodera rostochiensis* Woll.) in the genus *Lycopersicon*. Euphytica 20:93–101

Ernst K, Kumar A, Kriseleit D, Kloos DU, Phillips MS, Ganal M (2002) The broad-spectrum potato cyst nematode resistance gene (Hero) from tomato is the only member of a large gene family of NBS-LRR genes with an unusual amino acid repeat in the LRR region. Plant J 31:127–136

Flis B, Hennig J, Strzelczyk-Zyta D, Gebhardt C, Marczewski W (2005) The *Ry-f sto* gene from *Solanum stoloniferum* for extreme resistant to Potato virus Y maps to potato chromosome XII and is diagnosed by PCR marker GP122718 in PVY resistant potato cultivars. Mol Breed 15:95–101

Folkertsma RT, Spassova MI, Prins M, Stevens MR, Hille J, Goldbach RW (1999) Construction of a bacterial artificial chromosome (BAC) library of *Lycopersicon esculentum* cv. Stevens and its application to physically map the *Sw-5* locus. Mol Breed 5:197–207

Fraser R, Loughlin S (1980) Resistance to tobacco mosaic virus in tomato: effects of the *Tm-1* gene on virus multiplication. J Gen Virol 48:87–96

Fulladolsa A, Navarro F, Kota R, Severson K, Palta J, Charkowski A (2015) Application of marker assisted selection for *Potato Virus Y* resistance in the University of Wisconsin potato breeding program. Am J Pot Res 92:444–450

Fuller V, Lilley C, Urwin P (2008) Nematode resistance. New Phytol 180:27–44

Ganal MW, Simon R, Brommonschenkel S, Arndt M, Phillips MS, Tanksley SD, Kumar A (1995) Genetic mapping of a wide spectrum nematode resistance gene (*Hero*) against *Globodera rostochiensis* in tomato. Mol Plant Microb Interact 8:886–891

Gebhardt C, Mugniery D, Ritter E, Salamini F, Bonnel E (1993) Identification of RFLP markers closely linked to the *H1* gene conferring resistance to *Globodera rostochiensis* in potato. Theor Appl Genet 85:541–544

Gebhardt C, Valkonen JPT (2001) Organization of genes controlling disease resistance in the potato genome. Ann Rev Phytopath 39:79–102

Ghislain M, Byarugaba A, Magembe E, Njoroge A, Rivera C, Roman M, Tovar J, Gamboa S, Forbes G, Krueze J, Barekye A, Kiggundu A (2019) Stacking three late blight resistance genes from wild species directly into African highland potato varieties confers complete field resistance to local blight races. Plant Biotech J 17:1119–1129

Gill U, Scott JW, Shekasteband R, Ogundiwin E, Schuit C, Francis DM, Sim S-C, Smith H, Hutton SF (2019) *Ty-6*, a major begomovirus resistance gene on chromosome 10, is effective against Tomato yellow leaf curl virus and Tomato mottle virus. Theor Appl Genet 132:1543–1554

Grube RC, Radwanski ER, Jahn M (2000) Comparative genetics of disease resistance within the Solanaceae. Genetics 155:873–887

Hall T (1980) Resistance at the *TM-2* locus in the tomato to tomao mosaic virus. Euphytica 29:189–197

Hämäläinen JH, Sorri VA, Watanabe KN, Gebhardt C, Valkonen JPT (1998) Molecular examination of a chromosome region that controls resistance to potato Y and A potyviruses in potato. Theor Appl Genet 96:1036–1043

Hämäläinen JH, Watanabe KN, Valkonen JPT, Arihara A, Plaisted RL, Pehu E, Miller L, Slack SA (1997) Mapping and marker-assisted selection for a gene for extreme resistance to potato virus Y. Theor Appl Genet 94:192–197

Hameed A, Zaidi S, Shakir S, Mansoor S (2018) Applications of new breeding technologies for potato improvement. Front Plant Sci 9:925

Hanson PM, Bernacchi D, Green S, Tanksley SD, Muniyappa V, Padmaja AS, Chen H-m, Kuo G, Fang D, Chen J-t (2000) Mapping a wild tomato introgression associated with tomato yellow leaf curl virus resistance in a cultivated tomato line. J Amer Soc Hort Sci 125:15–20

Hanssen IM, Lapidot M, Thomma BPHJ (2010) Emerging viral diseases of tomato crops. Mol Plant Microb Interact 5:539–548

Hirakawa H, Shirasawa K, Miyatake K, Nunome T, Negoro S, Ohyama A, Yamaguchi H, Sato S, Isobe S, Tabata S, Fukuoka H (2014) Draft genome sequence of eggplant (*Solanum melongena* L.): the representative *Solanum* species indigenous to the Old World. DNA Research

Hulbert SH, Webb CA, Smith SM, Sun Q (2001) Resistance gene complexes: evolution and utilization. Annu Rev Phytopathol 39:285–312

Hutton SF, Scott JW, Schuster DJ (2012) Recessive resistance to *Tomatto yellow leaf curl virus* from the tomato cultivar Tyking is located in the same rregion as *Ty-5* on chromosome 4. HortSci 47:324–327

Ishibashi K, Masuda K, Naito S, Meshi T, Ishikawa M (2007) An inhibitor of viral RNA replication is encoded by a plant resistance gene. Proc Natl Acad Sci U S A 104:13833–13838

Jacobs JME, van Eck HJ, Horsman K, Arens PFP, Verkerk-Bakker B, Jacobsen E, Pereira A, Stiekema WJ (1996) Mapping of resistance to the potato cyst nematode *Globodera rostochiensis* from the wild potato species *Solarium Vernei*. Mol Breed 2:51–60

Jansky S, Charkowski A, Douches D, Gusmini G, Richael C, Bethke P, Spooner D, Novy R, De Jong H, De Jong W, Bamberg J, Thompson A, Bizimungu B, Holm D, Brown C, Haynes K, Sathuvalli V, Veilleux R, Miller CJ, Bradeen J, Jiang J (2016) Reinventing potato as a diploid inbred line-based crop. Crop Sci 56:1412–1422

Janssen G, Norel A, Janssen R, Hoogendoorn J (1997) Dominant and additive resistance to the root-knot nematodes *Meloidogyne chitwoodi* and *M. fallax* in Central American *Solanum* species. Theor Appl Genet 94:692–700

Ji Y, Schuster DJ, Scott JW (2007) *Ty-3*, a begomovirus resistance locus near the Tomato yellow leaf curl virus resistance locus *Ty-1* on chromosome 6 of tomato. Mol Breed 20:271–284

Ji Y, Scott JW, Hanson P, Graham E, Maxwell DP (2007) Sources of resistance, inheritance, and location of genetic loci conferring resistance to members of the tomato-infecting begomoviruse. In: Czosnek H (ed) Tomato yellow curl virus disease. Springer, Dordrecht, pp 343–362

Ji Y, Scott JW, Schuster DJ, Maxwell DP (2009) Molecular mapping of *Ty-4*, a new Tomato yellow leaf curl virus resistance locus on chromosome 3 of tomato. J Am Soc Hort Sci 134:281–288

Jia H, Zhang Y, Orbovic V, Xu J, White F, Jones J, Wang N (2017) Genome editing of the disease susceptibility gene CsLOB1 in citrus confers resistance to citrus canker. Plant Biotech J 15:817–823

Jiang W, Zhou H, Bi H, Fromm M, Yang B, Weeks D (2013) Demonstration of CRISPR/Cas9/sgRNA-mediated targeted gene modification in Arabidopsis, tobacco, sorghum and rice. Nucl Acids Res 41:188

Jones J, Zitter T, Momol T, Miller S (2014) Compendium of tomato diseases and pests, 2nd edn. APS Press, St. Paul

Kaloshian I, Yaghoobi J, Liharska T, Hontelez J, Hanson D, Hogan P, Jesse T, Wijbrandi J, Simons G, Vos P, Zabel P, Williamson V (1998) Genetic and physical localization of the root-knot nematode resistance locus *Mi* in tomato. Mol Gen Genet 257:376–385

Kanyuka K, Bendahmane A, Rouppe van der Voort JNAM, Vossen EAGvd, Baulcombe DC (1999) Mapping of intra-locus duplications and introgressed DNA: aids to map-based cloning of genes from complex genomes illustrated by physical analysis of the Rx locus in tetraploid potato. Theor appl genet 98:679–689

Kasai K, Morikawa Y, Sorri VA, Valkonen JPT, Gebhardt C, Watanabe KN (2000) Development of SCAR markers to the PVY resistance gene Ry(adg) based on a common feature of plant disease resistance genes. Genome 43:1–8

Kil E-J, Kim S, Lee Y-J, Byun H-S, Park J, Seo H, Kim C-S, Shim J-K, Lee J-H, Kim J-K, Lee K-Y, Choi H-S, Lee S (2016) *Tomato yllow leaf curl virus* (TTYLCV-IL): a seed-transmissible geminivirus in tomatoes. Scienitfic Reports 6:19013

Klein-Lankhorst R, Rietveld P, Machiels B, Verkerk R, Weide R, Gebhardt C, Koornneef M, Zabel P (1991) RFLP markers linked to the root knot nematode resistance gene *Mi* in tomato. Theor Appl Genet 81:661–667

Kreike CM, de Koning JRA, Vinke JH, van Ooijen JW, Gebhardt C, Stiekema WJ (1993) Mapping of loci involved in quantitatively inherited resistance to the potato cyst-nematode *Globodera rostochiensis* pathotype Ro1. Theor Appl Genet 87:464–470

Kreike CM, de Koning JRA, Vinke JH, van Ooijen JW, Stiekema WJ (1994) Quantitatively-inherited resistance to *Globodera pallida* is dominated by one major locus in *Solanum spegazzinii*. Theor Appl Genet 88:764–769

Lanfermeijer F, Dijkhuis J, Sturre M, de Haan P, Hille J (2003) Cloning and characterization of the durable tomato mosaic virus resistance gene *Tm-2(2)* from *Lycopersicon esculentum*. Plant Mol Biol 52:1037–1049

Lanfermeijer F, Warmink J, Hille J (2005) The products of the broken *Tm-2* and the durable *Tm-22* resistance genes from tomato differ in four amino acids. J Exper Bot 56:2925–2933

Lapidot M, Karniel U, Gelbart D, Fogel D, Evenor D, Kutsher Y, Makhbash Z, Nahon S, Shlomo H, Chen L, Reuveni M, Levin I (2015) A novel route controlling begomovirus resistance by the messenger RNA surveillance factor pelota. PLoS Genet 11:e1005538

Leister D, Ballvora A, Salamini F, Gebhardt C (1996) A PCR-based approach for isolating pathogen resistance genes from potato with potential for wide application in plants. Nat Genet 14:421–429

Leisner CP, Hamilton JP, Crisovan E, Manrique-Carpintero NC, Marand AP, Newton L, Pham GM, Jiang J, Douches DS, Jansky SH, Buell CR (2018) Genome sequence of M6, a diploid inbred clone of the high-glycoalkaloid-producing tuber-bearing potato species *Solanum chacoense*, reveals residual heterozygosity. Plant J 94:562–570

Malnoy M, Viola R, Jung M-H, Koo O-J, Kim S, Kim J-S, Velasco R, Kanchiswamy C (2016) DNA-free genetically edited grapevine and apple protoplast using CRISPR/Cas9 ribonucleoproteins. Front Plant Sci 7:1904

Meiyalaghan S, Paget M, Thompson S, Thomson S, Baldwin S, Anderson J, Genet R, Lewthwaite S (2018) High resolution DNA melting markers for identification of *H1*-linked resistance to potato cyst nematode. Mol Breed 38:79

Meshi T, Motoyoshi F, Adachi A, Watanabe Y, Takamatsu N, Okada Y (1988) Two concomitant base substitutions in the putative replicase genes of tobacco mosaic virus confer the ability to overcome the effects of a tomato resistancegene, *Tm-1*. EMBO J 7:1575–1581

Messeguer R, Ganal M, De Vincente MC, Young ND, Bolkan H, Tanksley SD (1991) High resolution RFLP map around the root knot nematode resistance gene (Mi) in tomato. Theor Appl Genet 82:529–536

Milczarek D (2012) A Multiplex PCR method of detecting markers linked to genes conferring resistance to *Globodera rostochiensis*. Am J Pot Res 89:169–171

Milczarek D, Flis B, Przetakiewicz A (2011) Suitability of molecular markers for selection of potatoes resistant to *Globodera* spp. Am J Pot Res 88:245–255

Milligan SB, Bodeau J, Yaghoobi J, Kaloshian I, Zabel P, Williamson VM (1998) The root knot nematode resistance gene Mi from tomato is a member of the leucine zipper, nucleotide binding, leucine-rich repeat family of plant genes. Plant Cell 10:1307–1319

Mori K, Sakamoto Y, Mukojima N, Tamiya S, Nakao T, Ishii T, Hosaka K (2011) Development of a multiplex PCR method for simultaneous detection of diagnostic DNA markers of five disease and pest resistance genes in potato. Euphytica 180:347–355

Motoyoshi F, Oshima N (1977) Expression of genetically controlled resistance to tobacco mosaic virus infection in isolated tomato leaf mesophyll protoplasts. J Gen Virol 34:499–506

Nadakuduti S, Buell C, Voytas D, Starker C, Douches D (2018) Genome editing for crop improvement—applications in clonally propagated polyploids with a focus on potato (*Solanum tuberosum* L.). Front Plant Sci 9:1607

Nekrasov V, Wang C, Win J, Lanz C, Weigel D, Kamoun S (2017) Rapid generation of a transgene-free mildew resistant tomato by genome deletion. Sci Rep 7:482

Nie X, Lalany F, Dickison V, Wilson D, Singh M, De Koeyer D, Murphy A (2016) Detection of molecular markers linked to *Ry* genes in potato germplasm for marker-assisted selection for extreme resistance to PVY in AAFC's potato breeding program. Can J Plant Sci 96:737–742

Nombela G, Williamson VM, Muniz M (2003) The root-knot nematode resistance gene *Mi-1.2* of tomato is responsible for resistance against the whitefly *Bemisia tabaci*. Mol Plant Microb Interact 16:645–649

Ortega F, Lopez-Vizcon C (2012) Application of molecular Marker-Assisted Selection (MAS) for disease resistance in a practical potato breeding programme. Potato Res 55:1–13

Ottoman RJ, Hane DC, Brown CR, Yilma S, James SR, Mosley AR, Crosslin JM, Vales MI (2009) Validation and implementation of Marker-Assisted Selection (MAS) for PVY resistance (*Ryadg* gene) in a tetraploid potato breeding program. Am J Pot Res 86:304–314

Paal J, Henselewski H, Muth J, Meksem K, Menendez CM, Salamini F, Ballvora A, Gebhardt C (2004) Molecular cloning of the potato Gro1-4 gene conferring resistance to pathotype Ro1 of the root cyst nematode *Globodera rostochiensis*, based on a candidate gene approach. Plant J 38:285–297

Palukaitis P (2012) Resistance to viruses of potato and their vectors. Plant Pathol J 28:248–258

Park J, Yang H, De Jong W, Wang X (2018) An evaluation of two *H1*-linked markers and their suitability for selecting *Globodera rostochiensis* resistant potatoes in the New York breeding program. Am J Pot Res 95:170–177

Paulson RE, Webster JM (1972) Ultrastructure of the hypersensitive reaction in roots of tomato, *Lycopersicon esculentum* L., to infection by the root-knot nematode *Meloidogyne incognita*. Physiol Plant Pathol 2:227–234

Perez de Castro A, Miguel Blanca J, Jose Diez M, Nuez Vinals F (2007) Identification of a CAPS marker tightly linked to the tomato yellow leaf curl disease resistance gene *Ty-1* in tomato. Eur J Plant Pathol 117:347–356

Querci M, Baulcombe DC, Goldbach RW, Salazar LF (1995) Analysis of the resistance-breaking determinant of potato virus X (PVX) strain HB on different potato genotypes expressing extreme resistance to PVX. Phytopathology 85:1003–1010

Quirin EA, Mann H, Meyer RS, Traini A, Chiusano ML, Litt A, Bradeen JM (2012) Evolutionary meta-analysis of Solanaceous resistance gene and *Solanum* resistance gene analog sequences and a practical framework for cross-species comparisons. Mol Plant-Microbe Interact 25:603–612

Ramakrishnan A, Ritland C, Blas Sevillano R, Riseman A (2015) Review of potato molecular markers to enhance trait selection. Am J Potato Res 92:455–472

Razaimi R, Bougouffa S, Morton M, Lightfoot D, Alam I, Essack M, Arold S, Kamau A, Schmockel S, Pailles Y, Shahid M, Michell C, Al-Babili S, Ho Y, Tester M, Bajic V, Negroa S (2018) The genome sequence of the wild tomato *Solanum pimpinellifolium* provides insights into salinity tolerance. Front Plant Sci 9:1402

Reem N, Van Eck J (2019) Application of CRISPR/Cas9-mediated gene editing in tomato. Method Mol Biol 1917:171–182

Ritter E, Debener T, Barone A, Salamini F, Gebhardt C (1991) RFLP mapping on potato chromosomes of two genes controlling extreme resistance to potato virus X (PVX). Mol Gen Genet 227:81–85

Rossi M, Goggin FL, Milligan SB, Kaloshian I, Ullman DE, Williamson VM (1998) The nematode resistance gene Mi of tomato confers resistance against the potato aphid. Proc Natl Acad Sci U S A 95:9750–9754

Rouppe van der Voort J, Wolters P, Folkertsma R, Hutten R, Pv Z, Vinke H, Kanyuka K, Bendahmane A, Jacobsen E, Janssen R (1997) Mapping of the cyst nematode resistance locus Gpa2 in potato using a strategy based on comigrating AFLP markers. Theor Appl Genet 95:874–880

Rybicki E (2015) A top ten list for economically important plant viruses. Arch Virol 160:17–20

Sagredo BD, Mathias MR, Barrrientos CP, Acuna IB, Kalazich JB, Santos Rojas J (2009) Evaluation of scar RYSC3 marker of the *Ryadg* gene to select resistant genotypes to potato virus Y (PVY) in the INIA potato breeding program. Chilean J Ag Res 69:305–315

Sato M, Nishikawa K, Komura K, Hosaka K (2006) *Potato virus Y* resistance gene, *Rychc*, mapped to the distal end of potato chromosome 9. Euphytica 149:367–372

Scheben A, Wolter F, Batley J, Puchta H, Edwards D (2017) Towards CRISPR/Cas crops—bringing together genomics and genome editing. New Phytol 216:682–698

Scott JW, Hutton SF (2015) Fla. 8638B and Fla. 8624 tomato breeding lines with *Begomovirus* resistance genes *ty-5* plus *Ty-6* and *Ty-6,* respectively. HortSci 50:1405–1407

Smith P (1944) Embryo culture of a tomato species hybrid. Proc Am Soc Hortic Sci 44:413–416

Song Y-S, Hepting L, Schweizer G, Hartl L, Wenzel G, Schwarzfischer A (2005) Mapping of extreme resistance to PVY (*Rysto*) on chromosome XII using anther-culture-derived primary dihaploid potato lines 111:879–887

Song Y-S, Schwarzfischer A (2008) Development of STS markers for selection of extreme resistance (*Rysto*) to PVY and maternal pedigree analysis of extremely resistant cultivars. Am J Pot Res 85:159–170

Spassova MI, Prins TW, Folkertsma RT, Klein-Lankhorst RM, Hille J, Goldbach RW, Prins M (2001) The tomato gene *Sw5* is a member of the coiled coil, nucleotide binding, leucine-rich repeat class of plant resistance genes and confers resistance to TSWV in tobacco. Mol Breed 7:151–161

Stevens MR, Lamb EM, Rhoads DD (1995) Mapping the *Sw-5* locus for tomato spotted wilt virus resistance in tomatoes using RAPD and RFLP analyses. Theor Appl Genet 90:451–456

Stevens MR, Scott SJ, Gergerich RC (1992) Inheritance of a gene for resistance to tomato spotted wilt virus (TSWV) from *Lycopersicon peruvianum* Mill Euphytica 59:9–17

Stevenson WR, Loria R, Franc GD, Weingartner DP (2001) Compendium of potato diseases, 2nd edn. APS Press, St. Paul

Strachan SM, Armstrong MR, Kauer A, Wright KM, Lim TY, Baker K, Jones J, Bryan G, Blok V, Hein I (2019) Mapping the *H2* resistance effective against *Globodera pallida* pathotype Pa1 in tetraploid potato. Theor Appl Genet (Published on line January 21, 2019)

Szajko K, Chrzanowska M, Witek K, Strzelczyk-Zyta D, Gebhardt C, Hennig J, Marczewski W (2008) The novel gene *Ny-1* on potato chromosome IX confers hypersensitive resistance to *Potato virus Y* and is an alternative to *Ry* genes in potato breeding for PVY resistance. Theor Appl Genet 116:297–303

Szajko K, Strzelczyk-Zyta D, Marczewski W (2014) *Ny-1* and *Ny-2* genes conferring hypersensitive response to potato virus Y (PVY) in cultivated potatoes: mapping and marker-assisted selection validation for PVY resistance in potato breeding. Mol Breed 34:267–271

Tanksley SD, Ganal MW, Prince JP, Vicente MCd, Bonierbale MW, Broun P, Fulton TM, Giovannoni JJ, Grandillo S, Martin GB (1992) High density molecular linkage maps of the tomato and potato genomes. Genetics 132:1141–1160

Valkonen JPT (2007) Viruses: economic losses and biotechnological potential. In: Vreugdenhil D, Bradshaw J, Gebhardt C, Govers F, Mackerron DKL, Taylor MA, Ross HA (eds) Potato biology and biotechnology advances and perspective. Elsevier, Oxford, pp 619–641

van Eck L, Bradeen J (2019) Hunting for novel disease resistance genes: observations and opportunities from the *Rosaceae*. Acta Hort 1232:125–134

Verlaan MG (2013) Characterization of major resistance genes to Tomatto Yellow Leaf Curl Virus. Wageningen University

Verlaan MG, Hutton SF, Ibrahem RM, Kormelink R, Visser RGF, Scott JW, Edwards JD, Bai Y (2013) The tomato yellow leaf curl virus resistance genes *Ty-1* and *Ty-3* Are allelic and code for DFDGD-Class RNA–dependent RNA polymerases. PLoS Genet 9: e1003399

Verlaan MG, Szinay D, Hutton SF, De Jong H, Kormelink R, Visser RGF, Scott JW, Bai Y (2011) Chromosomal rearrangements between tomato and *Solanum chilense* hamper mapping and breeding of the TYLCV resistance gene *Ty-1*. Plant J 68:1093–1103

Vossen EAGvd, Rouppe van der Voort JNAM, Kanyuka K, Bendahmane A, Sandbrink H, Baulcombe DC, Bakker J, Stiekema WJ, Klein Lankhorst RM (2000) Homologues of a single resistance-gene cluster in potato confer resistance to distinct pathogens: a virus and a nematode. Plant j 23:567–576

Wang Y, Cheng X, Shan Q, Zhang Y, Liu J, Gao C (2014) Simultaneous editing of three homoeoalleles in hexaploid bread wheat confers heritable resistance to powdery mildew. Nat Biotech 32:947–951

Weber H, Pfitzner A (1998) Tm-22 resistance in tomato requires recognition of the carboxy terminus of the movement protein of tomato mosaic virus. Mol Plant Microb Interact 11:498–503

Williamson VM (1998) Root-knot nematodie resistance genes in tomato and their potential for future use. Ann Rev Phytopath 36:277–293

Williamson VM, Ho JY, Wu FF, Miller N, Kaloshian I (1994) A PCR-based marker tightly linked to the nematode resistance gene, Mi in tomato. Theor Appl Genet 87:757–763

Witek K, Jupe F, Witek A, Baker D, Clark M, Jones J (2016) Accelerated cloning of a potato late blight-resistance gene using RenSeq and SMRT sequencing. Nat Biotech 34:656–660

Yamafuji R, Watanabe Y, Meshi T, Okada Y (1991) Replication of TMV-L and Lta1 RNAs and their recombinants in TMV-resistant *Tm-1* tomato protoplasts. Virology 183:99–105

Yamaguchi H, Ohnishi J, Saito A, Ohyama A, Nunome T, Miyatake K (2018) An NB-LRR gene, *TYNBS1*, is responsible for resistance mediated by the *Ty-2 Begomovirus* resistance locus of tomato. Theor Appl Genet 131:1345–1362

Yan Z, Perez de Castro A, Diez M, Hutton SF, Visser R, Wolters A-M, Bai Y, Li J (2018) Resistance to tomato yellow leaf curl virus in tomato germplasm. Front Plant Sci 9:1198

Yang X, Caro M, Hutton SF, Scott JW, Guo Y, Wang X, Rashid M, Szinay D, de Jong H, Visser R, Bai Y, Du Y (2014) Fine mapping of the tomato yellow leaf curl virus resistance gene *Ty-2* on chromosome 11 of tomato. Mol Breed 34:749–760

Zamir D, Ekstein-Michelson I, Zakay Y, Navot N, Zeidan M, Sarfatti M, Eshed Y, Harel E, Pleban T, van-Oss H, Kedar N, Rabinowitch H, Czosnek H (1994) Mapping and introgression of a tomato yellow leaf curl virus tolerance gene, *TY-1*. Theor Appl Genet 88:141–146

Zhang L-H, Mojtahedi H, Kuang H, Baker B, Brown C (2007) Marker-assisted selection of Columbia root-knot nematode resistance introgressed from *Solanum bulbocastanum*. Crop Sci 47:2021–2026

Zhou X, Liu J, Bao S, Yang Y, Zhuang Y (2018) Molecular cloning and characterization of a wild eggplant *Solanum aculeatissimum* NBS-LRR gene, involved in plant resistance to *Meloidogyne incognita*. Int J Mol Sci 19:583

6 Pepper and Eggplant Genetic Resources

Laura Toppino, Jaime Prohens, Giuseppe Leonardo Rotino, Mariola Plazas, Mario Parisi, Carolina Carrizo García, and Pasquale Tripodi

Abstract

Peppers (*Capsicum* spp.) and eggplants (*Solanum* spp.) are vegetable crops from the Solanaceae family that have spread worldwide and are important as a source of health-related compounds for humans. Several pepper and eggplant species have been domesticated, and a large number of wild relatives exist for both crops. Over 25,000 domesticated and wild accessions are stored in the major germplasm banks, representing the readily available genepool for genetic improvement of the cultivated forms. In the present chapter, we revisit the origin, domestication, and spread of peppers and eggplants, describing the main botanical characteristics, agronomic and qualitative properties, and providing novel insight into the recent classification and phylogenetic relationships. Potentiality and constraints of the use of genetic resources in intraspecific and interspecific hybridization, description of breeding programs carried out in the past years for agronomic traits, improvement of biotic (fungal, bacterial, nematodes, and viruses) and abiotic stresses (heat, drought, and salinity), tolerance, and genetic studies for the identification of genes underlying traits of interest, are widely discussed.

6.1 Pepper

6.1.1 Introduction

Pepper (*Capsicum* spp.) had its origin in the tropical and subtropical World's regions (Hunziker 2001) in a wide area that includes Mesoamerican and Andean countries. The genus includes over 40 species (Carrizo Garcia et al. 2016; Barboza et al. 2019) with different chromosome number and ploidy level (Tripodi and Kumar 2019). The most widely consumed species are the cultivated *C. annuum* and four domesticated ones including *C. frutescens*, *C. chinense*, *C. baccatum*, and *C. pubescens*. Collection of genetic resources are available in the form of seeds in various germplasm banks constituted across the globe for which international initiatives promoting the collection, maintenance,

L. Toppino · G. L. Rotino
CREA Research Centre for Genomics and Bioinformatics, 26836 Montanaso Lombardo, LO, Italy

J. Prohens · M. Plazas
Instituto de Conservación y Mejora de la Agrodiversidad Valenciana, Universitat Politècnica de València, 46022 Valencia, Spain

M. Parisi · P. Tripodi (✉)
CREA Research Centre for Vegetable and Ornamental Crops, 84098 Pontecagnano Faiano, SA, Italy
e-mail: pasquale.tripodi@crea.gov.it

C. Carrizo García
Instituto Multidisciplinario de Biología Vegetal, Córdoba, Argentina

© Springer Nature Switzerland AG 2021

D. Carputo et al. (eds.), *The Wild Solanums Genomes*, Compendium of Plant Genomes,
https://doi.org/10.1007/978-3-030-30343-3_6

and distribution are ongoing. The largest genebanks are represented by the United States Department of Agriculture (USDA) and the World Vegetable Center (AVRDC) (Taiwan), holding about 15,000 wild and cultivated species. Several other gene banks such as the Banco de Germoplasma Hortalicas (BGH) in Brazil, the Chile Pepper Institute of the New Mexico State University (NMSU), the Leibniz Institute of Plant Genetics and Crop Plant Research (IPK) of Germany, and the Centre for Genetic Resources of Netherland (CGN-WUR), provide more than a thousand of accessions each. Efforts to explore the genetic diversity of these collections have been carried out using different molecular methods of investigation. Nicolai et al. (2013) by employing microsatellites markers determined the genetic structure of over 1300 accessions retrieved from a broad range of countries. Colonna et al. (2019), report genotyping by sequencing, for investigating genomic diversity of 373 accessions belonging to 11 *Capsicum* species. Lee et al. (2016) analyzed 3821 accessions representing a large part of the Asian germplasm using 48 transcriptome-based SNPs. To date, the latter investigation is the largest in terms of accessions studied. Although the degree of overlap in terms of inventoried accessions between these collections is not currently known, international efforts (e.g., G2P-SOL; http://www.g2p-sol.eu/, DivSeek; https://divseekintl.org/) have been undertaken with the aims to a better management of these genetic resources and the dissection of their genetic potentialities, in order to face the actual challenges of agriculture (i.e., demographic growth, climate changes, etc.).

Pepper represents a rich source of health beneficial compounds, including vitamins, sugars, organic acids, flavonoids, and isoprenoids which provide a high nutritional value and a strong antioxidant capacity. The genus is best known for the presence of Capsaicinoids, a group of metabolites responsible for the pungency of fruits and produced in the placenta cells covering seeds. These compounds, which give a uniqueness to this crop among vegetables, are synthesized through the parallel biosynthetic pathways of phenylalanine and valine/leucine (Aza-González et al. 2011). Over 20 capsaicinoids are recognized, among them, capsaicin is the most represented accounting for over 80% of total content in spicy types.

The pungency sensation is due to the stimulation of receptors responding to harmful stimuli (Jordt and Julius 2002) and is under control of *Pun1*, a dominant gene mapped on chromosome 2 encoding for a putative acyltransferase 3 (*AT3*) (Stewart et al. 2005). Recessive alleles at *Pun1* locus leading to non-pungency have been identified in *C. annuum* (*pun1*), *C. chinense* ($pun1^2$), and *C. frutescens* ($pun1^3$) (Stewart et al. 2007). Molecular characterization has shown a 2.5-kb deletion located in the genomic interval between the promoter and the first exon of the *pun1* gene affecting the AT3 biosynthesis. For the gene $pun1^2$, a four base pair deletion in the first exon leads to a frameshift mutation resulting in the production of an AT3 truncated protein. These genetic disruptions lead to a reduced level of capsaicin levels up to 70% (Stewart et al. 2005). Stellari et al. (2010) reported a second gene, *Pun2*, regulating the pungency in the wild species *C. chacoense*. The *Pun2* locus has been mapped on chromosome 7, and the recessive allele *pun2* coding for loss of pungency has a recessive epistatic interaction with *Pun1*. Recent mapping studies allowed to identify in a cultivated pepper core collection, five candidate genes involved in capsaicinoid biosynthesis (Han et al. 2018). Moreover, additional candidate genes underlying capsaicinoid biosynthesis have been identified in panels of pungent and non-pungent cultivars and their related crossing progenies using a combination of GBS and transcriptome sequencing approaches (Park et al. 2018). The variation of the level of capsaicinoids is due to the complex interaction of the genotype (G) and environment (E). Although results evidenced how the response to environmental fluctuation depends on the type of plant material tested, the genotypic component is reported to play a major role in the accumulation and content of capsaicinoids (Zwedie and Bosland 2000; Gurung et al. 2011; Tripodi et al. 2018). However, more data and experiments are required to draw better conclusions.

Fresh peppers afford great quantities of ascorbic acid (vitamin C), which could reach up to five-folds the recommended daily dosage, particularly in local varieties (Mennella et al. 2018). The large variety of colors is due to carotenoids such as β-carotene, zeaxanthin, violaxanthin, and lutein, which provide yellow and orange pigments. Moreover, capsanthin and capsorubin are metabolites exclusive of peppers, which provide red fruit color. All these compounds protect cells against oxidative damage as a result of interacting with oxygen molecules and scavenging peroxyl radicals, ensuring beneficial effects on human health and preventing common degenerative diseases (Gomez-Garcia and Ochoa-Alejo 2013).

6.1.2 Domesticated and Wild Species, Provenance, Botanical Description, Phylogenesis

The knowledge on *Capsicum* is strongly centered around the most important domesticated species, namely *C. annuum*, *C. chinense*, and *C. frutescens*. However, the genus includes the other two domesticated species, *C. baccatum* and *C. pubescens*, as well as ca. 36 wild species (Fig. 6.1). The wild species have received so little attention that currently new species are being described (e.g., Barboza et al. 2019, 2020) while the identity of some collections is still under study (Barboza, pers. comm.). Besides, except for a few cases [e.g., *C. baccatum* (Eshbaugh 1970), *C. eximium* (Eshbaugh 1982), *C. lanceolatum* (Bosland and González 2000)], the only available information regarding the wild species is limited to the original descriptions, which use to be succinct in former times [e.g., *C. mirabile*, *C. campylopodium* (Martius 1846)].

The diversity of *Capsicum* was first thoroughly analyzed from a phylogenetic perspective a few years ago, including 34 out of the 35 species recognized by then, when hypotheses about species affinities and character evolution were proposed (Carrizo García et al. 2016). Later on, six new species were described, while the phylogenesis of the genus was revisited (Barboza et al. 2019; Carrizo García et al. *in prep*). The latest integral *Capsicum* phylogenetic reconstruction based on preliminary RAD-sequencing data has provided a slightly new evolutionary scenario, mostly congruent as regards the previous circumscription of clades (Carrizo García et al. *in prep*). Nine clades are recognized within *Capsicum*, two of them monotypic, with only one species position (*C. coccineum*) weakly resolved and not assigned to any clade (Fig. 6.1; Carrizo García et al. *in prep*). *Capsicum* clades can be in turn grouped into two strongly divergent lineages, one formed only by the Andean clade and the other one including the remaining ones. Most *Capsicum* clades can be associated with a particular geographic area of Central-South America, where their species are native to (Fig. 6.1), except for the Annuum and Baccatum clades that include several widespread species currently cultivated in different regions of the entire world. Wild Capsicums grow in a wide variety of environments, from rain forests, such as the Atlantic Forest of SE Brazil, the Yungas in Bolivia and NW Argentina, and the montane forests of Peru, to drier areas, such as those found in the Brazilian Caatinga ecoregion or the dry Chaco ecoregion in Bolivia and Argentina. Even so, each species is usually limited to certain conditions, which in some cases determine very narrow areas of distribution (e.g., *C. tovarii*, *C. friburgense*, *C. pereirae*), turning them into vulnerable species.

For most audiences, *Capsicum* species diversity is mostly unknown and commonly limited to those best-known, i.e., the cultivated *C. annuum* and its close allies, which are very similar to each other (except for the fruits). In general, *Capsicum* plants develop as (perennial) herbs to shrub that can be up to 4–5 m tall (Barboza et al. 2011; Carrizo García et al. 2013). Flowers are small-sized (usually ca. 1–1.5 cm wide or less, larger by exception, as in some cultigens), in general solitary (Fig. 6.2a, h, i, k, m), and less frequently 2–4 per node (Fig. 6.2b, d, f, g) or grouped in multi-flowered fascicles (Fig. 6.2e). One key feature to distinguish *Capsicum*, shared only with its sister genus *Lycianthes*, regards the calyx developed as an entire plate or cup-shaped

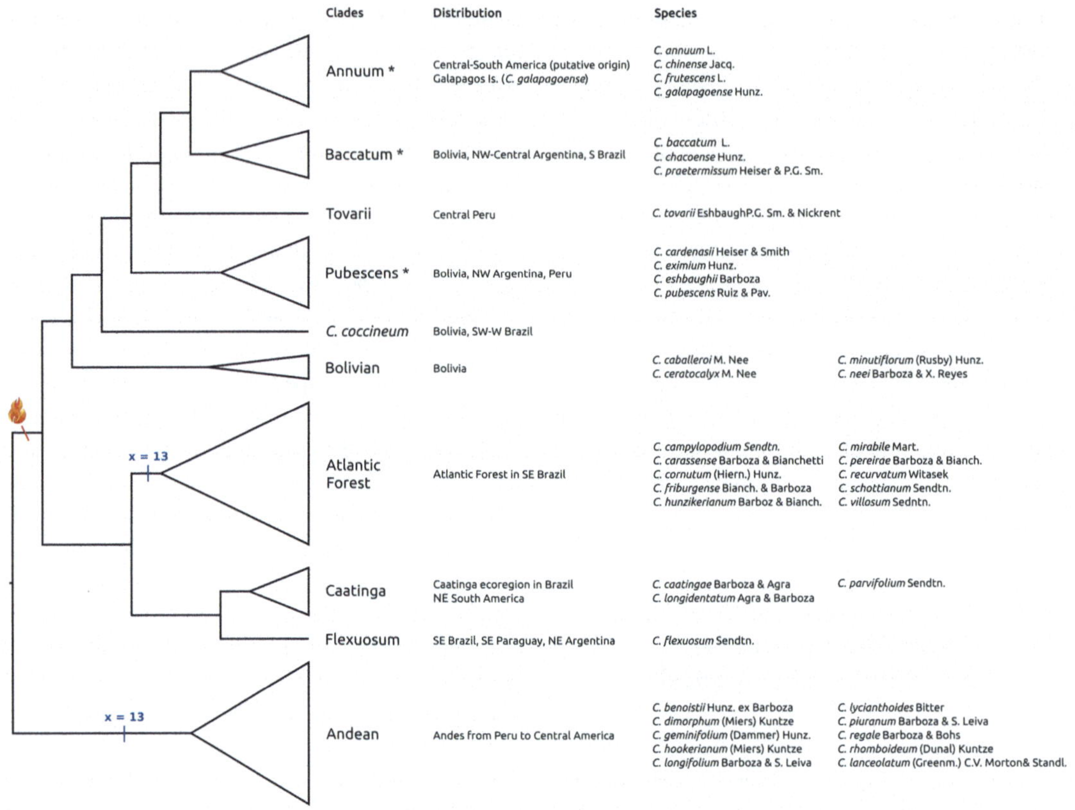

Fig. 6.1 Phylogenetic reconstruction of Capsicum, species grouping, and distribution. The terminal triangles represent multi-species clades (*C. coccineum* position is weakly resolved, not assigned to any clade). The putative origins of pungency (red) and the base chromosome number x=13 (blue) are marked. Species included in each clade and their geographic origin and distribution are listed on the right. Asterisks point to clades including domesticated species

structure, with or without tooth-like appendages (Figs. 6.2 and 6.3). The presence, number (most frequently five), length, and texture of the teeth are diagnostic characters useful to distinguish species, e.g., slender (Fig. 6.2m), short, and thick (Fig. 6.2i), barely noticeable (Fig. 6.2f), or completely absent (Fig. 6.3c). Corollas are mostly pentamerous, though more than five petals are often observed among cultigens (Fig. 6.2), and vary from stellate or rotaceous-stellate (Fig. 6.2a–f) to wide-campanular (Fig. 6.2g–j), exceptionally tubular, or tubular-urceolate (Fig. 6.2k, l). The pure white corollas of the domesticated *C. annuum* and allies are the best known (Fig. 6.2m), showing variable pigmentation only in some ornamentals. By contrast, a range of colors and spot patterns are found among the wild species. Yellow corollas are dominant in the Andean (except in *C. lanceolatum*, Fig. 6.2g–j) and Bolivian clades (Fig. 6.2b), purple and lilac is the common color in the Pubescens clade (Fig. 6.2f, k), and a broad variety of spotted corollas, i.e., green, golden-yellow, garnet spots on mostly white petals, are generally found in the Caatinga, Atlantic Forest, Flexuosum, and Baccatum clades (Fig. 6.2a, d), with some exceptions (Fig. 6.2c, e, l, m). Droplets of nectar may be perceptible on the petals, close to the throat of the corolla (Fig. 6.2a, f), due to particular ducts that convey the nectar to the outside, by capillarity, from the nectary at the base of the ovary (Vogel 1998). This attribute would lead to generalistic pollination (Vogel 1998), a relevant feature for breeding purposes.

Fig. 6.2 Flower diversity in *Capsicum*. A. *C. flexuosum*. B. *C. minutiflorum*. C. *C. praetermissum*. D. *C. recurvatum*. E. *C. caatingae*. F. *C. pubescens*. G. *C. rhomboideum*. H. *C. lanceolatum*. I-J. *C. lycianthoides*. K. *C. cardenasii*. L. *C. friburgense*. M. *C. chacoense*. Observe the differences in the position of the flowers, in the calyx teeth, and in the corolla shape and pigmentation. Photographs by C. Carrizo García (A-D, F-K, M) and M. Sterpetti (E, L)

Fig. 6.3 Fruit diversity in wild *Capsicum* species. A. *C. galapagoense*. B. *C. baccatum* var. *baccatum*. C. *C. flexuosum*. D. *C. rhomboideum*. E. *C. schottianum*. F. *C. recurvatum*. Photographs by C. Carrizo García

Capsicum fruits are berries, either sweet or hot. The diversity of fruit size, shape, and color among *Capsicum* cultigens (e.g., Tripodi and Greco 2018), as well as in the level of pungency, highly exceeds that registered among the wild Capsicums. The wild fruits are small rounded to ovoid berries (Fig. 6.3), ranging from 3–4 to 11 mm long, (oranged) red to garnet in the species of most clades (Fig. 6.3a–d), yellowish golden in the species of the Caatinga and Atlantic Forest clades (Fig. 6.3e, f), and singularly purple in *C. regale* (Barboza et al. 2020). The fruits of the wild Capsicums are also pungent in variable degrees, although no case has yet been recorded as pungent as the super-hot domesticated chiles. All species of the Andean clade, *C. longidentatum* (Barboza et al. 2011), and some populations of *C. chacoense*, *C. baccatum*, *C. eximium* (Tewksbury et al. 2006), and *C. flexuosum* (Garcés-Claver et al. 2007), have naturally non-pungent fruits. Fruit pungency would be a derived trait within the genus, which may have appeared after the divergence of the Andean clade (Carrizo García et al. 2016; Fig. 6.1). Another singular fruit feature is the development of noticeable giant cells in the pericarp, that give the typical blistered appearance to its inner surface; those cells are absent only in *C. baccatum* var. *umbilicatum* and across the Andean clade.

Seed shape, size, and color are variable features. According to their shape, two main seed types can be recognized, either discoidal and compressed or angular prismatic. Seed color ranges from pale creamy to brown, blackish-brown, or black (Carrizo García et al. 2016), while seed size ranges between ca. 2–6 mm wide. The seeds of certain species and clades are recognizable by a particular combination of these features, such as large discoid, curled, blackish-brown seeds in *C. pubescens* or small, prismatic and black in *C. flexuosum,* the Andean and Atlantic Forest clades.

Apart from the above-mentioned morphological attributes, another interesting feature is the presence of two basic chromosome numbers, i.e., $x = 12$ and $x = 13$ (Moscone et al. 2007; Carrizo García et al. 2016), a phenomenon known as disploidy. The basic chromosome number $x = 12$ characterizes most clades, including all the domesticated species and their closest allies (Carrizo García et al. 2016). The number $x = 13$ has only been recorded in two clades, the Andean and Atlantic Forest ones (Carrizo García et al.

2016; Barboza et al. 2019, 2020). Nevertheless, because these two clades are the more speciose, including altogether nearly half of *Capsicum* species, both basic chromosome numbers are almost evenly present in the genus. Based on phylogenetic and cytogenetic evidence, two independent origins have been hypothesized for the base chromosome number $x = 13$ (Fig. 6.1), appearing as derived states within the genus (Carrizo García et al. 2016), since $x = 12$ is found in the sister genus *Lycianthes* and also in other Solanoideae genera (e.g., Chiarini et al. 2010, and references therein).

The clades resolved within *Capsicum* after different approaches seem to be consistently circumscribed, except for the former Bolivian clade (Barboza et al. 2019; Carrizo García et al. 2016; Carrizo García et al. *in prep*). Different combinations of morphological, chemical, and cytological features are useful to characterize them. For instance, mostly yellow corollas, non-pungent fruits, without giant cells in the pericarp, and small, prismatic, black seeds are recorded among the species of the Andean clade, while generally stellate corollas, with different patterns of colorful spots, yellowish golden fruits, and black seeds are observed in the Atlantic Forest clade. A general trend of clades' diversification can be drawn across the genus, although some of them may vary the position (and therefore their relationships) according to the approach followed and the data used (cf. Carrizo García et al. 2016; Barboza et al. 2019). The latest phylogenetic hypothesis, obtained using genome-wide DNA sequence data, has contributed to strongly solve most previous controversial issues (Carrizo García et al. *in prep*). There is strong support to sustain that: all the domesticated species are resolved among the most derived clades (i.e., Annuum, Baccatum, and Pubescens); the Annuum and Baccatum clades are sister groups; *C. tovarii* and *C. flexuosum* are consistently resolved as monotypic lineages; the Andean clade forms the most diverging lineage that splits at the base of the genus (Fig. 6.1). Important unresolved issues involve species/clades centered in Bolivian territory. Since the knowledge about *Capsicum* diversity is still in progress, which evidences how little attention have received the wild species, more progresses would be possible if extensive field expeditions and new collections are done to gain a thoroughly understanding of the genus.

6.1.3 Interspecific Breeding in *Capsicum*

The main objectives in of pepper breeding regard the selection of high yielding varieties, good adaptation to drought and salinity cultivation conditions, resistances to various pests and diseases, qualitative traits in terms of the content of health-related compounds (capsaicinoids, vitamins, carotenoids, and flavonoids), marketability of the fruits for both industry and fresh consumption following the standard required by the market.

Despite the large number of domesticated and wild species, breeding activities in *Capsicum* occurred mainly at intraspecific level. At the same time, less use of interspecific hybridization has been made due to pre-zygotic (e.g., pollen rejection) and post-zygotic incompatibility (embryo abortion) as well as for the divergence of chromosome number between the cultivated pepper and many wild relatives. As a general rule, crosses of species of the same clade have a major probability to produce fertile hybrids, although the success depends on the accession used. Reciprocal crosses within the Annuum complex are possible between *C. annuum*, *C. chinense*, and *C. frutescens,* although a percentage of unsuccess can be observed (Manzur et al. 2015). Embryo rescue technique has been used to overcome the genetic barriers occurring in interspecific breeding, and its use is reported for the cross *C. annuum* X *C. baccatum* (Yoon et al. 2004; Manzur et al. 2015; Cremona et al. 2018). However, embryo rescue must be considered as a first step devoted to avoiding the embryo abortion since, in some cases, the sterility of hybrids occurred, and successive backcrosses were required (Yoon et al. 2004). The need of facilities for in vitro culture as well as the non-applicability of embryo rescue to all interspecific crosses makes these techniques not exploitable in many breeding programs. Therefore, the use of bridge species has been demonstrated as a valid approach to facilitate introgression breeding from *C. baccatum* to *C. annuum* (Yoon and Park 2005; Manzur et al. 2015). Two strategies were proposed: (1) a first cross *C. annuum* (female parent) × *C. chinense* (male parent), and then a subsequent cross of the hybrid as female parent to *C. baccatum* (male); (2) a first cross between *C. baccatum* (female) × *C. chinense* (male) followed by the cross of the obtained hybrid to *C. annuum* as male parent. As reported by Yoon and Park (2005) and Manzur et al. (2015), the first combination leads to a variable pollen viability or hybrid sterility and allows the recovery of only 25% of the cultivated genome. Instead, within the second scheme, a higher number of fertile combinations and a greater pollen viability have been observed, increasing the percentage of coverage to 50% of the cultivated genome. Overall, it must be recognized that the cross's success is genotype-dependent, and the use of the first or second bridge species strategy depends on the trait to introgress.

Interspecific breeding has been carried out to introgress qualitative traits and determine loci involved in the variation of size and shape. *Capsicum frutescens* and *C. chinense* were successfully crossed to *C. annum* in order to develop interspecific mapping population for the identification of quantitative trait loci (QTLs) related to fruit quality. In the cross between the large blocky cultivar Maor with the small-fruited accessions 'BG 2816' (*C. frutescens*) (Rao et al. 2003), eight QTLs involved in the variation of agronomic traits were detected. Experimental mapping population using the accessions 'PI 159,234' and 'PI 152225' of *C. chinense* allowed to identify two major QTLs on chromosome 10 (*fs 10.1*) and chromosome 3 (*fs 3.1*) that explained about 60 and 40% of phenotypic variation for the size and shape of the fruit, respectively (Ben Chaim et al. 2003a, b). Moreover, two major QTLs for fruit size and shape explaining over the 60% of variation were identified on the chromosomes 2 and 4 using lines carrying introgressions of *C. chinense* ('PI 152225') and *C. frutescens* ('BG 2816') into the *C. annuum* inbred line '100/63' (Zygier et al. 2005).

6.1.4 Potentiality of *Capsicum* Genetic Resources in Breeding for Disease Resistance

Several pathogens affect the productivity and quality of pepper causing severe damage to the cultivations, including fungal and bacterial diseases, nematodes, and insects. Fungal diseases include pathogens damaging the vegetative parts of the plants and the fruits such as powdery mildew (*Leveillula taurica*) and antrachnose (*Colletotrichum* spp.), as well as soil-borne pathogens such as *Phytophthora capsici* (Phytophthora crown and root rot), *Verticillium dahliae* (Verticillum wilt), *Fusarium* spp, (Fusarium wilt) and *Rhizoctonia solani* (Rhizoctonia root rot). Root-knot nematodes are also very destroying pests in pepper.

6.1.4.1 Fungal and Bacterial Diseases

Powdery mildew, casual agent *Leveillula taurica*, is a dangerous disease for cultivated pepper occurring in warm and humid conditions. Source of resistances were found in the secondary and tertiary *Capsicum* genepool. Different accessions of *C. frutescens* ('IHR 703'), *C. chinense* ('KC616'), *C. baccatum* ('CNPH' genotypes), and *C. pubescens* ('KC' accessions) were found to be from moderate to highly resistant being used as parent line in genetic improvement programs (Anand et al. 1987; De Souza and Café-Filho 2003; McCoy and Bosland 2019). Genetic mapping has been instead mainly performed within the *C. annuum* background using an intraspecific double haploid (DH) population derived from the cross 'H3' (resistant) × 'Vania' (susceptible), allowing to identify QTLs in five chromosomal regions (Chr 5, 6, 9, 10, 12) with additive and epistatic interaction (Lefebvre et al. 2003). Recently, the gene *PMR1* located on chromosome 4 has been mapped in intraspecific

C. annuum F_2 and F_3 segregant populations using the resistant accessions' 'VK515R' and 'Singang' (Jo et al. 2017).

Fruit rot antrachnose is due to different *Colletotrichum* species, which cause typical lesions on the fruits characterized by concentric rings making the fruits not marketable (Yumi Baba et al. 2019). Source of resistance to *C. truncatum*, *scovillei*, and *C. capsici* was found in *C. chinense* ('PBC932') and *C. baccatum* ('PBC80', 'PBC81') (Mongkolporn and Taylor 2018). These genotypes have been used for genetic mapping (Voorrips et al. 2004; Kim et al. 2010; Ying et al. 2015) allowing to identify about 30 QTLs distributed on six chromosomal regions and QTLs with major effects on chromosomes 5, 8 and 9.

Among soil-borne disease, *Phytophthora capsici* is undoubtedly the most destructive and widely spread pathogen in pepper, causing root deterioration, lesions on the stem, wilting, stunting, and necrosis of the plant (Quirin et al. 2005). Breeding efforts have been carried out within the *C. annuum* genepool due to the lack of source of resistances in the other *Capsicum* species (Barchenger et al. 2018). Since 2003, different mapping studies have been carried out, most of which using the highly resistant cultivated pepper genotype 'Serrano Criollo de Morelos CM334' (Thabuis et al. 2004; Quirin et al. 2005; Naegele et al. 2014a). All investigations reported the chromosome 5 as the main genetic region carrying QTLs with effect on the resistances to Phytophthora root rot. More recently, Siddique et al. 2019 using 188 $F_{7:8}$ RILs [*C. annuum* CM334 (as resistant parent) x ECW30R (as susceptible line)] detected 117 significant SNPs across the genome associated with *Phytophthora capsici* resistance, including SNPs on chromosomes 5, 7, and 11 that colocalized with the QTLs identified here and in previous studies. Verticillium wilt represents a serious threat to pepper production requiring every year costly and non-environmentally friendly strategies for its management including soil fumigants and fungicides (Vasileva et al. 2019). Sources of resistance have been found in an extensive study involving the screening of about 400 pepper genotypes against two *V. dahliae* strains (Gurung et al. 2015). Several *C. annuum* accessions from Mexico showed a different degree of resistance, while, among wilds and domesticated, two *C. frutescens* (PI 281396, PI281397) accessions exhibiting a low level of vascular discoloration, due to the fungal, attacks were identified. These genotypes can be considered as robust sources of resistance and easily used in breeding programs. In addition, specific PCR based markers have been developed to assist breeding for *Verticillium* (Barchenger et al. 2017).

Fusarium spp. and *Rhizoctonia solani* represent polifagous and destructive pathogens for which yield losses could reach peaks of 80% (Loganathan et al. 2013). For the former, germplasm screening allowed to identify different resistant accessions in the cultivated background (Anaya-López et al. 2011; Maruti et al. 2014), for the latter, two *C. baccatum* genotypes ('PI 439410' and 'PI 5556119') were found to the highly resistant, while the *C. annuum* accessions 'Long Chili' (a Korean hybrid) and 'PI 167061', exhibited over than 65% of survival plants post-inoculation. Although various disease screening studies have been carried out, allowing the identification of accessions useful in breeding programs, the genetic basis of the resistances has not yet been identified. Nevertheless, it has been recently found that the expression of the *CaChi2* gene confers a level of resistance to *Fusarium oxysporum* (Ferniah et al. 2018).

Bacterial wilt in pepper is caused by 3 species of *Ralstonia solanacearum*, (Elphinstone 2005; Safni et al. 2014) grouped in species complex and classified into 'races' (Buddenhagen et al. 1962; Hayward 1964; He et al. 1983) and 'biovars' (Hayward 1964, 1991).

Sources of resistance were found in *C. baccatum* and *C. chinense* accessions (Lebeau et al. 2011) as well as in cultivated *annuum* genotypes ('Perennial', 'Narval', 'MC4', 'CA8', 'PI 3222719', 'LS2341'). The genetic basis has been investigated by Mimura et al. (2009), which identified a major QTL on chromosome 1 in a *C. annuum* intraspecific DH population obtained crossing the resistant cultivar "LS2341' to 'Yolo

Wonder' (susceptible cultivar). More recently, a major QTL (*qRRs-10.1*) was identified on chromosome 1 in the same species using SLAF-BSA and QTL mapping by Du et al. (2019). Different species of *Xanthomonas* cause bacterial spot including *X. vesicatoria and euvesicatoria, X. perforans, and X. gardneri, and.* Currently, seven races virulent of *X. campestris* pv. *vesicatoria* infecting pepper have only been differentiated based on resistance conferred by six genes of resistance, four of which are dominant (*Bs1-Bs4*), and two recessives (*bs5-bs6*). The genes *Bs1* and *Bs3* were found in the *C. annuum* accessions 'PI 163192' and 'PI 271322', *Bs2* in *C. chacoense* 'PI 260435' and *Bs4* in *C. pubescens* 'PI 235047' (Kim and Hartmann 1985; Hibberd et al. 1987). The recessive genes *bs5* and *bs6* were discovered in the *C. annuum* accessions 'PI 271322', 'Pep13', and 'PI 163192' (Jones et al. 1998; Csillery et al. 2004). Although no QTL studies are reported, genetic mapping allowed to detect markers linked to the genes that can facilitate introgression breeding (Tai et al. 1999; Pierre et al. 2000; Romer et al. 2010).

6.1.4.2 Viral Diseases

The number of virus species infecting pepper crops and their incidences has increased considerably in the past 30 years due to a combination of several factors, which increased the enlargement of viruses and vectors such as the enlargement of commercial trades and climate changes. It is possible to count about 50 species causing damages to pepper (Kormelink 2011) for which preventing strategies such as the use of genetic resistances are essential for the control.

Among insect-transmitted viruses, it is possible to count potyviruses, cucumoviruses, and tospoviruses, which are more frequent and severe in open fields, although it is possible to observe them frequently in protected crops. Among Tospoviruses, TSWV (*Tomato spotted wilt orthotospovirus*) is the most dangerous and spread dangerous. Hypersensitive response-based resistance to TSWV has been identified in the *C. chinense* accessions 'PI 152225', 'PI 159234', and 'PI 159236', which are the most adopted in introgression breeding programs (Black et al. 1991; Boiteux 1995). From these resistance sources, the dominant gene *Tsw* has been mapped in the long arm of chromosome 10 (Jahn et al. 2000), and markers tightly linked were identified in derived F_2 populations (Moury et al. 2000). Several other sources of resistances carrying the *Tsw* gene have been found in wild and domesticated species (Cebolla-Cornejo et al. 2003; Di Dato et al. 2015). However, the occurrence of resistance-breaking (RB) strains able to overcome *Tsw* have been reported in different countries (Roggero et al. 2002; Ferrand et al. 2015; Jiang et al. 2017) and for which efforts to identify the source of germplasm resistant are ongoing (Parisi et al. 2015; Almási et al. 2016; Soler et al. 2015). Interesting results are reported by Soler et al. (2015), which identified in the *C. baccatum* accession 'PIM26-1a' good level of tolerance to both wild type and RB-TSWV isolates.

Potyviruses is another big family, including nine viral species of which PVY (*Potato virus Y*) is the most widespread. Several recessive (*pvr1*, *pvr2*^{1}, *pvr2*^{2}, *pvr3*, *pvr5*, *pvr6*, *pvr 8*) and dominant (*Pvr4*, *Pvr7*) genes are reported and named according to their chronological order of identification (Dogimont et al. 1996; Kyle and Palloix 1997). Resistances have been reported in different *C. annuum*, *C. frutescens* and in *C. chinense* cultivars (Kyle and Palloix 1997; Yeam et al. 2005; Di Dato et al. 2015; Parisi et al. 2020). The chromosomal positions of *pvr* genes were retrieved through allelisms tests involving the wild *C. chinense* accessions PI 159236 and PI 152225, allowing to position *pvr2* on chromosome 4 and *Pv4 and Pvr7* on chromosome 10 (Murphy et al. 1998; Grube et al. 2000). A new *PepMoV* resistance colocalizing at the *Pvr4* and *Pvr7* loci, has been recently reported (Venkatesh et al. 2018). Considering *pvr6*, it was mapped on chromosome 3 using intraspecific *C. annuum* F_2 lines (Caranta et al. 1996). QTL studies have been carried out in the genetic background of *C. annuum* using DHs derived from the cross CM334 × Perennial (Caranta et al. 1997; Quenoille et al. 2014) evidencing main QTLs closely linked to *pvr2* and *pvr 6*.

Aphids *Myzus persicae* and *Aphis gossypii* mainly transmit *Cucumber mosaic virus*

(CMV) of the *Cucumovirus* genus. In wild peppers, sources of resistance were identified *in C. frutescens* ('BG2814-6' and 'PBC8688') and *C. baccatum* ('PI 439381-1-3') (Grube et al. 2000; Suzuki et al. 2003; Guo et al. 2017), while within cultivated pepper, 'Vania' and 'Perennial' resulted the most used in breeding, exhibiting a resistance from moderate to high (Caranta et al. 1997, 2002). Mechanism of defense included partial resistance to initial virus infection, inhibition of virus replication, and inhibition of movement through cells.

Pepper resistance against CMV is given by two single gene. The first one, *Cmr1*, a dominant gene located on chromosome 2 and able to cover a broad range of CMS strains (Kang et al. 2010), has been extensively used in breeding until overcome by the isolate CMV-P1. The second one, *cmr2*, a recessive gene located on chromosome 8 and conferring resistance to CMV-P1, was recently discovered by Choi et al. (2018). Moreover, different QTLs have been identified in intraspecific segregating populations in *C. annuum* (Ben Chaim et al. 2001; Caranta et al. 1997, 2002) and *C. frutescens* (Guo et al. 2017) leading to the identification of major QTLs on chromosomes 5 and 11.

Regarding tobamoviruses, PMMoV (*Pepper mild mottle virus*), TMV (*Tobacco mosaic virus*), and ToMV (*Tomato mosaic virus*) are the most common in pepper. Several alleles (L^1-L^4) conferring resistant to pathotypes infecting pepper have been identified. The alleles L^1-L^3 are specific for certain pathotypes and have been identified in both cultivated and domesticated pepper (Lefebvre et al. 1995; Tomita et al. 2008). The allele L^4, which has a broad spectrum of resistance against PMMoV, TMV, and ToMV (Yang et al. 2009), have been identified from the wild *C. chacoense* (Matsunaga et al. 2003).

For viral diseases, several molecular markers linked to the genes have been detected. Di Dato et al. (2015) carried out a comprehensive study analyzing markers for different pathogens in nine *Capsicum* species and verifying the correspondence marker/resistant phenotype based on historical data. The markers' transferability across species resulted to be conditioned by the breaking of the linkage between markers and target genes. As a consequence, allelic markers such as those used to detect resistance to potyviruses resulted to be robust due to the absence of recombination. Therefore, assisted breeding could benefit from the work carried out during the last 20 years for to develop molecular markers linked to pathogen resistance.

6.1.4.3 Nematodes

Root-knot nematodes (RKN) are widespread and represent the major pathogens in vegetable production regions of the world (Barbary et al. 2015). It is known that 90 species are described as belonging to the genus *Meloidogyne.* Among them, *M. incognita* and *M. javanica* are the most important species for the sweet pepper. However, the species *M. enterolobii* has gained importance (Brito et al. 2007; Pinheiro et al. 2015). They cause damage by feeding and by inducing large galls or 'knots' throughout the root system of infected plants, which can alter the uptake of water and nutrients. The nematode infection of roots increases the incidence and severity of fungal wilt diseases, negatively influencing yield and quality of crops (Katsantonis et al. 2003).

Genetic control is the most sustainable way to manage RKN since it poses no risk to human health; it is relatively of low cost and does not pollute the environment (Barbary et al. 2015).

In *C. annuum*, many genes showing resistance to RKN is associated with several dominant genes (the *Me* genes) that are thought to act independently in gene-for-gene interactions (Djian-Caporalino et al. 1999). Six existing *Me* genes have reported: some specific to certain *Meloidogyne* species (*Me4*, *Mech1*, and *Mech2*), whereas others are effective against a wide range of *M. arenaria*, *M. javanica*, and *M. incognita*, (*Me1*, *Me3*, and *Me7*). Useful markers tightly linked to *Me1* were recently developed by Wang and collaborators (2018). Comparative mapping data suggest that the three clusters of R genes conferring resistance to nematodes are conserved within and between solanaceous species crops (eggplant, pepper, and tomato) (Djian-Caporalino et al. 2007).

Cultivated and wild *Capsicum* germoplasm has been screened by several researchers to identify suitable sources of resistance for *annuum* breeding programs. Then, different accessions of *C. chinense* ('PI 441641', '201-16', '201-21') showed resistance to *M. hapla*, as well as *C. baccatum* ('PI 439381' and 'PI 267729'), *C. frutescens* ('589-20'). Resistances to *M. incognita* were detected in accessions of *C. chinense* ('PA-353', 'PA-398', 'PA-426', '201-26', '547-7'), *C. chacoense* ('528-8', '529-8', '46-530/7'), *C. annuum* ('PR205' accession carrying *CaMi* gene), 'Carolina Cayenne' (carrying *N* gene and an additional recessive gene), and *C. frutescens* ('28-201' and 'Santanka XS') (Di Vito et al. 1991, 1993; Chen et al. 2007; Fery and Dukes 1996; Thies and Fery 2002). The last accession was also resistant to *M. arenaria* and *javanica* and heat-stable resistant (Thies et al. 1997). Finally, some genotypes (*C. chinense* 201-16 and '201-21' accessions and *C. frutescens* '589-20' accession) were the most promising for their multiple resistances to different species of RKN (*M. hapla*, *incognita*, *arenaria*, and *javanica*) (Di Vito et al. 1991). Recently, a good source of resistance to *M. enterolobii*, which is the most aggressive species, was detected in 'UENF 1730' accession and in BRS Nandaia (*C. chinense)* (Pinheiro et al. 2015, 2020). In addition, *C. annuum* accessions 'PM217' and 'PM702' were found as resistant to *M. chitwoodi* (Djian-Caporalino et al. 2007).

6.1.5 Genetic Resources Of *Capsicum* in Breeding for Tolerance/Resistance to Abiotic Stresses

Climate change and global warming, increasing salinity in water and soil, and the need to provide water resources represent limiting factors to the cultivation of pepper. The optimal temperature for pepper ranges from 20 to 30 °C, with impacts on plant growth and reproduction below and above these values (Berke et al. 2005; Guo et al. 2014). Sources of tolerance to heat stress and low temperature have been detected in the genetic resource of *Capsicum*. Habanero pepper (*C. chinense*) is reported to activate mechanisms able to reduce leaf temperature and increase CO_2 assimilation when temperature increase occurs (Garruna-Hernandez et al. 2014). Usman et al. (2014) reported different accessions of *C. annuum* and *C. frutescens*, selected from the AVRDC genebank, tolerant to high temperature measured based on membrane thermostability phenotyping parameters. *Capsicum annuum* ornamental peppers are reported to tolerate well low temperatures in mature plants (Stommel and Bosland 2007). Different ornamental cultivars have been identified as a good source of resistance/tolerance to heat ('Explosive Ember', 'Chilly Chili', 'Medusa', 'Thai Hot', 'Treasures Red') and cold ('Black Pearl', 'Red Missile', and 'Salsa Yellow') (Gajanayake et al. 2011). Genomic approaches involving the sequencing of the seedling transcriptome of resistant (*C. annuum* 'R597') and susceptible (*C. annuum* 'S590') pepper lines, allowed to identify about 4000 differentially expressed genes annotated as heat shock proteins (highly expressed in 'S590'), heat shock transcription factors, hormone, as well as calcium and kinase signaling (highly expressed in 'R597') (Li et al. 2015).

Drought stress in pepper severely influence the physiology of the plant, causing disturbance of turgor pressure in cells and affecting main biological processes, which reduce plant growth and crop yield. Okunlola et al. (2017) showed that moderate and severe drought induces significant reduction not only of leaf relative water content and growth rate but also of metabolites (carotenoid, chlorophyll), demonstrating how pepper plants need more attention at vegetative stage than flowering or fruiting stages during drought stresses. The authors evidenced how *C. chinense* under different drought treatments increased the phenol content at vegetative stage concerning *C. annuum* and *C. frutescens*, also reducing the capsaicin content at fruiting stage. These mechanisms emphasize the high tolerance to drought stress of *C. chinense*. Moreover, it was demonstrated a relationship between the content of capsaicinoids and the tolerance to drought. Under drought conditions, high pungent cultivars showed a higher water retention and a minor yield reduction respect to those with low or medium pungency, regardless of the species (i.e., *C. annuum* cv 'Perennial' and *C. chinense* 'BGH1719') (Phimchan and Techawongstien 2012). Recently, correlations between the content of antioxidants in seeds and water stress-tolerant traits in seedlings of *C. annuum* have been reported (Sahytia et al. 2018). The genotypes grown under water deficit exhibited high total phenolics, high DPPH scavenging activity, and proline accumulation. The latter is an osmolyte allowing to maintain turgor pressure in the cell by drawing extracellular water as a strategy employed by plants to achieve tolerance toward drought (Sahytia et al. 2018). Furtermore, tolerances to drought stress in diverse *C. frutescens* and *C. chinense* cultivars revealed mechanisms due adaptation to zones of origin (Bernau et al. 2020).

The development of salt-tolerance cultivars is highly demanded in pepper. It has been demonstrated how stressed plants drastically reduce both water use efficiency (WUE) and nitrogen use efficiency (NUE) (Huez-Lopez et al. 2011). In particular, the excess of sodium and chloride ions in the cells reduces plant growth increasing the production of reactive oxygen species (Munns and Tester 2008). Several *C. annuum* cultivars were reported to be highly tolerant to salinity treatments ('Early Jalapeno', 'AZ-20', 'Ancho', and 'Cayenne'), while Habanero was reported to have a low survival under high salt concentrations (Niu and Rodriguez 2010; Niu et al. 2010). Although several studies have been carried out in order to identify the genes responsible for the resistance/tolerance to salinity in pepper, as reviewed by Chhapekar et al. (2018), to date, there are not extensive studies performed to develop germplasm resources resistance to salinity and no exotic libraries have been developed.

6.2 Eggplant

6.2.1 Introduction

Humans have domesticated three eggplants species, of which the most important is the common or brinjal eggplant (*Solanum melongena*), whose center of origin is southeast Asia (Meyer et al. 2012). The two other cultivated species are the scarlet eggplant (*S. aethiopicum*) and the gboma eggplant (*S. macrocarpon*), which were domesticated in Africa (Lester and Daunay 2003). While the common eggplant is a well-known vegetable in most regions of the world, scarlet and gboma eggplants are less known and are mostly cultivated in sub-Saharan Africa (Schippers 2000) and occasionally in some other parts of the world, such as Southeast Asia, and in the case of the scarlet eggplant also in the Caribbean, Brazil and some parts of the south of Italy (Sunseri et al. 2010). The three cultivated species are diploid ($2n = 2x = 24$).

It is unknown when the domestication of eggplants took place. The oldest written record referring to common eggplant is from 300 BC in Sankskrit documents from the Indo-Burma region (Daunay and Janick 2007). Some centuries later, eggplants are mentioned in Chinese documents (Wang et al. 2008). The westward migration of common eggplant took place through Persia, and its expansion throughout the Mediterranean basin took place with the Muslim expansion in the Middle East, North of Africa, and Spain of the seventh and eighth centuries (Daunay and Janick 2007). From there, it spread to other places such as the rest of Europe and America and Oceania with European colonization (Nuez et al. 2002). Much less is known of the domestication and spread of scarlet and gboma eggplants, but it has been suggested that the presence of the scarlet eggplant in the Caribbean and Brazil probably was a result of the slave trade from Africa (Plazas et al. 2014).

The number of accessions of cultivated eggplants conserved in genebanks, according to AVGRIS and Genesys databases, is of 6632 (Taher et al. 2017). Of these, 5665 are of common eggplant, 798 of scarlet eggplant, and 169 of gboma eggplant. Regarding wild eggplant relatives, 1304 accessions are known to be stored in germplasm banks, with the species with higher representation being *S. incanum* (167 accessions), and *S. torvum* (132 accessions). A gap analysis study performed by Syfert et al. (2016)

revealed that eggplant wild relatives are underrepresented in germplasm collections and that efforts should be devoted to their collection and conservation. The largest collections of eggplant genetic resources, according to Genesys, are those of the World Vegetable Center (Taiwan), the Plant Genetic Resources Conservation Unit at the University of Georgia, USDA-ARS (USA), the Center for Genetic Resources (The Netherlands), and the NI Vavilov Research Institute of Plant Genetic Resources (Russia) (Taher et al. 2017). Also, important collections not reporting to Genesys are those of the Institute of Vegetables and Flowers (China), the National Bureau of Plant Genetic Resources (India) and the gene bank of Institute of Vegetable and Floriculture Science, Naro (Japan). As occurs with pepper, the G2P-SOL project; http://www.g2p-sol.eu/) is aimed at a better organization, detection of duplicates, and enhancement of the eggplant genetic resources at a global level.

6.2.2 Domesticated Eggplants and Their Wild Relatives

The three domesticated eggplants (*S. melongena*, *S. aethiopicum* and *S. macrocarpon*) belong to *Solanum* subgenus *Leptostemonum*. They are related to a large number of phenotypically variable (Figs. 6.4 and 6.5) African and Asian wild species of this subgenus, collectively known as the 'spiny' solanums group, due to the presence of prickles in many species of the group (Lester and Daunay 2003; Knapp et al. 2013; Vorontsova et al. 2013; Vorontsova and Knapp 2016). The subgenus *Leptostemonum* is the largest major clade in genus *Solanum* and it contains around 450 species from the Americas, where the greatest diversity exists, and from the Old World, mostly from Africa and from Oceania (Vorontsova et al. 2013; Vorontsova and Knapp 2016). Within this subgenus, *S. melongena* and *S. macrocarpon*, together with their respective wild ancestors *S. insanum* and *S. dasyphyllum* (Lester and Daunay 2003; Knapp et al. 2013) are considered as members of section *Melongena* (D'Arcy 1972), while *S. aethiopicum* and its wild ancestor *S. anguivi* (Lester and Daunay 2003) are included within section *Oliganthes*. However, recent molecular studies (Acquadro et al. 2017; Gramazio et al. 2017a) suggest that the genetic distances between the three species are similar and that *S. macrocarpon* and *S. melongena* should be included in separate sections. The three cultivated species are partially interfertile (Oyelana and Ugborogho 2008), and therefore, each of them is a source of variation for traits of interest for the others (Prohens et al. 2012). In the same way, wild relatives of potential interest for common eggplant breeding might also be useful for the genetic improvement of the scarlet and gboma eggplants. However, most studies on the taxonomy relationships and interest of wild eggplant relatives for breeding purposes have been performed focusing on the common eggplant (*S. melongena*). Therefore, the subsequent discussion on genepools and the eggplant complex will be in relation to the common eggplant *S. melongena*.

At present, it is considered that the primary genepool of common eggplant contains a single wild species (*S. insanum*), which is the wild ancestor of eggplant (Knapp et al. 2013; Ranil et al. 2017). Hybrids and segregating generations between *S. melongena* and *S. insanum* are completely fertile (Kouassi et al. 2016; Plazas et al. 2016). Indeed, intermediate forms, frequently weedy, exist between both species (Knapp et al. 2013; Ranil et al. 2017) and evidence of natural hybridization and introgression between the two species has been described (Davidar et al. 2015; Mutegi et al. 2015). The proximity of both species at the morphological level is high, and several authors considered *S. insanum* as part of *S. melongena* (Lester and Hasan 1990; Daunay et al. 2001), or as a botanical variety of *S. melongena* (*S. melongena* var. *insanum*) (Isshiki et al. 1994). In this respect, Lester and Hasan (1990) and Daunay et al. (2001) considered two groups of *S. melongena* (groups E and F), which correspond to what is currently considered *S. insanum* (Knapp et al. 2013; Ranil et al. 2017).

The secondary genepool is constituted by many wild species of African and Asian origin within the subgenus *Leptostemonum* (Syfert et al.

Fig. 6.4 Flower diversity in eggplant wild relatives. A. *S. insanum*. B. *S. anguivi*. C. *S. campylacanthum*. D. *S. dasyphyllum*. E. *S. incanum*. F. *S. pyracanthos*. G. *S. linnaeanum*. H. *S. tomentosum*. I. *S. elaeagnifolium*. J. *S. sisymbriifolium*. K. *S. torvum*. Photographs by M. Plazas and J. Prohens

Fig. 6.5 Fruit diversity in eggplant wild relatives. A. *S. insanum*. B. *S. anguivi*. C. *S. campylacanthum*. D. *S. dasyphyllum*. E. *S. incanum*. F. *S. linnaeanum*. G. *S. tomentosum*. H. *S. elaeagnifolium*. I. *S. torvum*. Photographs by M. Plazas and J. Prohens

2016). These last authors listed 48 species from the secondary genepool belonging to the Eggplant clade (nine species), Climbing clade (three species), and Anguivi grade (36 species). Of these, the closest to eggplant is the Eggplant clade which includes, in addition to the cultivated eggplant *S. melongena* and the primary genepool species *S. Insanum*, the following nine wild species (Syfert et al. 2016): *S. agnewiorum*, *S. aureitomentosum*, *S. campylacanthum*, *S. cerasiferum*, *S. incanum*, *S. lichtensteinii*, *S. linnaeanum*, *S. rigidum*, and *S. umtuma*. The wild *S. campylacanthum, S. incanu*m, and *S. lichtensteinii* were considered by Lester and Hasan (1990) and Daunay et al. (2001) as different forms of *S. incanum sensu lato* (Knapp et al. 2013) and included by these authors as part of the 'eggplant complex' of closest species to eggplant. The Anguivi grade is the next closest group to cultivated eggplant (Vorontsova et al.

2013). Apart from African species, this group includes some Asian species, such as *S. violaceum* (Aubriot et al. 2016), and two endemisms from the Canary Islands (*S. lidii* and *S. vespertilio*) (Prohens et al. 2007), as well as the wild ancestors of the scarlet and gboma eggplants (Plazas et al. 2014). Finally, the Climbing clade is the phylogenetically least close to *S. melogena* among the secondary genepool species (Vorontsova et al. 2013) and includes the wild species *S. stipitatostellatum*, *S. richardii*, and *S. zanzibarense*. Some of these secondary genepool species are known to give partially fertile hybrids (with a different range of fertility levels among species) with *S. melongena* and therefore can be effectively used in eggplant breeding (Prohens et al. 2013; Rotino et al. 2014; Liu et al. 2015; Kouassi et al. 2016; Gramazio et al. 2017b). Although for some species considered as part of this secondary genepool, the development of hybrids with eggplant has not been reported, such as in the case of the wild species of the Climbing clade, they were considered as part of the secondary genepool based on phylogenetic proximity either to cultivated eggplant or to wild species with which hybrids with eggplant can be obtained (Syfert et al. 2016). Other species of *Solanum* subgenus *Leptostemonum* of African and Asian origin (Vorontsova et al. 2013; Aubriot et al. 2016; Vorontsova and Knapp 2016) might also be incorporated in the future to this secondary genepool when more information on their crossability of phylogenetic relationships becomes available.

The tertiary genepool includes New World species of *Solanum* subgenus *Leptostemonum* with which hybridization is rarely successful (Rotino et al. 2014; Kouassi et al. 2016; Plazas et al. 2016) and in which special techniques such as embryo rescue or somatic hybridization have to be used to obtain hybrids, which generally have a high degree of sterility (Gleddie et al. 1986; Bletsos et al. 1998). Although few species of particular interest for eggplant breeding, such as *S. elaeagnifolium*, *S. sisymbriifolium*, *S. torvum*, or *S. viarum* (Daunay and Hazra 2012; Rotino et al. 2014) are recognized as members of the eggplant tertiary genepool (Syfert et al. 2016; García-Fortea et al. 2019), many other species unexplored for potential sexual or somatic hybridization with eggplant both from the Americas and Australia may in a future be included as members of this tertiary genepool.

A recent gap analysis of eggplant wild relatives performed by Syfert et al. (2016) revealed that many eggplant wild relatives are poorly represented in germplasm banks and that specific *ex situ* and *in situ* conservation actions are needed for the conservation of wild eggplant relatives. In this respect, hot spots for diversity of wild eggplant relatives have been identified in eastern and southern Africa, where many wild relatives of eggplant are naturally present (Syfert et al. 2016; Vorontsova and Knapp 2016). The evaluation of the extinction risk of crop wild relatives (CWRs) revealed that the species *S. lidii*, an endemism from the Canary Islands (Prohens et al. 2007; Gramazio et al. 2020) is critically endangered, other nine (*S. agnewiorum*, *S. aldabrense*, *S. deflexicarpum*, *S. inaequiradians*, *S. litoraneum*, *S. malindeiense*, *S. setaceum*, *S. torreanum*, and *S. vespertilio*) are threatened, three (*S. nigriviolaceum*, *S. platacanthum*, and *S. rigidum*) are near threatened, and one (*S. ruvu*) is probably extinct in the wild (Syfert et al. 2016). Unfortunately, to our knowledge, none of these species, except *S. lidii* and *S. vespertilio* (Kaushik et al. 2016), has been evaluated for eggplant breeding.

6.2.3 Interspecific Hybridization and Introgression Breeding in Eggplant

Common eggplant exhibits a wide diversity in phenotypic, physiological, and biochemical characteristics, (Portis et al. 2015; Kaushik et al. 2016), and several varieties of eggplant improved with regard to the fruit size, weight, and shape, and some resistance to diseases and pests were developed mainly through the classical breeding methods by exploiting the intraspecific variability (Kalloo 1993; Rotino et al. 2014; Miyatake et al. 2016). However, despite its large morphological diversity, its genetic diversity is narrow, particularly in the modern cultivars (Muñoz-Falcón et al. 2009). In Europe and some Asian countries, and mostly where the cultivation system became more intensive, the release of F_1 hybrids has contributed to the loss of eggplant landraces thus, unavoidably, causing genetic erosion in *S. melongena* (Daunay et al. 1997; Daunay and Hazra 2012). The limited genetic diversity of cultivated eggplant (*S. melongena*) contrasts with the large morphological and genetic variation present in the wild eggplant relatives (Meyer et al. 2012; Vorontsova et al. 2013; Vorontsova and Knapp 2016). Wild and allied species of eggplant are an important reserve of potential genetic variability and allelic variation for the traits underlying many agronomic and qualitative features of plant and fruit as well as for the content of nutritional and functional compounds. They could represent a source for broadening the genetic background of eggplant (*Solanum melongena*) for developing a new generation of eggplant cultivars with dramatically improved yield and quality, as well as for addressing the challenges posed by adaptation to the climate change. Contrarily to many other crops, in which wild species have mainly been used as source for introgression breeding (Hajjar and Hodgkin 2007; Warschefsky et al. 2014), eggplant breeders have neglected their potential for long, and wild relatives have not made a relevant contribution to the development of new eggplant cultivars (Daunay and Hazra 2012; Rotino et al. 2014).

The use of conventional breeding methods to introgress the traits of interest from allied species was used only sporadically, as many wild relatives displayed partial cross-compatibility with the cultivated species, thus hampering their employment in crop improvement (Ano et al. 1991; Blestos et al. 1998). Moreover, the capability of eggplant to cross to species of other genera or subgenera resulted very low (Daunay et al. 1991) and attempts at more distant crosses as that of *S. melongena* with S. *lycopersicon* also resulted in sterile hybrids (Rao 1979). Although interspecific hybrids between eggplant and wild relatives have been obtained through sexual crosses using 27 species (Daunay and Hazra 2012; Rotino et al. 2014; Devi et al. 2015; Plazas et al. 2016), most of the studies have been conducted for taxonomic purposes and preliminary breeding works and have not undertaken the development of backcross generations. Moreover, these hybrids displayed different degrees of viability and fertility, and only a limited number of the employed species were able to develop fertile or partially fertile F_1 progenies that were successfully backcrossed to the cultivated *S. melongena*. Indeed, interspecific crosses with *Solanum* species display stepwise crossability patterns ranging from the development of seedless fruits from crosses, fruits containing shriveled seeds, non-germinating seeds, early death of the embryos, abnormal hybrid parents, hybrids with virtually sterile, partially fertile or fertile pollen, hybrids not able to yield progenies and hybrids yielding viable progenies but with variable levels of fertility (Daunay and Hazra 2012). Cultivated eggplant is self-compatible (Daunay et al. 1991); therefore, the frequent infertility of interspecific hybrids between *S. melongena* and others *Solanum* spp. could be attributable to a lack of affinity of the genomes involved in the cross, causing difficulty in a correct pairing of chromosomes at the meiosis and the formation of irregular and sterile microspores and egg-cells.

Different strategies were followed to combine the cultivated eggplant with *S. sisymbriifolium* and *S. torvum*, two wild species of eggplant possessing a large number of resistance traits to severe diseases. However, crosses gave no hybrids at all and only partially fertile hybrids were obtained through embryo rescue (Bletsos et al. 1998; Cao et al. 2009; Afful et al. 2018; Çürük and Dayan 2018), but no backcross generations were ever produced (Sihachakr et al. 1988, 1989; Kumchai et al. 2013). Regarding *S. sisymbriifolium*, no viable sexual interspecific hybrids were obtained, and tetraploid somatic hybrids resulted sterile (Gleddie et al. 1986; Bletsos et al. 1998). Reduced fertility or infertility is a common phenomenon in the interspecific hybrids of eggplant with other *Solanum* species (e.g., *S. macrocarpon*, *S. aethiopicum*, *S. linnaeanum*, *S. torvum*) belonging to the Melongena section as well (Daunay et al. 1991;

Nee 1999; Bletsos et al. 2004; Kameswara Rao 2011). Sometimes, the sterile interspecific hybrids were directly commercially employed as rootstock (e.g., cv 'Taibyo VF' from a cross between eggplant and *S. grandiflorum*; Hasnumnahar et al. 2012) especially in Japan in fungal and bacterial wilt infested fields. Also, alloplasmic male-sterile *S. melongena* lines containing the cytoplasm of wild relatives of eggplant have been obtained (Khan et al. 2015). The employment of different accessions of the same wild relative with various lines of eggplant sometimes improved the chance to get fertile interspecific hybrids. Interspecific crossing barriers may also be broken down by making several attempts of crossing to the same flowers along different days (Rotino et al. 2014).

Many papers reported unidirectional crossability of *S. melongena* also with some of its closer wild relatives, whose crosses yielded fertile progeny only when genotypes were used as female parents (Chopde and Wanjari 1974; Rajasekaran 1970a, b; Wanjari 1976; Behera and Singh 2002; Sharma et al. 1980, 1984) or, on the contrary, as male (Omidiji 1979; Rajasekaran 1970c; Sharma et al. 1980, 1984). The information on both interspecific and intraspecific crossability is important in developing a comprehensive breeding strategy. The crossability and hybridization studies of different *S. melongena* accessions with its related species were carried out with inconsistent results for long (Gowda and Seenappa 1991; Behera and Singh 2002). A study involving thirteen cultivated genotypes of eggplant as female parents and four wild *Solanum* species (*S. incanum, S. aethiopicum, S. integrifolium*, and *S. indicum*) as pollen donors, elucidated the crossability relationship of *S. melongena* genotypes with the wild species, determined by the occurrence of fruit set (%), number of seeds per fruit and germination (%) of F_1 seed (Devi et al. 2015). A first systematic attempt to obtain multiple sets of eggplant materials containing introgressions from different wild relatives resulted in the development of over 50 combinations of hybrids and 30 first backcross (BC_1) generations (Kouassi et al. 2016; Plazas et al. 2016) obtained combining six *S. melongena* varieties and 14 wild species from different genepools, including the primary (*S. incanum* and *S. insanum*), secondary (*S. anguivi, S. dasyphyllum, S. lichtensteinii, S. linnaeanum, S. pyracanthos, S. tomentosum*, and *S. violaceum*), and tertiary (*S. elaeagnifolium, S. sisymbriifolium*, and *S. torvum*) genepools. Fruit set, hybrid seed, and seed germination were obtained between *S. melongena* and all wild species of the primary and secondary genepools. The highest fruit set percentage and quantity of seeds per fruit were obtained with the primary genepool species *S. insanum* as well as with some secondary genepool species, like *S. anguivi, S. dasyphyllum, S. incanum*, or *S. lichtnsteinii*. However, some differences among species were observed depending on the direction of the hybridization. This intense introgressiomics approach has allowed to conclude that eggplant is amenable to interspecific hybridization with a large number of wild species, including tertiary genepool materials. These hybrid materials displayed a wide variability and are the starting point for introgression breeding in eggplant, and some cases might also be useful as rootstocks for eggplant grafting. (Knapp et al. 2013; Plazas et al. 2016).

Efforts have been made to transfer alien genes controlling important traits from wild and allied relatives into the cultivated genepool of eggplant, and using unconventional (somatic hybridization, embryo rescue, and genetic transformation) breeding methods (Daunay and Hazra 2012; Rotino et al. 2014). Fertility restoration by doubling the F_1 hybrid's ploidy level has been successfully reported in the sexual hybrid between eggplant and *S. aethiopicum* (Isshiki and Taura 2003) and *S. violaceum* (Isshiki and Kawajiri 2002). Relevant knowledge acquired on eggplant regeneration from different organs, tissues, cells, and protoplasts has resulted in the development of somatic hybrids (Rajam et al. 2008) thus providing the opportunity to apply this technology for introgression of alien genes for the genetic improvements of eggplant (Sihachakr et al. 1994; Collonnier et al. 2001a; Kashyap et al. 2003; Rajam and Kumar 2006) overcoming sexual barriers or improving the fertility of interspecific hybrids. However, also with these techniques, the obtained hybrids often resulted in highly sterile, preventing their practical utilization (Gleddie et al. 1986; Sihachakr et al. 1988; Guri and Sink 1988a, b; Sihachakr et al. 1989; Collonnier et al. 2003). Nonetheless, somatic hybridizations with *S. aethiopicum* gr. *gilo* and gr. *aculeatum* were two excellent examples of successful incorporation of somatic hybrids in breeding programs to introgress the traits of resistance to bacterial and fungal wilts from wild species into cultivated eggplant (Daunay et al. 1993; Rotino et al. 1998, 2001; Collonnier et al. 2001b). These results support the previous reports that the tetraploid status significantly improves the interspecific hybrid's fertility, which produces seeded fruits. However, in order to incorporate the somatic hybrid in a practical eggplant breeding program for introgression of traits of interest, two essential prerequisites still need to be accomplished beyond fertility: (1) occurrence of genetic recombination between the genomes of the two species and (2) reduction of the ploidy level to that of recurrent parent because generally the somatic hybrids possess the sum of the genomes of the involved species. Although first backcrossing of the dihaploids with the recurrent eggplants has been obtained with a particular difficulty, in subsequent backcross generations, morphology and fertility improved progressively (Rotino et al. 2005; Toppino et al. 2008a, b).

The development of populations of advanced backcross lines (ABLs) and introgression lines (ILs) in the eggplant genepool would make a major contribution to enhancing the use of CWRs in eggplant breeding (Prohens et al. 2017). Different kinds of introgression materials have been obtained with eggplant relatives, including eggplant breeding lines from the somatic hybrid with *S. aethiopicum* gr. *gilo* and *aculeatum* carrying an introgressed trait of resistance to *Fusarium oxysporum* (Toppino et al. 2008b). Backcrossed progenies from sexual cross with *S. linnaeanum* (=*S. sodomaeum*) displayed an improved resistance to *Verticillium* wilt (Acciarri et al. 2004; Liu et al. 2015). Backcrossed progenies obtained from the sexual hybrid of eggplant with *S. aethiopicum* gr. *aculeatum* showed resistance to *Fusarium* (Zhuang and Wang 2009). Interspecific hybridization has also been accomplished in the less cultivated *S. aethiopicum* Kumba group, and its BC_1 progenies have been analyzed given their possible exploitation for the improvement of eggplant (Prohens et al. 2012). Similarly, backcross progenies (BC_1) derived from the crossing between the *S. incanum* and eggplant have been used to construct an interspecific linkage map and development of introgression lines (Vilanova et al. 2010; Prohens et al. 2012, 2013) with improved tolerance to abiotic stresses. *Solanum tomentosum* has been successfully employed for alien genes introgression due to its partial fertility with cultivated species, and the hybrid obtained was then backcrossed with cultivated species for the improvement of agronomical and fruit quality traits (Toppino et al. 2017). Another interspecific hybrid using *S. viarum* was successfully backcrossed to eggplant, and after selfing for nine generations resulted in the identification of two lines having higher yield, good fruit quality, and high tolerance to the devastating insect fruit and shoot borer (Pugalendhi et al. 2010). Recently, a large set (>40) introgression lines carrying specific genomic fragments of the wild *S. incanum* in the genetic background of *S. melongena* was obtained using marker-assisted-selection. The materials obtained represent eggplant prebreeding materials of great interest to breeders (Gramazio et al. 2014). Moreover, a population of 90 IL lines of eggplant each carrying a fragment of the wild species *S. tomentosum* was obtained (Toppino et al. 2018) which could be of great interest for the disclosure of the genetic basis underlying the novel resistance traits to *Fusarium*, *Verticillium* and nematodes (*Meloidogyne* spp.). A RIL population of 170 progenies whose female parent is an IL from an interspecific somatic hybrid with *S. aethiopicum* has been recently developed and employed to anchor the new version of the eggplant genome to a genetic map (Rotino et al. 2017; Barchi et al. 2019). Introgression lines populations and molecular markers have accelerated the identification of alien segments introgressed from wild species and helped to assess their impact on phenotypic expression, opening new ways to solve constraints for accessing to the reservoir of alien alleles represented by the progenitors, allied and wild species of *S. melongena*. Further development of new tools and techniques will certainly improve the wild species' use in alien gene introgression in eggplant.

6.2.4 Potentiality of Genetic Resources of Eggplants in Breeding for Biotic Stress Tolerance and Resistance

In general, cultivated eggplant is susceptible to a variety of biotic and abiotic stresses, which may cause high yield and quality losses. Particularly, soil-borne diseases (bacterial and fungal wilts, nematodes) and insects are the most serious cause of yield losses both in greenhouse and in open-field cultivations (Sihachakr et al. 1994). Partial resistance/tolerance traits to most pathogens were found within the cultivated eggplant genepool, but their low efficacy, often, hindered an effective employment in breeding programs (Daunay et al. 1991). The narrow genetic base of eggplant is a severe constraint to the development of new highly performing cultivars, which must be able to contribute to increasing yield and performance under the environmental changes resulting from a climate-change scenario (Tilman et al. 2011; Ray et al. 2013). *Solanum melongena* landraces, allied and wild relatives, represent a great source of variation for many traits of agronomical interest and introgression breeding could significantly contribute to improve the resistance to pests and diseases, the quality and, more recently, the nutritional value of the fruits. Many wild and allied species have been reported to carry traits of resistance to most diseases and pests affecting eggplant, including fungi (*Leveillula taurica, Phomopsis vexans*, *Fusarium oxysporum*, and *F. solani*, *Verticillium dahliae* and V. *albo-atrum*, *Colletotrichum coccodes*, *Phytophthora parasitica* and *P. capsici*, *Cercospora solani*), bacteria (*Ralstonia solanacearum*), nematodes (*Meloidoygine* spp.), pests (*Leucinodes orbonalis, Epilachana vigintioctopunctata, Aphis gossypii, Tetranychus cinnabarinus, T. urticae*), viruses (*potato virus Y*, *eggplant mosaic virus*), and mycoplasma (Daunay and Hazra 2012; Rotino et al. 2014).

The allied species *S. aethiopicum* gr. *gilo* and gr. *aculeatum*, *S. incanum* and *S. macrocarpon* have been frequently subjected to evaluation and utilized as source of genetic variability because they are considered genetically closer to *S. melongena* (Rotino et al. 2014). Since they carry many relevant resistance traits; several studies have identified resistances to bacterial and several fungal wilts (*Phomopsis, Fusarium, Verticillium*) in *S. aethiopicum* gr. *gilo* (Collonnier et al. 2001a); resistance to fungi and some insects in *S. macrocarpon* (Schaff et al. 1982), *Verticillium* wilt and nematode resistance in *S. incanum* (Prohens et al. 2013; Robinson et al. 2001; Gisbert et al. 2011) and resistance to *Phomopsis* blight, mycoplasma (little leaf), bacterial wilt, and shoot and fruit borer in *S. aethiopicum* gr. *aculeatum* (Kalloo 1993). *Solanum linnaeanum* is considered a good source of resistance traits to *Verticillium* and other fungi (Daunay et al. 1991), as also to *Ralstonia* (Yamaguchi et al. 2010; Liu et al. 2015). New sources of resistance to *Phytophtora capsici* have been recently reported in *S. incanum*, *S. linnaeanum*, and *S. gilo* (Naegele et al. 2014b). A recent new source of resistance to nematodes (*Melodoygine* spp.) has been detected in *S. tomentosum* (Toppino et al. 2018). Among the wild species with great potential for eggplant crop improvement, *S. torvum* and *S. sisymbriifolium* have been traditionally considered as very interesting due to their multiple resistances to the most important eggplant fungal and bacterial diseases, like those caused by *Fusarium oxysporum*, *Verticillium dahliae* or *Ralstonia solanacearum*, as well as to nematodes (Bletsos et al. 2003; Gousset et al. 2005; Daunay and Hazra 2012; Nahar et al. 2014). *Solanum torvum* has been widely utilized as a rootstock (Bletsos et al. 2003), as grafted eggplant plantlets display a gained high tolerance to *Verticillium* and other abiotic stresses (high salinity, heavy metals).

These responses prompted the study to identify the genes of *S. torvum* involved in the tolerance to *Verticillium* and also to the heavy metal Cd by transcriptome analyses (Wang et al. 2010a; Yamaguchi et al. 2010; Xu et al. 2012), and the pathogenesis-related gene StDAHP was cloned following inoculation with *Verticillium* (Wang et al. 2010b).

Other tertiary genepool species that have recently raised interest for eggplant breeding are *S. elaeagnifolium*, which is an invasive weed highly tolerant to drought (Christodoulakis et al. 2009; Garcia-Fortea et al. 2019), and *S. bonariense*, a relative of *S. torvum* (Nurit-Silva et al. 2012) with high vigor and mostly unexplored. In most of the work aimed to assess the response of the allied species to the biotic and abiotic stresses, a single or a few accessions were often analyzed. It is worth pointing out that the allied and wild species probably display an even huge allelic variation for the useful traits searched. They have been mainly subjected to environmental selection, including the indirect effect of the human activities and not to the bottleneck of domestication. Therefore, the collections of wild and allied species need to be implemented and also to be better characterized for their biological and biochemical properties; concerning eggplant, only a few examples can be reported, like S. *torvum* (Gousset et al. 2005) and *S. aethiopicum* (Sunseri et al. 2010).

6.2.5 Potentiality of Genetic Resources of Eggplants in Breeding for Abiotic Stress Tolerance

As occurs with many domesticates, common eggplant is susceptible to several abiotic stresses, including drought, salinity, low and high temperatures, and soil toxicity (Savvas and Lenz 2000; Boyaci et al. 2009; Daunay and Hazra 2012; Wu et al. 2014; Díaz-Pérez and Eaton 2015; Plazas et al. 2019). Although genetic variation exists within the cultivated species (Behera et al. 2006; Liu et al. 2018), the genetic diversity of wild relatives is much higher (Vorontsova et al. 2013). In addition, wild eggplant relatives grow in a wide range of environments, and wild eggplant relatives from the Old World are frequently found in dry, semi-desertic, and even desertic environments (Vorontsova and Knapp 2016; Ranil et al. 2017) and therefore are an important source of variation for introgression of abiotic stress tolerance in eggplant (Prohens et al. 2017). In this respect, although the wild eggplant relatives have been barely explored for the genetic improvement of tolerance to stresses, the application of the Focused Identification of Germplasm Strategy (FIGS) is promising. FIGS is based on the assumption that wild species naturally thriving under stressful conditions carry adaptive genes for tolerance to the stresses they suffer (Street et al. 2016) and therefore could contribute to the identification of the most promising wild eggplant relatives for being incorporated in breeding programs for tolerance to abiotic stresses.

As in most vegetable crops, drought stress causes a reduction of growth and yield in eggplant, although compared to other vegetables of the same family such as tomato and pepper, it is considered more tolerant (Díaz-Pérez and Eaton 2015). Several studies reveal that intraspecific diversity exists for tolerance to drought in common eggplant (Bhatt et al. 2014; Fita et al. 2015; Saracanlao et al. 2016; Tani et al. 2018). Other studies found that in the scarlet eggplant, important differences can be found among accessions (Mibei et al. 2017; Sseremba et al. 2018). In this way, differences among cultivated eggplant genotypes suggest that selection of eggplant cultivars more tolerant to drought may be feasible. In addition to intraspecific diversity, some eggplant CWRs are very tolerant to drought, as they naturally grow under conditions where water availability is minimal (Knapp et al. 2013; Syfert et al. 2016; Vorontsova and Knapp 2016; Knapp et al. 2017). In this respect, some forms of the primary genepool of eggplant *S. insanum*, which can be quickly incorporated in breeding programs (Kouassi et al. 2016), are tolerant to drought (Ranil et al. 2017). Within the secondary genepool, *S. macrocarpon* and *S. linnaeanum* have been described as tolerant to drought (Daunay et al. 1991), while other species such as *S. incanum* are considered as very promising for improving the tolerance to drought in eggplant (Knapp et al. 2013; Gramazio et al. 2017b). Within the tertiary genepool, *S. elaeagnifolium*, which is an invasive weed highly tolerant to drought (Fita et al. 2015; Knapp et al. 2017) is of particular interest for improving the tolerance to drought in eggplant, particularly after backcrosses of the interspecific hybrid with eggplant have been obtained (García-Fortea et al. 2019).

Eggplant is susceptible to salinity, although less than pepper (Shahbaz et al. 2012). Salinity negatively affects plant growth yield and fruit quality (Savvas and Lenz 2000; Kirnak et al. 2002). However, differences among eggplant cultivars in tolerance to salinity have been described (Mustafa et al. 2017; Hannachi and Van Labeke 2018), opening the way to select varieties with a greater degree of tolerance. However, some wild species such as *S. insanum* (Brenes et al. 2020a), *S. linnaeanum* and *S. torvum* (Chen et al. 2012; Brenes et al. 2020b) have been found to be tolerant to salinity stress. The direct use of wild species tolerant to salinity as rootstocks may be a direct way to use CWRs for improving eggplant tolerance to salinity, as has been demonstrated for *S. torvum* (Colla et al. 2010).

Eggplant is susceptible to cold temperatures, which arrest growth and development and can also reduce pollen viability (Daunay and Hazra 2012). Differences among eggplant genotypes have been found for tolerance to low temperatures, opening the way for the development of more tolerant cultivars (Abak and Guler 1994; Boyaci et al. 2009). One of the main productive problems of growing eggplants at low temperatures is that low pollen viability results in decreased fruit set due to lack of pollination and fertilization (Donzella et al. 2000). The use of the intraspecific variation has allowed the selection and development of parthenocarpic cultivars which allow fruit set under low temperatures that reduce pollen viability (Kikuchi et al. 2008). Some wild eggplant species such as *S. aculeatissimum*, *S. grandiflorum*, or *S. mammosum* have been reported as tolerant to cold (Rotino et al. 2014; Yang et al. 2017) and could be of interest for breeding eggplants tolerant to low temperatures.

Although eggplant is a warm-loving crop, temperatures above 40 °C can arrest growth, induce flower buds abortion, reduction of pollen production and viability, and a decrease of yield and fruit quality (Wu et al. 2014). Little is known about genetic diversity for tolerance to high temperatures in eggplant, but it is known that some close wild species of eggplant, such as *S. incanum* (Knapp et al. 2013; Vorontsova et al. 2013) grow in areas where they are exposed to high temperatures during the summer season. These wild species might represent a genetic resource of interest for tolerance to high temperatures in eggplant.

Soil toxicity effects in eggplant have been barely studied, and little is known on intraspecific variation, although it is known that high soil concentrations of cadmium, chromium, lead, and nickel may have a toxic effect for eggplant plants (Pandey and Gopal 2010; Gautam et al. 2014a, b; Yuan et al. 2019). However, it has been found that grafting eggplant onto the eggplant wild relative *S. torvum* improves tolerance to cadmium toxicity by reducing the translocation to the aerial parts (Yamaguchi et al. 2011; Yuan et al. 2019).

6.2.6 Molecular Interspecific Mapping Populations

Both interspecific and intraspecific mapping populations have been developed in eggplant. The interspecific maps benefit from an enhanced frequency of marker polymorphism, but their relevance to marker-assisted crop breeding is limited due to the low transferability of the results to common eggplant; therefore, scientific attention moves toward the development and employment of intraspecific maps.

The allied species *S. linnaeanum* was efficiently crossed with eggplant to introgress the alien genes controlling *Verticillium* wilt following backcross breeding, which resulted in the development of several wilt tolerant lines of eggplant (Acciarri et al. 2004). The same species was used in the development of a mapping population of 58 F_2 individuals for construction of a molecular map (Doganlar et al. 2002) containing 233 markers distributed over 12 linkage groups, spanning 1480 cM and was used for comparative analysis between eggplant and tomato and then employed for the identification of QTLs for qualitative and quantitative traits (Frary et al. 2003). Subsequently, the previously mentioned map was improved by Wu et al. (2009) by adding total of 110 COSII markers (Conserved Ortholog Set, Wu et al. 2006) previously mapped in the tomato genome, and five tomato-derived markers were selected primarily in the regions with identified chromosome rearrangements between the genomes of eggplant and tomato. The obtained genetic map contains 347 markers, assigned to the 12 chromosomes in the haploid chromosome set of eggplant; it spans 1535 cm, with a framework marker density of about 6 cm, and the size of the chromosomes ranging from 105 to 159 cm. According to synteny between the tomato and eggplant genomes, the locations of an additional 522 COSII markers on the eggplant linkage map were deduced, bringing the total number of RFLP and COSII markers of known position in this interspecific eggplant population to 869. The improvement of the interspecific map previously established has been reported by Doganlar et al. (2014), mostly achieved by increasing the number of individuals used (from 58 to 108 F_2 individuals), as well as by adding AFLP and 117 previously unmapped RFLP and COSII markers. Overall, the newly developed map consists of 400 AFLPs, 348 RFLPs, and 116 COSII for a total of 864 markers, spanning 1518 cm, with the size of the chromosomes ranging from 93 to 152 cm.

Gramazio et al. (2014) developed a new interspecific (*S. melongena* × *S. incanum*) linkage map based on a first backcross (BC_1) generation (91 plants) toward the cultivated accession of eggplant 'AN-S-26', as a tool for introgressing *S. incanum* alleles involved in the biosynthesis of chlorogenic acid in the genetic background of *S. melongena*. The mapping population was genotyped with 243 molecular markers comprising 42 COSII, 99 SSRs, 88 AFLPs, nine CAPS, four SNPs, and the morphological marker PRICKLINESS. The linkage map covers 1085 cm, with linkage groups length comprised between 58.6 and 132.9 cm, a number of markers for each LG comprised between 16 and 27, and an average marker density of 4.46 cm. Chapter 10 summarizes results from a resequencing study of wild *S. incanum* and compares it to accessions of *S. melongena*.

Acknowledgements Authors are grateful to the European Union's Horizon 2020 Research and Innovation Programme under grant agreement No 677379 (G2P-SOL project: Linking genetic resources, genomes, and phenotypes of Solanaceous crops), and to Spanish Ministerio de Economía, Industria y Competitividad and Fondo Europeo de Desarrollo Regional (grant AGL2015-64755-R from MINECO/FEDER). Parts of the eggplant section were undertaken as part of the initiative 'Adapting Agriculture to Climate Change: Collecting, Protecting and Preparing Crop Wild Relatives', which is supported by the Government of Norway. The project is managed by the Global Crop Diversity Trust with the Millennium Seed Bank of the Royal Botanic Gardens, Kew, and implemented in partnership with national and international gene banks and plant breeding institutes around the world. For further information see the project website: http://www.cwrdiversity.org/. Mariola Plazas is grateful to Generalitat Valenciana and Fondo Social Europeo for a post-doctoral contract (APOSTD/2018/014).

Author Contributions PT drafted the manuscript outline. PT, CCG, and MPa contributed to the pepper subchapter 6.1; LT, JP, MPl, and GLR contributed to the eggplant subchapter 6.2. PT wrote paragraphs 6.1.1, 6.1.2, 6.1.4, and 6.1.6; CCG wrote paragraph 6.1.3; MPa and PT wrote conjunctly paragraph 6.1.5. JP and MPl wrote paragraphs 6.2.1, 6.2.2, 6.2.5; LT and GLR wrote paragraphs 6.2.3, 6.2.4, and 6.2.6. All authors revised and approved the manuscript.

Conflict of Interest The authors declare no conflicts of interest.

References

Abak K, Guler HY (1994) Pollen fertility and the vegetative growth of various eggplant genotypes under low temperature greenhouse conditions. Acta Hort 366:85–91

Acciarri N, Rotino GL, Sabatini E, Valentino D, Sunseri F, Mennella G, Tamietti G (2004) Improvement of eggplants for resistance to Verticillium. Proc Eucarpia Meet Genet Breed Capsicum Eggplant 12:178

Acquadro A, Barchi L, Gramazio P, Portis E, Vilanova S, Comino C, Plazas M, Prohens J, Lanteri S (2017) Coding SNPs analysis highlights genetic relationships and evolution pattern in eggplant complexes. PLoS ONE 12:

Afful NT, Nyadanu D, Akromah R, Amoatey HM, Annor C, Diawouh RG (2018) Evaluation of crossability studies between selected eggplant accessions with wild relatives S. *torvum, S. anguivi* and *S. aethiopicum* (Shum group). J Plant Breed Crop Sci 10:1–12

Almási A, Csilléry G, Salánki K, Nemes K, Palkovics L, Tóbiás I (2016) Comparison of wild type and resistance-breaking isolates of *Tomato spotted wilt virus* and searching for resistance on pepper. In: Proceedings of XVIth EUCARPIA *capsicum* and eggplant working group meeting Kecskemét, Hungary, pp 574–578

Anand N, Deshpande AA, Sridhar TS (1987) Resistance to powdery mildew in an accession of *Capsicum frutescens* and its inheritance pattern. Capsicum Eggplant Newsl 6:77–78

Anaya-López JL, González-Chavira MM, Villordo-Pineda E, Rodríguez-Guerra R, Rodríguez-Martínez R, Guevara-González RG, Guevara-Olvera L, Montero-Tavera V, Torres-Pacheco I (2011) Selection of chili pepper genotypes resistant to pathogenic wilt disease complex. Rev Mex De Cienc Agric 2:373–383

Ano G, Hebert Y, Prior P, Messiaen CM (1991) A new source of resistance to bacterial wilt of eggplants obtained from a cross: *Solanum aethiopicum* L. × *Solanum melongena* L. Agronomie 11:555–560

Aubriot X, Singh P, Knapp S (2016) Tropical Asian species show that the Old World clade of 'spiny solanums' (*Solanum* subgenus *Leptostemonum pro parte*: Solanaceae) is not monophyletic. Bot J Linnean Soc 181:199–223

Aza-González C, Núñez-Palenius HG, Ochoa-Alejo N (2011) Molecular biology of capsaicinoid biosynthesis in chili pepper (*Capsicum* spp.). Plant Cell Rept 30:695–706

Barbary A, Djian-Caporalino C, Palloix A, Castagnone-Sereno P (2015) Host genetic resistance to root-knot nematodes, *Meloidogyne* spp, in Solanaceae: from genes to the field. Pest Manag Sci 71:1591–1598

Barboza GE, Agra MF, Romero MV (2011) New endemic species of *Capsicum* (Solanaceae) from the Brazilian Caatinga: comparison with the re-circumscribed *C. parvifolium*. Syst Bot 36:768–781

Barboza GE, Carrizo García C, Leiva González S (2019) Four new Species of *Capsicum* (Solanaceae) from the tropical Andes and an update on the phylogeny of the genus. PLoS ONE 14:

Barboza GE, Carrizo García C, Scaldaferro M, Bohs L (2020) An amazing new *Capsicum* (Solanaceae) species from the Andean-Amazonian Piedmont. PhytoKeys, in press

Barchenger DW, Rodriguez K, Li Jiang, Hanson SF, Bosland PW (2017) Allele-specific CAPS marker in a *Ve1* homolog of *Capsicum* annuum for improved selection of *Verticillium dahliae* resistance. Mol Breed 37(11):134

Barchenger DW, Lamour KH, Bosland PW (2018) Challenges and strategies for breeding resistance in *Capsicum annuum* to the multifarious pathogen, *Phytophthora capsici*. Front Plant Sci 9:628

Barchi L, Pietrella M, Venturini L, Minio A, Toppino L, Acquadro A, Andolfo G, Aprea G, Avanzato C, Bassolino L, Comino C, Dal Molin A, Ferrarini A, Chappell Maor L, Portis E, Reyes-Chin-Wo S, Rinaldi R, Sala T, Scaglione D, Sonawane P, Tononi P, Almekias-Siegl E, Zago E, Ercolano MR, Aharoni A, Delledonne M, Giuliano G, Lanteri S, Rotino GL (2019) A chromosome-anchored eggplant genome sequence reveals key events in Solanaceae evolution. Sci Rep 9(1):1–13. https://www.nature.com/articles/s41598-019-47985-w

Behera TK, Singh N (2002) Inter-specific crosses between eggplant (*Solanum melongena* L.) with related *Solanum* species. Sci Hortic 95:165–172

Behera TK, Sharma P, Singh BK, Kumar G, Kumar R, Mohapatra T, Singh NK (2006) Assessment of genetic diversity and species relationships in eggplant (*Solanum melongena* L.) using STMS markers. Sci Hortic 107:352–357

Ben Chaim A, Grube RC, Lapidot M, Jahn M, Paran I (2001) Identification of quantitative trait loci associated with resistance to *Cucumber mosaic virus* in *Capsicum annuum*. Theor Appl Genet 102:1213–1220

Ben Chaim A, Borovsky Y, De Jong W, Paran I (2003a) Linkage of the *A* locus for the presence of anthocyanin and *fs10.1*, a major fruit-shape QTL in pepper. Theor Appl Genet 106:889–894

Ben Chaim A, Borovsky Y, Rao GU, Tanyolac B, Paran I (2003b) *fs3.1*: a major fruit shape QTL conserved in *Capsicum*. Genome 46:1–9

Berke T, Black LL, Talekar NS, Wang JF, Gniffke P, Green SK, Wang TC, Morris R (2005) Suggested cultural practices for chilli pepper. International Co-Operator's Guide

Bernau V, Barbolla LJ, McHale LK, Mercer KL (2020) Germination response of diverse wild and landrace chile peppers (Capsicum spp.) under drought stress simulated with polyethylene glycol. bio Rxiv preprint. https://doi.org/10.1101/2020.06.29.177386

Bhatt RM, Laxman RH, Singh TH, Divya MH, Rao AN (2014) Response of brinjal genotypes to drought and flooding stress. Vegetable Sci 41:116–124

Black LL, Hobbs HA, Gatti JM Jr (1991) Tomato spotted wilt virus resistance in *Capsicum chinense* PI 152225 and 159236. Plant Dis 75:863

Bletsos FA, Roupakias DG, Tsaktsira ML, Scaltsoyjannes AB, Thanassoulopoulos CC (1998) Interspecific hybrids between three eggplant (*Solanum melongena* L.) cultivars and two wild species (*Solanum torvum* Sw. and *Solanum sisymbriifolium* Lam.). Plant Breed 117:159–164

Bletsos F, Thanassoulopolous C, Roupakias D (2003) Effect of grafting on growth, yield, and Verticillium wilt of eggplant. Hortic Sci 38:183–186

Bletsos F, Roupakias D, Tsaktsira M, Scaltsoyjannes A (2004) Production and characterization of interspecific hybrids between three eggplant (*Solanum melongena* L.) cultivars and *Solanum macrocarpon*. L. Sci Hortic 101:11–21

Boiteux LS (1995) Allelic relationships between genes for resistance to *Tomato spotted wilt* tospovirus in *Capsicum* chinense. Theor Appl Genet 90:146–149

Bosland PW, González MM (2000) The rediscovery of *Capsicum lanceolatum* (Solanaceae), and the importance of nature reserves in preserving cryptic biodiversity. Biodivers Conserv 9:1391–1397

Boyaci HF, Oğuz A, Ünlü M, Denizer B, Abak K (2009) Growth, pollen quantity and quality and fruit characteristics of some parthenocarpic and non-parthenocarpic eggplants in unheated greenhouse. Acta Hort 807:239–244

Brenes M, Solana A, Boscaiu M, Fita A, Vicente O, Calatayud Á, Prohens J, Plazas M (2020a) Physiological and biochemical responses to salt stress in cultivated eggplant (*Solanum melongena* L.) and in *S. insanum*, a close wild relative. Agronomy 10:651

Brenes M, Pérez J, González-Orenga S, Solana A, Boscaiu M, Prohens J, Plazas M, Fita A, Vicente O (2020b) Comparative studies on the physiological and biochemical responses to salt stress of eggplant (*Solanum melongena*) and its rootstock *S. torvum*. Agriculture 10:328

Brito JA, Stanley JD, Kaur R, Cetintas R, Di Vito M, Thies JA, Dickson DW (2007) Effects of the *Mi1*, *N* and *Tabasco* genes on infection and reproduction of *Meloidogyne mayaguensis* on tomato and pepper genotypes. J Nematol 39:327–332

Buddenhagen I, Sequeira L, Kelman A (1962) Designation of races in *Pseudomonas solanacearum*. Phytopathology 52:726

Cao B, Jiang-Jun L, Yong W, Ya-Qing L (2009) Interspecific hybridization between *Solanum melongena* and *Solanum torvum*. Acta Hort Sin 36:209–214

Caranta C, Palloix A, Gebre-Selassie K, Lefebvre V, Moury B, Daubèze AM (1996) A complementation of two genes originating from susceptible *Capsicum annuum* lines confers a new and complete resistance to pepper veinal mottle virus. Phytopathology 86:739–743

Caranta C, Pflieger S, Lefebvre, V, Daube`ze AM, Thabuis A, Palloix A (2002) QTLs involved in the restriction of *Cucumber mosaic virus* (CMV) long-distance movement in pepper. Theor Appl Genet 104:586–591

Caranta C, Palloix A, Lefebvre V, Daube`ze AM (1997) QTLs for a component of partial resistance to *Cucumber mosaic virus* in pepper: Restriction of virus installation in host-cells. Theor Appl Genet 94:431–438

Carrizo García C, Palombo N, Weiss-Schneeweiss H, Barboza G. Red hot chiles or not quite: updating the evolutionary scenario of the genus *Capsicum* (Solanaceae) using RAD-seq data. In preparation

Carrizo García C, Sterpetti M, Volpi P, Ummarino M, Saccardo F (2013) Wild *Capsicums*: Identification and in situ analysis of Brazilian species. In: Lanteri S, Rotino GS (eds) Proceedings of XVth EUCARPIA *capsicum* and eggplant working group meeting. Torino, pp 205–213

Carrizo García C, Barfuss MHJ, Sehr EM, Barboza GE, Samuel R, Moscone EA, Ehrendorfer F (2016) Phylogenetic relationships, diversification and expansion of chili peppers (*Capsicum*, Solanaceae). Ann Bot 118:35–51

Cebolla-Cornejo J, Soler S, Gomar B, Soria MD, Nuez F (2003) Screening *Capsicum* germplasm for resistance to *Tomato spotted wilt virus* (TSWV). Ann Appl Biol 143:143–152

Chen R, Li H, Zhang L, Zhang J, Xiao J, Ye Z (2007) *CaMi*, a root-knot nematode resistance gene from hot pepper (*Capsicum annuum* L) confers nematode resistance in tomato. Plant Cell Rep 26:895–905

Chen G, Wang H, Gai JI, Zhu YL, Yang LF, Liu QQ, Zhang GC, Chen GH (2012) Construction and characterization of a full-length cDNA library and identification of genes involved in salinity stress in wild eggplant (*Solanum torvum* Swartz). Hort Environ Biotechnol 53:158–166

Chhapekar SS, Jaiswal V, Ahmad I, Gaur R, Ramchiary N (2018) Progress and prospects in *capsicum* breeding for biotic and abiotic stresses. In: Vats S (ed) Biotic and abiotic stress tolerance in plants. Springer, Singapore, pp 279–322

Chiarini FE, Moreno CN, Barboza GE, Bernadello G (2010) Karyotype characterization of Andean Solanoideae (Solanaceae). Caryologia 63:278–291

Choi S, Lee J-H, Kang W-H, Kim J, Huy HN, Park S-W, Son E-H, Kwon J-K and Kang B-C (2018) Identification of *Cucumber mosaic resistance* 2 (*cmr2*) That Confers Resistance to a New *Cucumber mosaic virus Isolate* P1 (CMV-P1) in Pepper (*Capsicum* spp). Front Plant Sci 9:1106. 103389/fpls201801106

Chopde PR, Wanjari KB (1974) Interspecific hybrids in solanum. Indian J Genet Plant Breed 34A:1318–1323

Christodoulakis NS, Lampri PN, Fasseas C (2009) Structural and cytochemical investigation of the leaf of silverleaf nightshade (*Solanum elaeagnifolium*), a drought-resistant alien weed of the Greek flora. Aust J Bot 57:432–438

Colla G, Rouphael Y, Leonardi C, Bie Z (2010) Role of grafting in vegetable crops grown under saline conditions. Sci Hortic 127:147–155

Collonnier C, Fock I, Kashyap V, Rotino GL, Daunay MC, Lian Y, Mariska IK, Rajam MV, Servaes A, Ducreux G, Sihachakr D (2001a) Applications of biotechnology in eggplant. Plant Cell Tiss Org Cult 65:91–107

Collonnier C, Mulya K, Fock II, Mariska II, Servaes A, Vedel F, Siljak-Yakovlev S, Souvannavong VV, Ducreux G, Sihachakr D (2001b) Source of resistance against *Ralstonia solanacearum* in fertile somatic hybrids of eggplant (*Solanum melongena* L.) with *Solanum aethiopicum* L. Plant Sci 160:301–313

Collonnier C, Fock I, Mariska I, Servaes A, Vedel F, Siljak-Yakovlev S, Souvannavong V, Sihachakr D (2003) GISH confirmation of somatic hybrids between *Solanum melongena* and *S. torvum*: assessment of resistance to both fungal and bacterial wilts. Plant Physiol Biochem 41:459–470

Colonna V, D'Agostino N, Garrison E, Albrechtsen A, Meisner J, Facchiano A, Cardi T, Tripodi P (2019) Genomic diversity and novel genome-wide association with fruit morphology in *Capsicum*, from 746 k polymorphic sites. Sci Rep 9:10067

Cremona S, Iovene M, Festa G, Conicella C, Parisi M (2018) Production of embryo rescued hybrids between the landrace "Friariello" (*Capsicum annuum* var *annuum*) and *C baccatum* var *pendulum* phenotypic and cytological characterization. Euphytica 214:129

Csillery G, Szarka E, Sardi E, Mityko J, Kapitany J, Nagy B, Szarka J (2004) The unity of plant defense: genetics, breeding and physiology. In: Proceeding 12th Eucarpia meeting on genetics and breeding of *capsicum* and eggplant, 17-19 May, Noordwijkerhout, the Netherlands, pp 147–153

Çürük S, Dayan A (2018) Production of diploid and amphidiploid interspecific hybrids of eggplant and *Solanum torvum*, and pollen fertility. J Anim Plant Sci 28:1485–1492

D'Arcy WG (1972) Solanaceae studies II: Typification and subdivisions of *Solanum*. Ann Missouri Bot Gard 59:262–278

Daunay MC, Hazra P (2012) Eggplant. In: Peter KV, Hazra P (eds) Handbook of vegetables. Studium Press, Houston, TX, pp 257–322

Daunay MC, Janick J (2007) History and iconography of eggplant. Chronica Hort 47:16–22

Daunay MC, Lester RN, Laterrot H (1991) The use of wild species for the genetic improvement of Brinjal (eggplant) (*Solanum melongena*) and tomato (*Lycopersicon esculentum*). In: Hawks JC, Lester RN, Nee M, Estrada N (eds) Solanaceae III, taxonomy, chemistry, evolution, vol 27. Royal Botanic Gardens Kew and Linnean Soc, London, pp 389–413

Daunay MC, Chaput MH, Sihachakr D, Allot M, Vedel F, Ducreux G (1993) Production and characterization of fertile somatic hybrids of eggplant (*Solanum melongena* L.) with *Solanum aethiopicum* L. Theor Appl Genet 85:841–850

Daunay MC, Lester RN, Ano G (1997) Les Aubergines. In: Charrier A, Jacquot M, Hamon S, Nicolas D (eds) L'amélioration des plantes tropicales, Repères. CIRAD-ORSTOM, Montpellier, pp 83–107

Daunay MC, Lester RN, Ano G (2001) Eggplant. In: Charrier A, Jacquot A, Hamon M, Nicolas D (eds) Tropical plant breeding. Science Publishers, Montpellier, France, pp 199–222

Davidar P, Snow AA, Rajkumar M, Pasquet R, Daunay MC, Mutegi E (2015) The potential for crop to wild hybridization in eggplant (*Solanum melongena*; Solanaceae) in southern India. Am J Bot 102:129–139

de Souza VL, Café-Filho AC (2003) Resistance to *Leveillula taurica* in the genus *Capsicum*. Plant Pathol 52:613–619

Devi CP, Munshi AD, Behera TK, Choudhary H, Gurung B, Saha P (2015) Cross compatibility in interspecific hybridization of eggplant, *Solanum melongena*, with its wild relatives. Sci Hortic 193:353–358

Di Dato F, Parisi M, Cardi T, Tripodi P (2015) Genetic diversity and assessment of markers linked to resistance and pungency genes in *Capsicum* germplasm. Euphytica 204:103–119

Di Vito M, Saccardo F, Zaccheo G (1991) Response of lines of *Capsicum* spp to Italian populations of four species of *Meloidogyne*. Nematol Medit 19:43–46

Di Vito M, Saccardo F, Zaccheo G (1993) Response of new lines of pepper to *Meloidogyne incognita*, *M javanica*, *M arenaria* and *M hapla*. Afro-Asian J Nematol 3:135–138

Díaz-Pérez JC, Eaton TE (2015) Eggplant (*Solanum melongena* L.) plant growth and fruit yield as affected by drip irrigation rate. HortScience 50:1709–1714

Djian-Caporalino C, Pijarowski L, Januel A, Lefebvre V, Daubèze A, Palloix A, Dalmasso A, Abad P (1999) Spectrum of resistance to root-knot nematodes and inheritance of heat-stable resistance in pepper (*Capsicum annuum* L). Theor Appl Genet 99:496–502

Djian-Caporalino C, Fazari A, Arguel MJ, Vernie T, VandeCasteele C, Faure I, Brunoud G, Pijarowski L, Palloix A, Lefebvre V, Abad P (2007) Root-knot nematode (*Meloidogyne* spp) *Me* resistance genes in pepper (*Capsicum annuum* L) are clustered on the P9 chromosome. Theor Appl Genet 114:473–476

Doganlar S, Frary A, Daunay M-CC, Lester RN, Tanksley SD (2002) A comparative genetic linkage map of eggplant (*Solanum melongena*) and its implications for genome evolution in the Solanaceae. Genetics 161:1697–1711

Doğanlar S, Frary A, Daunay M-C, Huvenaars K, Mank R, Frary A (2014) High resolution map of eggplant (*Solanum melongena*) reveals extensive chromosome rearrangement in domesticated members of the Solanaceae. Euphytica 1–11. 10.1007/s10681-014-1096-2

Dogimont C, Palloix A, Daubèze AM, Marchoux G, Gèbre-Selassie K, Pochard E (1996) Genetic analysis of broad spectrum resistance to potyviruses using doubled haploid lines of pepper (*Capsicum* annuum L). Euphytica 88:231–239

Donzella G, Spena A, Rotino GL (2000) Transgenic parthenocarpic eggplants: superior germplasm for increased winter production. Mol Breed 6:79–86

Du H, Wen C, Zhang X, Xu X, Yang J, Chen B, Geng S (2019) Identification of a major QTL (*qRRs-10.1*) that confers resistance to *Ralstonia solanacearum* in pepper (*Capsicum annuum*) using SLAF-BSA and QTL Mapping. Int J Mol Sci 20:5887

Elphinstone JG (2005) The current bacterial wilt situation: a global view in bacterial wilt disease and the ralstonia solanacearum species complex. In: Allen C, Prior P, Hayward AC (eds) Saint Paul, MN: APS Press, pp 9–28

Eshbaugh WH (1970) A biosystematic and evolutionary study of *Capsicum baccatum* (Solanaceae). Brittonia 22:31–43

Eshbaugh WH (1982) Variation and evolution in *Capsicum eximium* Hunz. Baileya 21:193–198

Ferniah RS, Kasiamdari RS, Priyatmojo A, Daryono BS (2018) Resistance response of chilli (*Capsicum annuum* L.) F1 to *Fusarium oxysporum* involves expression of the *CaChi2* gene. Trop Life Sci Res 29:29–37

Ferrand L, García ML, Resende RO, Balatti PA, Dal Bó E (2015) First report of a resistance-breaking isolate of tomato spotted wilt virus infecting sweet pepper harboring the *Tsw* gene in Argentina. Plant Dis 99:1869

Fery RL, Dukes PD (1996) The inheritance of resistance to the southern root-knot nematode in 'Carolina Hot' cayenne pepper. J Am Soc Hort Sc 121:1024–1027

Fita A, Fioruci F, Plazas-Ávila M, Rodríguez-Burruezo A, Prohens J (2015) Drought tolerance among accessions of eggplant and related species. Bull Univ Agric Sci Vet Med Cluj-Napoca: Hort 72:461–462

Frary A, Doganlar S, Daunay MC, Tanksley SD (2003) QTL analysis of morphological traits in eggplant and implications for conservation of gene function during evolution of solanaceous species. Theor Appl Genet 107:359–370

Gajanayake B, Trader BW, Reddy KR, Harkess RL (2011) Screening ornamental pepper cultivars for temperature tolerance using pollen and physiological parameters. Hort Sci 46:878–884

Garcés-Claver A, Fellman SM, Gil-Ortega R et al (2007) Identification, validation and survey of a single nucleotide polymorphism (SNP) associated with pungency in *Capsicum* spp. Theor Appl Genet 115:90–916

García-Fortea E, Gramazio P, Vilanova S, Fita A, Mangino G, Villanueva G, Arrones A, Knapp S, Prohens J, Plazas M (2019) First successful backcrossing towards eggplant (*Solanum melongena*) of a new World species, the silverleaf nightshade (*S. elaeagnifolium*) and characterization of interspecific hybrids and backcrosses. Sci Hortic 246:563–573

Garruna-Hernandez R, Orellana R, Larque-Saavedra A, Canto A (2014) Understanding the physiological responses of a tropical crop (*Capsicum chinense* Jacq.) at high temperature. PLoS One 9:e111402

Gautam M, Singh AK, Johri RM (2014a) Influence of Pb toxicity on yield, yield attributing parameters and photosynthetic pigment of tomato (*Solanum lycopersicum*) and eggplant (*Solanum melongena*). Indian J Agric Sci 84:808–815

Gautam M, Singh AK, Johri RM (2014b) Effect of chromium toxicity on growth, chlorophyll and some macronutrients of *Solanum lycopersicum* and *Solanum melongena*. Indian J Agric Sci 84:1115–1123

Gisbert C, Prohens J, Raigón MD, Stommel JR, Nuez F (2011) Eggplant relatives as sources of variation for developing new rootstocks: effects of grafting on eggplant yield and fruit apparent quality and composition. Sci Hortic 128:14–22

Gleddie S, Keller WA, Setterfield G (1986) Production and characterization of somatic hybrids between *Solanum melongena* L. and *S. sisymbriifolium* Lam. Theor Appl Genet 71:613–621

Gomez-Garcia MR, Ochoa-Alejo N (2013) Biochemistry and molecular biology of carotenoid biosynthesis in chili peppers (*Capsicum* spp.). Int J Mol Sci 14:19025–19053

Gousset C, Collonnier C, Mulya K, Mariska I, Rotino GL, Besse P, Servaes A, Sihachakr D (2005) *Solanum torvum*, as a useful source of resistance against bacterial and fungal diseases for improvement of eggplant (*S. melongena L.)* Plant Sci 168:319–327

Gowda PHR, Seenappa K (1991) Occurrence of natural inter-specific hybrids of *S. incanum* L. Crop Res 4:352–354

Gramazio P, Prohens J, Plazas M, Andújar I, Herraiz FJ, Castillo E, Knapp S, Meyer RS, Vilanova S (2014) Location of chlorogenic acid biosynthesis pathway and polyphenol oxidase genes in a new interspecific anchored linkage map of eggplant. BMC Plant Biol 14:350

Gramazio P, Prohens J, Borràs D, Plazas M, Herraiz FJ, Vilanova S (2017a) Comparison of transcriptome-derived simple sequence repeat (SSR) and single nucleotide polymorphism (SNP) markers for genetic fingerprinting, diversity evaluation, and establishment of relationships in eggplants. Euphytica 213:264

Gramazio P, Prohens J, Plazas M, Mangino G, Herraiz FJ, Vilanova S (2017b) Development and genetic characterization of advanced backcross materials and an introgression line population of *Solanum incanum* in a *S. melongena* background. Front Plant Sci 8:1477

Gramazio P, Jaén-Molina R, Vilanova S, Prohens J, Marrero Á, Caujapé-Castells J, Anderson GJ (2020) Fostering conservation via integrated use of conventional approaches and high-throughput SPET genotyping: a case study using the endangered Canarian endemics *Solanum lidii* and *S. vespertilio* (Solanaceae). Front Plant Sci 11:757

Grube RC, Zhang Y, Murphy JF, Loaiza-Figueroa F, Lackney VK, Provvidenti R, Jahn MK (2000) New source of resistance to *Cucumber mosaic virus* in *Capsicum frutescens*. Plant Dis 84:885–891

Guo W, Chen R, Du XZ, Yin Y, Gong Z, Wang G (2014) Reduced tolerance to abiotic stress in transgenic

Arabidopsis overexpressing a *Capsicum annuum* multiprotein bridging factor. BMC Plant Biol 14:138
Guo G, Wang S, Liu J, Pan B, Diao W, Ge W et al (2017) Rapid identification of QTLs underlying resistance to Cucumber mosaic virus in pepper (*Capsicum frutescens*). Theor Appl Genet 130:41–52
Guri A, Sink KC (1988a) Organelle composition in somatic hybrids between an atrazine resistant biotype of *Solanum nigrum* and *Solanum melongena*. Plant Sci 58:51–58
Guri A, Sink KC (1988b) Interspecific somatic hybrid plants between eggplant (*Solanum melongena*) and *Solanum torvum*. Theor Appl Genet 76:490–496
Gurung T, Techawongstien S, Suriharn B, Techawongstien S (2011) Impact of environments on the accumulation of Capsaicinoids in *Capsicum* spp. Hort Sci 46:1576–1581
Gurung S, Short DPG, Hu X, Sandoya GV, Hayes RJ, Subbarao KV (2015) Screening of wild and cultivated *capsicum* germplasm reveals new sources of *verticillium* wilt resistance. Plant Dis 99:1404–1409
Hajjar R, Hodgkin T (2007) The use of wild relatives in crop improvement: a survey of developments over the last 20 years. Euphytica 156:1–13
Han K, Lee HY, Ro NY, Hur OS, Lee JH, Kwon JK, Kang BC (2018) QTL mapping and GWAS reveal candidate genes controlling capsaicinoid content in *Capsicum*. Plant Biotechnol J 16:1546–1558
Hannachi S, Van Labeke MC (2018) Salt stress affects germination, seedling growth and physiological responses differentially in eggplant cultivars (*Solanum melongena* L.). Sci Hortic 228:56–65
Hasnumnahar MST, Khan MD, Isshiki S (2012) Inheritance analysis of fertility restoration genes (*Rf*) in a male sterile system of eggplant using cytoplasm of *Solanum grandiflorum*. Aust J Crop Sci 6:475–479
Hayward AC (1964) Characteristics of *Pseudomonas solanacearum*. J App Bacteriol 27:265–277
Hayward AC (1991) Biology and epidemiology of bacterial wilt caused by *Pseudomonas solanacearum*. Ann Rev Phytopathol 29:65–87
He LY, Sequeira L, Kelman A (1983) Characteristic of *Pseudomonas solanacearum* from China. Plant Dis 67:1357–1361
Hibberd AM, Bassett MJ, Stall RE (1987) Allelism tests of three dominant genes for hypersensitive resistance to bacterial spot of pepper. Phytopathology 77:1304–1307
Huez-Lopez MA, Ulery AL, Samani Z, Picchioni G, Flynn RP (2011) Response of chille pepper (*Capsicum annuum* L.) to salt stress and organic and inorganic nitrogen sources: I. growth and yield. Trop Subtrop Agroecosyst 14:757–763
Hunziker AT (2001) Genera Solanacearum: the genera of Solanaceae illustrated, arranged according to a new system Gantner, Ruggell
Isshiki S, Kawajiri N (2002) Effect of cytoplasm of *Solanum violaceum* Ort. on fertility of eggplant (*S. melongena* L.). Sci Hortic 93:9–18
Isshiki S, Taura T (2003) Fertility restoration of hybrids between *Solanum melongena* L. and S. *aethiopicum* L. Gilo Group by chromosome doubling and cytoplasmic effect on pollen fertility. Euphytica 134:195–201
Isshiki S, Okubo H, Fujieda K (1994) Phylogeny of eggplant and related *Solanum* species constructed by allozyme variation. Sci Hortic 59:171–176
Jahn M, Paran I, Hoffmann K, Radwanski ER, Livingstone KD, Grube RC, Aftergoot E, Lapidot M, Moyer J (2000) Genetic mapping of the *Tsw* locus for resistance to the *Tospovirus Tomato spotted wilt virus* in *Capsicum* spp. and its relationship to the *Sw-5* gene for resistance to the same pathogen in tomato. Mol Plant Microbe Interact 13(6):673–682
Jiang L, Huang Y, Sun L, Wang B, Zhu M, Li J, Huang C, Liu Y, Li F, Liu Y, Dong J, Zhang Z, Tao X (2017) Occurrence and diversity of *Tomato spotted wilt virus* isolates breaking the *Tsw* resistance gene of *Capsicum* chinense in Yunnan, southwest China. Plant Pathol 66:980–989
Jo J, Venkatesh J, Han K, Lee H-Y, Choi GJ, Lee HJ, Choi D, Kang B-C (2017) Molecular mapping of *PMR1*, a Novel Locus conferring resistance to powdery mildew in pepper (*Capsicum annuum*). Front Plant Sci 8:2090. 103389/fpls201702090
Jones JB, Stall RE, Bouzar H (1998) Diversity among *Xanthomonas* pathogenic on pepper and tomato. Annu Rev Phytopathol 36:41–58
Jordt SE, Julius D (2002) Molecular basis for species-specific sensitivity to 'hot' chili peppers. Cell 108:421–430
Kalloo G (1993) Eggplant (*Solanum melongena*). In: Kalloo G (ed) Genetic improvement of vegetable crops. Pergamon Press, Oxford, pp 587–604
Kameswara Rao C (2011) Use of Brinjal (*Solanum melongena* L.) in alternative systems of medicine in India. Foundation for Biotechnology Awareness and Education Bangalore India
Kang WH, Hoang N, Yang HB, Kwon JK, Jo SH, Seo JK, Kim KH, Choi D, Kang BC (2010) Molecular mapping and characterization of a single dominant gene controlling CMV resistance in peppers (*Capsicum annuum* L) Theor Appl Genet 120:1587–1596
Kashyap V, Kumar SV, Collonnier C, Fusari F, Haicour R, Rotino GL, Sihachakr D, Rajam M (2003) Biotechnology of eggplant. Sci Hortic 97:1–25
Katsantonis D, Hillocks RJ, Gowen S (2003) Comparative effect of root-knot nematode on severity of verticillium and fusarium wilt in cotton. Phytoparasitica 31:154–162
Kaushik P, Prohens J, Vilanova S, Gramazio P, Plazas M (2016) Phenotyping of eggplant wild relatives and interspecific hybrids with conventional and phenomics descriptors provides insight for their potential utilization in breeding. Front Plant Sci 7:677
Khan MMR, Hasnunnahar M, Iwayoshi M, Ogura-Tsujita Y, Isshiki S (2015) Pollen degeneration in three functional male-sterile lines of eggplant with wild *Solanum* cytoplasms. Hortic Environ Biotechnol 56:350–357

Kikuchi K, Honda I, Matsuo S, Fukuda M, Saito T (2008) Stability of fruit set of newly selected parthenocarpic eggplant lines. Sci Hortic 115:111–116

Kim BS, Hartmann RW (1985) Inheritance of a gene (*Bs3*) conferring hypersensitive resistance to *Xanthomonas campestris* pv *vesicatoria* in pepper (*Capsicum* annuum). Plant Dis 69:233–235

Kim S, Kim KT, Kim DH et al (2010) Identification of quantitative trait loci associated with anthracnose resistance in chili pepper (*Capsicum* spp). Korean J Hortic Sci Technol 28:1014–1024

Kirnak H, Tas I, Kaya C, Higgs D (2002) Effects of deficit irrigation on growth, yield, and fruit quality of eggplant under semi-arid conditions. Austral J Agric Res 53:1367–1373

Knapp S, Vorontsova MS, Prohens J (2013) Wild relatives of the eggplant (*Solanum melongena* L.: Solanaceae): New understanding of species names in a complex group. PLOS ONE 8:e57039

Knapp S, Sagona E, Carbonell AKZ, Chairini F (2017) A revision of the *Solanum elaeagnifolium* clade (Elaeagnifolium clade; subgenus *Leptostemonum*, *Solanaceae*). PhytoKeys 84:1–104

Kormelink R (2011) The molecular biology of tospoviruses and resistance strategies. In: Elliott RM, Plyusin A (eds) Bunyaviridae: molecular and cellular biology. Plenum Press, New York, pp 163–191

Kouassi B, Prohens J, Gramazio P, Kouassi AB, Vilanova S, Galán-Ávila A, Herraiz FJ, Kouassi A, Segui-Simarro JM, Plazas M (2016) Development of backcross generations and new interspecific hybrid combinations for introgression breeding in eggplant (*Solanum melongena*). Sci Hortic 213:199–207

Kumchai J, Wei YC, Lee CY, Chen FC, Chin SW (2013) Production of interspecific hybrids between commercial cultivars of the eggplant (*Solanum melongena* L.) and its wild relative *S. torvum*. Genet Mol Res 12:755–764

Kyle MM, Palloix A (1997) Proposed revision of nomenclature for potyvirus resistance genes in *Capsicum*. Euphytica 97:183–188

Lebeau A, Daunay MC, Frary A, Palloix A, Wang JF, Dintinger J, Chiroleu F, Wicker E, Prior P (2011) Bacterial wilt resistance in tomato, pepper, and eggplant: genetic resources respond to diverse strains in the *Ralstonia solanacearum* species complex. Phytopathology 101:154–165

Lee HY, Ro NY, Jeong HJ, Kwon JK, Jo J, Ha Y, Jung A, Han JW, Venkatesh J, Kang BC (2016) Genetic diversity and population structure analysis to construct a core collection from a large *Capsicum* germplasm. BMC Genet 17:142

Lefebvre V, Palloix A, Caranta C, Pochard E (1995) Construction of an intra-specific integrated linkage map of pepper using molecular markers and doubled-haploid progenies. Genome 38:112–121

Lefebvre V, Daubeze AM, Voort RJ, Peleman J, Bardin M, Palloix A (2003) QTLs for resistance to powdery mildew in pepper under natural and artificial infections. Theor Appl Genet 107:661–666

Lester RN, Daunay MC (2003) Diversity of African vegetable *Solanum* species and its implications for a better understanding of plant domestication. Schiftren Genetischen Ressourcen 22:137–152

Lester RN, Hasan SMZ (1990) Origin and domestication of the brinjal eggplant, *Solanum melongena*, from *Solanum incanum*, in Africa and Asia. In: Hawkes JG, Lester RN, Nee M, Estrada N (eds) Solanaceae III: Taxonomy, chemistry and evolution. Royal Botanic Gardens Kew, Kew, UK, pp 369–387

Li T, Xu X, Li Y, Wang H, Li Z, Li Z (2015) Comparative transcriptome analysis reveals differential transcription in heat-susceptible and heat-tolerant pepper (*Capsicum annuum* L.) cultivars under heat stress. J Plant Biol 58:411–424

Liu J, Zheng Z, Zhou X, Feng C, Zhuang Y (2015) Improving the resistance of eggplant (*Solanum melongena*) to Verticillium wilt using wild species *Solanum linnaeanum*. Euphytica 201:463–469

Liu J, Yang Y, Zhou X, Bao S, Zhuang Y (2018) Genetic diversity and population structure of worldwide eggplant (*Solanum melongena* L.) germplasm using SSR markers. Genet Resour Crop Evol 65:1663–1670

Loganathan M, Venkataravanappa V, Saha S, Sharma BK, Tirupathi S, Verma MK (2013) Morphological, cultural and molecular characterizations of *Fusarium* wilt infecting tomato and chilli. National Symposium on Abiotic and Biotic Stress Management in Vegetable Crops April 12-14, Indian Society of Vegetable Science, IIVR, Varanasi

Manzur JP, Fita A, Prohens J, Rodríguez-Burruezo A (2015) Successful wide hybridization and introgression breeding in a diverse set of common peppers (*Capsicum annuum*) using different cultivated Ají (*C baccatum*) accessions as donor parents. PLoS ONE 10(12):

Martius KFP von (1846) Flora brasiliensis 10. F. Fleischer (Publisher)

Maruti TB, Tembhurne BV, Chavan RL, Amaresh YS (2014) Reaction of chilli (*Capsicum annuum* L) genotypes and hybrids against Fusarium wilt (*Fusarium solani*). J Spices Aromatic Crops 23:186–191

Matsunaga H, Saito T, Hirai M, Nunome T, Yoshida T (2003) DNA markers linked to *Pepper mild mottle virus* (PMMoV) resistant locus (L 4) in *Capsicum*. J Jpn Soc Hort Sci 72:218–220

McCoy JE, Bosland PW (2019) Identification of resistance to powdery mildew in chile pepper. Hort Sci 54:4–7

Mennella G, D'Alessandro A, Francese G, Fontanella D, Parisi M, Tripodi P (2018) Occurrence of variable levels of health-promoting fruit compounds in horn-shaped Italian sweet pepper varieties assessed by a comprehensive approach. J Sci Food Agric 98:3280–3289

Meyer RS, Karol KG, Little DP, Nee MH, Litt A (2012) Phylogeographic relationships among Asian eggplants and new perspectives on eggplant domestication. Mol Phylogenet Evol 63:385–701

Mibei EK, Owino WO, Ambuko J, Giovannoni JJ, Onyango AN (2017) Metabolomic analyses to evaluate the effect of drought stress on selected African Eggplant accessions. J Sci Food Agric 98:205–216

Mimura Y, Kageyama T, Minamiyama Y, Masashi H (2009) QTL analysis for resistance to *Ralstonia solanacearum* in *Capsicum* accession 'LS2341'. J Japan Soc Hort Sci 78:307–313

Miyatake K, Saito T, Negoro S, Yamaguchi H, Nunome T, Ohyama A, Fukuoka H (2016) Detailed mapping of a resistance locus against *Fusarium* wilt in cultivated eggplant (*Solanum melongena*). Theor Appl Genet 129:357–367

Mongkolporn O, Taylor PWJ (2018) Chili anthracnose: *Colletotrichum* taxonomy and pathogenicity. Plant Pathol 67:1255–1263

Moscone EA, Scadalferro MA, Gabriele M (2007) The evolution of chili peppers (*Capsicum* - Solanaceae): a cytogenetic perspective. Acta Hortic 745:137–169

Moury B, Pflieger S, Blattes A, Lefebvre V, Palloix A (2000) A CAPS marker to assist selection of *Tomato spotted wilt virus* (TSWV) in pepper. Genome 43:943–951

Munns R, Tester M (2008) Mechanisms of salinity tolerance. Annu Rev Plant Biol 59:651–681

Muñoz-Falcón JE, Prohens J, Vilanova S, Nuez F (2009) Diversity in commercial varieties and landraces of black eggplants and implications for broadening the breeders' gene pool. Ann Appl Biol 154:453–465

Murphy JF, Blauth JR, Livingstone KD, Lackney VK, Jahn MK (1998) Genetic mapping of the *pvr1* locus in *Capsicum* spp and evidence that distinct potyvirus resistance loci control responses that differ at the whole plant and cellular levels. Mol Plant-Microbe Interact 11:943–951

Mustafa Z, Ayyub CM, Amjad M, Ahmad R (2017) Assessment of biochemical and ionic attributes against salt stress in eggplant (*Solanum melongena* L.) genotypes. J Animal Plant Sci 27:503–509

Mutegi E, Snow AA, Rajkumar M, Pasquet R, Ponniah H, Daunay MC, Davidar P (2015) Genetic diversity and population structure of wild/weedy eggplant (*Solanum insanum*, Solanaceae) in southern India: Implications for conservation. Am J Bot 102:140–148

Naegele RP, Ashrafi H, Hill TA, Reyes Chin-Wo S, Van Deynze AE, Hausbeck MK (2014a) QTL mapping of fruit rot resistance to the plant pathogen *Phytophthora capsici* in a recombinant inbred line *Capsicum annuum* population. Phytopathology 104:479–483

Naegele RP, Boyle S, Quesada-Ocampo LM, Hausbeck MK (2014b) Genetic diversity, population structure, and resistance to *Phytophthora capsici* of a worldwide collection of eggplant germplasm. PloS one 9(5):e95930

Nahar K, Matsumoto I, Taguchi F, Inagaki Y, Yamamoto M, Toyoda K, Mukaihara T (2014) *Ralstonia solanacearum* type III secretion system effector Rip36 induces a hypersensitive response in the non-host wild eggplant *Solanum torvum*. Mol Plant Pathol 15:297–303

Nee M (1999) Synopsis of *Solanum* in the New World. In: Nee M, Symon DE, Lester RN, Jessop JP (eds) Solanaceae IV: advances in biology and utilization. Royal Botanic Gardens, Kew, pp 285–333

Nicolaı M, Cantet M, Lefebvre V, Sage-Palloix AM, Palloix A (2013) Genotyping a large collection of pepper (*Capsicum* spp.) with SSR loci brings new evidence for the wild origin of cultivated *C. annuum* and the structuring of genetic diversity by human selection of cultivar types. Genet Resour Crop Evol 60:2375–2390

Niu G, Rodríguez DS (2010) Rapid screening for relative salt tolerance among chili pepper genotypes. Hort Science 45:1192–1195

Niu G, Rodríguez DS, Call E, Bosland PW, Ulery A, Acosta E (2010) Responses of eight chili peppers to saline water irrigation. Sci Hortic 126:215–222

Nuez F, Prohens J, Valcárcel JV, Fernández de Córdova P (2002) Colección de Semillas de Berenjena del Centro de Conservación y Mejora de la Agrodiversidad Valenciana. Instituto Nacional de Investigación y Tecnología Agraria y Alimentaria, Madrid, Spain

Nurit-Silva K, Costa-Silva R, Basílio IJ, Agra MDF (2012) Leaf epidermal characters of Brazilian species of *Solanum* section Torva as taxonomic evidence. Botany 90:806–814

Okunlola GO, Olatunji OA, Akinwale RO, Tariq A, Adelusi AA (2017) Physiological response of the three most cultivated pepper species (*Capsicum* spp.) in Africa to drought stress imposed at three stages of growth and development. Sci Hortic 224:198–205

Omidiji MO (1979) Crossability relationship between some species of *Solanum*, *Lycopersicon* and *Capsicum* cultivated. In: Nigeria JG, Hawkes RN, Lester AD (eds) Skelding the biology and taxonomy of the solanaceae

Oyelana OA, Ugborogho RE (2008) Phenotypic variation of F1 and F2 populations of three species of *Solanum* L. (Solanaceae). Afr J Biotechnol 7:2359–2367

Pandey VK, Gopal R (2010) Nickel toxicity effects on growth and metabolism of eggplant. Intl J Vegetable Sci 16:351–360

Parisi M, Di Dato F, Minutolo M, Festa G, Alioto D (2015) Screening *Capsicum* spp for tolerance to a resistance-breaking strain of *Tomato spotted wilt virus* by artificial inoculation. J Plant Pathol 97:S57

Parisi M, Alioto D, Tripodi P (2020) Overview of biotic stresses in pepper (*Capsicum* spp.): Sources of genetic resistance, molecular breeding and genomics. Int J Mol Sci 21, 2587

Park M, Lee JH, Han K, Jang S, Han J, Lim JH, Jung JW, Kang BC (2018) A major QTL and candidate genes for capsaicinoid biosynthesis in the pericarp of *Capsicum chinense* revealed using QTL-seq and RNA-seq. Theor Appl. https://doi.org/10.1007/s00122-018-3238-8

Phimchan P, Techawongstien S (2012) Impact of drought stress on the accumulation of capsaicinoids in *Capsicum* cultivars with different initial capsaicinoid levels. Hort Sci 47:1204–1209

Pierre M, Noel L, Lahaye T, Ballvora A, Veuskens J, Ganal M, Bonas U (2000) High-resolution genetic mapping of the pepper resis-tance locus *Bs3* governing recognition of the *Xanthomonas campestris* pv *vesicatora AvrBs3* protein. Theor Appl Genet 101:255–263

Pinheiro JB, Boiteux LS, Almeida MRA, Pereira RB, Galhardo LCS, Carneiro RMDG (2015) First Report of *Meloidogyne enterolobii* in *Capsicum* Rootstocks Carrying The *Me*1 and *Me3*/*Me7* Genes in Central Brazil. Nematropica 45:184–188

Pinheiro JB, da Silva GO, Macêdo AG, Biscaia D, Ragassi CF, Ribeiro C, de Carvalho SI, Reifschneider FJB (2020) New resistance sources to root-knot nematode in *Capsicum* pepper. Horticultura Bras 38:33–40

Plazas M, Andújar I, Vilanova S, Gramazio P, Herraiz FJ, Prohens J (2014) Conventional and phenomics characterization provides insight into the diversity and relationships of hypervariable scarlet (*Solanum aethiopicum* L.) and gboma (*S. macrocarpon*) eggplant complexes. Front Plant Sci 5:318

Plazas M, Vilanova S, Gramazio P, Rodríguez-Burruezo A, Fita A, Herraiz FJ, Ranil R, Fonseka R, Niran L, Fonseka H, Kouassi B, Kouassi A, Kouassi A, Prohens J (2016) Interspecific hybridization between eggplant and wild relatives from different genepools. J Am Soc Hortic Sci 141:34–44

Plazas M, Nguyen HT, González-Orenga S, Fita A, Vicente O, Prohens J, Boscaiu M (2019) Comparative analysis of the responses to water stress in eggplant (*Solanum melongena*) cultivars. Plant Physiol Biochem 143:72–82

Portis E, Cericola F, Barchi L, Toppino L, Acciarri N, Pulcini L, Sala T, Lanteri S, Rotino GL (2015) Association mapping for fruit, plant and leaf morphology traits in eggplant. PLoS ONE 10:

Prohens J, Anderson GJ, Herraiz FJ, Bernardello G, Santos-Guerra A, Crawford D, Nuez F (2007) Genetic diversity and conservation of two endangered eggplant relatives (*Solanum vespertilio* Aiton and *Solanum lidii* Sunding) endemic to the Canary Islands. Genet Resour Crop Evol 54:451–464

Prohens J, Plazas M, Raigón MD, Seguí-Simarro JM, Stommel JR, Vilanova S (2012) Characterization of interspecific hybrids and backcross generations between two cultivated eggplants (*Solanum melongena* and *S. aethiopicum* Kumba group) and implications for eggplant breeding. Euphytica 186:517–538

Prohens J, Whitaker BD, Plazas M, Vilanova S, Hurtado M, Blasco M, Gramazio P, Stommel JR (2013) Genetic diversity in morphological charactes and phenolic acids content resulting from an interspecific cross between eggplant (*Solanum melongena*) and its wild ancestor (*S. incanum*). Ann Appl Biol 162:242–257

Prohens J, Gramazio P, Plazas M, Dempewolf H, Kilian B, Díez MJ, Fita A, Herraiz FJ, Rodríguez-Burruezo A, Soler S, Knapp S, Vilanova S (2017) Introgressiomics: a new approach for using crop wild relatives in breeding for adaptation to climate change. Euphytica 213:158

Pugalendhi L, Veeraragavathatham D, Natarjan S, Praneetha S (2010) Utilizing wild relative (*Solanum viarum*) as resistant source to shoot and fruit borer in brinjal (*Solanum melongena* Linn.). Electron J Plant Breed 1:643–648

Quenouille J, Paulhiac E, Moury B, Palloix A (2014) Quantitative trait loci from the host genetic background modulate the durability of a resistance gene: a rational basis for sustainable resistance breeding in plants. Heredity 112:579–587

Quirin EA, Ogundiwin EA, Prince JP, Mazourek M, Briggs MO, Chlanda TS, Kim K-T, Falise M, Kang B-C, Jahn MM (2005) Development of sequence characterized amplified region (SCAR) primers for the detection of *Phyto52*, a major QTL for resistance to *Phytophthora capsici* Leon In pepper. Theor Appl Genet 110:605–612

Rajam MV, Kumar SV (2006) Eggplant. In: Pua EC, Davey MR (eds) Biotechnology in agriculture and forestry, vol 59. Transgenic crops IV. Springer, Berlin, pp 201–219

Rajam MV, Rotino GL, Sihachakr D, Souvannavong V, Mansur E, Kumar PA (2008) Eggplant 2. In: Kole KC, Hall TC (eds) Compendium of transgenic crop plants: transgenic vegetable crops. Blackwell Publ, USA, pp 47–72

Rajasekaran S (1970a) Cytogenetic studies on the F1 hybrid of *Solanum macrocarpon* L. (*S. melongena* var. 'bulsarensis Arkigar') × *S. melongena* L. Auara 2:21–28

Rajasekaran S (1970a) Sterility in an inter-varietal hybrid *Solanum melongena* L. × *S. melongena* var. 'bulsarensis Argikar'. Madras Agric J 57:194–196

Rajasekaran S (1970b) Ovule sterility of the F1 hybrid *Solanum melongena* var bulsarensis Argikar. Jpn J Genet 45:163–166

Ranil RHG, Prohens J, Aubriot X, Niran HML, Plazas M, Fonseka RM, Vilanova S, Fonseka HH, Gramazio P, Knapp S (2017) *Solanum insanum* L. (subgenus *Leptostemonum* Bitter, Solanaceae), the neglected wild progenitor of eggplant (*S. melongena* L.): a review of taxonomy, characteristics and uses aimed at its enhancement for improved eggplant breeding. Genet Resour Crop Evol 64:1707–1722

Rao NN (1979) The barriers to hybridization between *Solanum melongena* and some other species of Solanum. In: Hawkes JG, Lester RN, Skelding AD (eds) The Biology and taxonomy of the Solanaceae. Press, London, Acad, pp 605–614

Rao GU, Chaim AB, Borovsky E, Paran I (2003) Mapping of yield related QTLs in pepper in an interspecific cross of *Capsicum annuum* and *C. frutescens*. Theor Appl Genet 106:1457–1466

Ray D, Muelle N, West P, Foley J (2013) Yield trends are insufficient to double global crop production by 2050. PLoS ONE 8:

Robinson RW, Shail JW, Gao YX, Doganlar S (2001) Interspecific hybridization of eggplant for Verticillium wilt resistance and other useful traits. Solanaceae V.

Advances in taxonomy and utilization. Nijmegen University Press, Nijmegen, pp 279–291

Roggero P, Masenga V, Tavella L (2002) Field isolates of *Tomato spotted wilt virus* overcoming resistance in pepper and their spread to other hosts in Italy. Plant Dis 86:950–954

Romer P, Jordan T, Lahaye T (2010) Identification and application of a DNA-based marker that is diagnostic for the pepper (*Capsicum annuum*) bacterial spot resistance gene *Bs3*. Plant Breed 129:737–740

Rotino GL, Perri E, D'Alessandro A, Mennella G (1998) Characterization of fertile somatic hybrids between eggplant (*S. melongena* L.) and S. *integrifolium*. Proc. Eucarpia Meet Genet Breed capsicum eggplant 10:213–217

Rotino GL, Mennella G, Fusari F, Vitelli G, Tacconi MG, D'Alessandro A, Acciarri N (2001) Towards introgression of resistance to *Fusarium oxysporum* f. sp. *melongenae* from *Solanum integrifolium* into eggplant. In: Proceedings of the 11th Eucarpia meeting on genetics and breeding of capsicum and eggplant. Antalya, Turkey, pp 303–307

Rotino GL, Sihachakr D, Rizza F, Valè G, Tacconi MG, Alberti P, Mennella G (2005) Current status in production and utilization of dihaploids from somatic hybrids between eggplant (*Solanum melongena* L.) and its wild relatives. Acta Physiol Plant 27:723–733

Rotino GL, Sala T, Toppino L (2014) "Eggplant," In: Pratap A, Kumar J (eds) Alien gene transfer in crop plants, vol 2. Springer, New York, pp 381–409

Rotino GL, Lanteri S, Sala T, Toppino L, Acquadro A, Barchi L, Portis E, Rinaldi R, Scaglione D, Dal Molin A, Minio A, Ferrarini A, Tononi P, Zamperin G, Fantini E, Pietrella M, Giuliano G, Delledonne M (2017) An Eggplant (*Solanum melongena* L.) High Quality Genome Draft. In: Proceedings of the plant and animal genome XXIInd Conference. 10-15 January PAG San Diego, California, USA

Safni I, Cleenwerck I, De Vos P, Fegan M, Sly L, Kappler U (2014) Polyphasic taxonomic revision of the *Ralstonia solanacearum* species complex: proposal to emend the descriptions of *Ralstonia solanacearum* and *Ralstonia syzygii* and reclassify current *R syzygii* strains as *Ralstonia syzygii* subsp *syzygii* subsp nov *R solanacearum* phylotype IV strains as *Ralstonia syzygii* subsp *indonesiensis* subsp nov *banana* blood disease bacterium strains as *Ralstonia syzygii* subsp *celebesensis* subsp nov and *R solanacearum* phylotype I and III strains as *Ralstonia pseudosolanacearum* sp nov. Int J Syst Evol Microbiol 64:3087–3103. 101099/ijs0066712-0

Sahytia UL, Krishna MSR, Deepthi SR, Shiva Prasad G, Kasim PD (2018) Seed antioxidants interplay with drought stress tolerance indices in Chilli (*Capsicum annuum* L.) Seedlings. BioMed Res Int 2018:1605096

Saracanlao RJR, Ocampo ETM, Canama AO, Manaday SJB, Maghirang RG, Delfin EF (2016) SSR-based genetic relationship in eggplant (*Solanum melongena*) genotypes with varying morphological response to drought. Philippine J Crop Sci 41:57–64

Savvas D, Lenz F (2000) Effects of NaCl or nutrient-induced salinity on growth, yeildm and composition of eggplants grown in rockwool. Sci Hortic 84:37–47

Schaff DA, Jelenkovic G, Boyer CD, Pollack BL (1982) Hybridization and fertility of hybrid derivatives of *Solanum melongena* L. and *Solanum macrocarpon* L. Theor Appl Genet 62:149–153

Schippers RR (2000) African indigenous vegetables—an overview of the cultivated species. Natural Resources Institute, Chatham, UK

Shahbaz M, Ashraf M, Al-Qurainy F, Harris PJC (2012) Salt tolerance in selected vegetable crops. Crit Rev Plant Sci 31:303–320

Sharma DR, Chowdhury JB, Ahuja U, Dhankhar BS (1980) Interspecific hybridization in genus *Solanum*: a cross between *S. melongena* and *S. khasianum* through embryo culture. Z Pfl anzenzüchtg 85:248–253

Sharma DR, Sareen PK, Chowdhury JB (1984) Crossability and pollination in some non-tuberous *Solanum* species. Indian J Agric Sci 54:514–517

Siddique MI, Lee HY, Ro NY, Kan K, Venkatesh J, Solomon AM, Patil AS, Changkwian A, Kwon JK, Kang BC (2019) Identifying candidate genes for *Phytophthora capsici* resistance in pepper (*Capsicum annuum*) via genotyping-by-sequencing-based QTL mappingand genome-wide association study. Sci Rep 9:9962

Sihachakr D, Haicour R, Serraf I, Barrientos E, Herbreteu C, Ducreux G, Rossignol L, Souvannavong V (1988) Electrofusion for the production of somatic hybrids plants of *Solanum melongena* L. and *Solanum khasianum* C.B. Clark. Plant Sci 57:215–223

Sihachakr D, Haicour R, Barientos E, Ducreux G, Rossignol L (1989) Somatic hybrid plants produced by electrofusion between *Solanum melongena* L. and *Solanum torvum* SW. Theor Appl Genet 77:1–6

Sihachakr D, Daunay MC, Serraf I (1994) Somatic hybridization of eggplant (*Solanum melongena* L.) with its close and wild relatives. In: Bajaj YPS (ed) Biotechnology in agriculture and forestry, somatic hybridization in crop improvement, vol I. Springer, Berlin, Heidelberg, pp 255–278

Soler S, Debreczeni DE, Vidal E, Aramburu J, López C, Galipienso L, Rubio L (2015) A new *Capsicum baccatum* accession shows tolerance to wild-type and resistance-breaking isolates of *Tomato spotted wilt virus*. Annals Appl Biol 167:343–353

Sseremba G, Tongoona P, Elebu J, Danquah EY, Kizito EB (2018) Heritability of drought resistance in *Solanum aethiopicum* Shum group and combining ability of genotypes for drought tolerance and recovery. Sci Hortic 240:213–220

Stellari GM, Mazourek M, Jahn MM (2010) Contrasting modes for loss of pungency between cultivated and wild species of *Capsicum*. Heredity 104:460–471

Stewart C Jr, Kang B, Mazourek M, Liu K, Moore SL, Paran I, Jahn MM (2005) The *Pun1* gene for pungency in pepper encodes a putative acyltransferase. Plant J 42:675–688

Stewart C, Mazourek M, Stellari GM, O'Connell M, Jahn M (2007) Genetic control of pungency in *C. chinense* via the *Pun1* locus. J Exp Botany 58:979–991

Stommel JR, Bosland P (2007) in Flower breeding and genetics: Issues, challenges, and opportunities for the 21st century. In: Anderson NO (ed) Ornamental pepper *Capsicum annuum*. Springer, Dordrecht, The Netherlands. pp 561–599

Street K, Bari A, Mackay M, Amri A (2016) How the focused identification of germplasm strategy (FIGS) is used to mine plant genetic resources for adaptive traits. In: Maxted N, Dulloo ME, Ford-Lloyd B (eds) Enhancing crop genepool use: capturing wild relative and landrace diversity for crop improvement. CABI, Wallingford, UK, pp 54–65

Sunseri F, Polignano GB, Alba V, Lotti C, Visignano V, Mennella G, D'Alessandro A, Bacchi M, Riccardi P, Fiori MC, Ricciardi L (2010) Genetic diversity and characterization of African eggplant germplasm collection. Afr J Plant Sci 4:231–241

Suzuki K, Kuroda T, Miura Y, Murai J (2003) Screening and field trials of virus resistant sources in *Capsicum* spp. Plant Dis 87:779–783

Syfert MM, Castañeda-Álvarez N, Khoury CK, Särkinen T, Sosa CC, Achicanoy HA, Bernau V, Prohens J, Daunay MC, Knapp S (2016) Crop wild relatives of the brinjal eggplant (*Solanum melongena*): poorly represented in genebanks and many species at risk of extinction. Am J Bot 103:635–651

Taher D, Solberg SØ, Prohens J, Chou Y, Rakha M, Wu T (2017) World Vegetable Center eggplant collection: Origin, composition, seed dissemination and utilization in breeding. Front Plant Sci 8:1484

Tai T, Dahlbeck D, Stall RE, Peleman J, Staskawicz BJ (1999) High-resolution genetic and physical mapping of the region containing the *Bs2* resistance gene of pepper. Theor Appl Genet 99:1201–1206

Tani E, Kizis D, Markellou E, Papadakis I, Tsamadia D, Leventis G, Makrogianni D, Karapanos I (2018) Cultivar-dependndt responses of eggplant (*Solanum melongena* L.) to simultaneous *Verticillium dahliae* infection and drought. Front Plant Sci 9:1181

Tewksbury JJ, Manchego C, Haak DC, Levey DJ (2006) Where did the Chili get its spice? Biogeography of capsaicinoid production in ancestral wild chili species. J Chem Ecol 32:547–564

Thabuis A, Palloix A, Servin B Daubèze AM, Signoret P, Hospital F, Lefebvre V (2004) Marker-assisted introgression of 4 *Phytophthora capsici* resistance QTL alleles into a bell pepper line: validation of additive and epistatic effects. Mol Breed 14:9–20

Thies JA, Fery RL (2002) Heat stability of resistance to southern root-knot nematode in bell pepper genotypes homozygous and heterozygous for the *N* gene. J Am Soc Hortic Sci 127:371–375

Thies JA, Mueller JD, Fery RL (1997) Effectiveness of resistance to southern root-knot nematode in 'Carolina Cayenne' pepper in greenhouse, microplot, and field tests. J Am Soc Hortic Sci 122:200–204

Tilman D, Balzer C, Hill J, Befort BL (2011) Global food demand and the sustainable intensification of agriculture. Proc Natl Acad Sci USA 108:20260–20264. 10.1073/pnas.1116437108

Tomita R, Murai J, Miura Y, Ishihara H, Liu S, Kubotera Y, Honda A, Hatta R, Kuroda T, Hamada H, Sakamoto M, Munemura I, Nunomura O, Ishikawa K, Genda Y, Kawasaki S, Suzuki K, Meksem K, Kobayashi K (2008) Fine mapping and DNA fiber FISH analysis locates the tobamovirus resistance gene L3 of *Capsicum chinense* in a 400-kb region of R-like genes cluster embedded in highly repetitive sequences. Theor Appl Genet 117:1107–1118

Toppino L, Mennella G, Rizza F, D'Alessandro A, Sihachakr D, Rotino GL (2008a) ISSR and isozyme characterization of androgenetic dihaploids reveals tetrasomic inheritance in tetraploid somatic hybrids between *Solanum melongena* and *Solanum aethiopicum* group gilo. J Hered 99:304–315

Toppino L, Valè G, Rotino GL (2008b) Inheritance of *Fusarium* wilt resistance introgressed from *Solanum aethiopicum* Gilo and Aculeatum groups into cultivated eggplant (*S. melongena*) and development of associated PCR-based markers. Mol Breed 22:237–250

Toppino L, Ribolzi S, Bassolino L, Rotino GL (2017) Development of an introgression population of eggplant from the wild species *S. indicum*. In: Proceedings of the joint congress SIBV-SIGA (Italian Society of Agricultural Genetics) Annual Congress. Pisa, 19-22 Settembre 2017

Toppino L, Ribolzi S, Shaaf S, Bassolino L., Carletti G, Fadda S., Rossini L., Boyaci HF, Caliskan S, Unlu A, Rotino GL (2018) Development of an introgression lines population and genetic mapping of novel traits linked to key breeding traits in eggplant. In: Proceedings of the 62th SIGA Congress Verona. Italy, 25/28 Sept 2018

Tripodi P, Greco B (2018) Large scale phenotyping provides insight into the diversity of vegetative and reproductive organs in a wide collection of wild and domesticated peppers (*Capsicum* spp.). Plants 7:103

Tripodi P, Kumar S (2019) The capsicum crop: an introduction. In: Ramchiary N, Kole C (eds) The capsicum genome. Compendium of plant genomes. Springer, Cham. https://doi.org/10.1007/978-3-319-97217-6_1

Tripodi P, Cardi T, Bianchi G, Migliori CA, Schiavi M, Rotino GL, Lo Scalzo R (2018) Genetic and environmental factors underlying variation in yield performance and bioactive compound content of hot pepper varieties (*Capsicum annuum*) cultivated in two contrasting Italian locations. Eur Food Res and Technol 244:1555–1567

Usman MG, Rafii MY, Ismail MR, Malek MA, Latif MA (2014) Heritability and genetic advance among chili pepper genotypes for heat tolerance and morphophysiological characteristics. Sci World J Article ID:308042, 14

Vasileva K, Todorova V, Masheva S (2019) Evaluation of collection of pepper (*Capsicum* spp.) resources for resistance to *Verticillium dahliae* Kleb. Bulg J Agric Sci 2019:1030–1038

Venkatesh J, An J, Kang WH, Jahn M, Kang BC (2018) Fine mapping of the dominant potyvirus resistance gene *Pvr7* reveals a relationship with *Pvr4* in *Capsicum annuum*. Phytopathology 108:142–148

Vilanova S, Blasco M, Hurtado M, Munoz-Falcon JE, Prohens J, Nuez F (2010) Development of a linkage map of eggplant based on a *S. incanum* × *S. melongena* backcross generation. In: Prohens J, Rodriguez-Burruezo A (eds) Advances in genetics and breeding of capsicum and eggplant. Editorial de la Universitat Politecnica de Valencia, Spain, pp 435–439

Vogel S (1998) Remarkable nectaries: Structure, ecology, organoleptic perspectives III. Nectar ducts. Flora 193:113–131

Voorrips RE, Finkers R, Lia Sanjaya (2004) QTL mapping of anthracnose (*Colletotrichum* spp) resistance in a cross between *Capsicum annuum* and *C chinense*. Theor Appl Genet 109:1275–1282

Vorontsova MS, Knapp S (2016) A revision of the "spiny solanums", *Solanum* subgenus *Leptostemonum* (Solanaceae), in Africa and Madagascar. Syst Bot Monograph 99:1–432

Vorontsova MS, Stern S, Bohs L, Knapp S (2013) African spiny *Solanum* (subgenus *Leptostemonum*): a thorny phylogenetic tangle. Bot J Linn Soc 173:176–193

Wang JX, Gao TG, Knapp S (2008) Ancient Chinese literature reveals pathways of eggplant domestication. Ann Appl Biol 102:891–897

Wang Z, Chao X, DengWei J, LePing H, QuanSheng H, Qing Y (2010a) Cloning and expression analysis of Verticillium wilt pathogenesis-related gene *StDAHP* in *Solanum torvum*. China Biotechnol 30:48–53 b

Wang Z, Guo JL, Zhang F, Huang QS, Huang LP, Yang Q (2010b) Differential expression analysis by cDNA-AFLP of *Solanum torvum* upon *Verticillium dahliae* infection. Russ J Plant Physiol 57:676–684

Wang X, Fazari A, Cao Y, Zhang Z, Palloix A, Mao S, Zhang B, Djian-Caporalino C, Wang L (2018) Fine mapping of the root-knot nematode resistance gene *Me1* in pepper (*Capsicum annuum* L.) and development of markers tightly linked to *Me1*. Mol Breed 38:39

Wanjari KB (1976) Cytogenetic studies on F1 hybrids between *Solanum melongena* L and S. *macrocarpon* L. Hortic. Res. 15:77–83

Warschefsky E, Penmetsa RV, Cook D, von Wettberg EJB (2014) Back to the wilds: tapping evolutionary adaptations for resilient crops through systematic hybridization with crop wild relatives. Am J Bot 101:1791–1800

Wu FN, Mueller LA, Crouzillat D, Petiard V, Tanksley SD (2006) Combining bioinformatics and phylogenetics to identify large sets of single copy, orthologous genes (COSII) for comparative, evolutionary and systematic studies: a test case in the euasterid plant clade. Genetics 174:1407–1420

Wu FN, Eannetta NT, Xu YM, Tanksley SD (2009) A detailed synteny map of the eggplant genome based on conserved ortholog set II (COSII) markers. Theor Appl Genet 118:927–935. 10.1007/s00122-008-0950-9

Wu X, Yao X, Chen J, Zhu Z, Zhang H, Zha D (2014) Brassinosteroids protect photosynthesis and antioxidant system of eggplant seedlings from high-temperature stress. Acta Physiol Plant 36:251–261

Xu J, Sun J, Du L, Liu X (2012) Comparative transcriptome analysis of cadmium responses in *Solanum nigrum* and *Solanum torvum*. New Phytol 196: 110–124

Yamaguchi H, Fukuoka H, Arao T, Ohyama A, Nunome T, Miyatake K, Negoro S (2010) Gene expression analysis in cadmium-stressed roots of a low cadmium-accumulating solanaceous plant, *Solanum torvum*. J Exp Bot 61:423–437. 10.1093/jxb/erp313

Yamaguchi N, Mori S, Baba K, Kaburagi-Yada S, Arao T, Kitajima N, Hokura A, Terada Y (2011) Cadmium distribution in the root tissues of solanaceous plants with contrasting root-to-shoot Cd translocation efficiencies. J Exp Bot 71:198–206

Yang HB, Liu WY, Kang WH, Jan M, Kang BC (2009) Development of SNP markers linked to the *L* locus in *Capsicum* spp by a comparative genetic analysis. Mol Breed 24:433

Yang X, Liu F, Zhang Y, Wang L, Cheng Y (2017) Cold-responsive miRNAs and their target genes in the wild eggplant species *Solanum aculeatissimum*. BMC Genom 18:1000

Yeam I, Kang BC, Lindeman W, Frantz JD, Faber N, Jahn MM (2005) Allele-specific CAPS markers based on point mutations in resistance alleles at the *pvr1* locus encoding *eIF4E* in *Capsicum*. Theor Appl Genet 112:178–186

Ying SC, Li MS, Hai ZZ, Alain P, Hao WL, Xi ZB (2015) Resistances to anthracnose (*Colletotrichum acutatum*) of *Capsicum* mature green and ripe fruit are controlled by a major dominant cluster of QTLs on chromosome P5. Sci Hortic 181:81–88

Yoon B, Park HG (2005) Trispecies bridge crosses (*Capsicum annuum* x *C chinense*) x *C baccatum*, as an alternative for introgression of anthracnose resistance from *C baccatum* into *C annuum*. J Kor Soc Hort Sci 46:5–9

Yoon JB, Do JW, Yang DC, Park HG (2004) Interspecific cross compatibility among five domesticated species of *Capsicum* genus. J Kor Soc Hort Sci 45:324–329

Yuan HH, Sun LZ, Tai PD, Liu W, Li XJ, Hao L (2019) Effects of grafting on root-to-shoot cadmium translocation in plants of eggplant (*Solanum melongena*) and tomato (*Solanum lycopersicum*). Sci Total Environ 652:

Yumi Baba V, Constantino LV, Tiemi Ivamoto S, Paladini Moreira AF, Bervelieri Madeira T, Nixdorf SL, Rodrigues R, Azeredo Gonçalves LS (2019) *Capsicum-Colletotrichum* interaction: identification of resistance sources and quantification of secondary metabolites in unripe and ripe fruits in response to anthracnose infection. Sci Hortic 246:469–477

Zewdie Y, Bosland PW (2000) Evaluation of genotype environment and genotype-by-environment interaction for capsaicinoids in *Capsicum annuum* L. Euphytica 111:185–190

Zhuang Y, Wang S (2009) Identification of interspecific hybrid and its backcross progenies of *Solanum aethiopicum* group Aculeatum and *S. melongena*. Jiangsu J Agric Sci 27:137–140

Zygier S, Ben Chaim A, Efrati A, Kaluzky G, Borovsky Y, Paran I (2005) QTLs mapping for fruit size and shape in chromosomes 2 and 4 in pepper and a comparison of the pepper QTL map with that of tomato. Theor Appl Genet 111:437–445

Knowledge on the Genomes of Wild Tomato Species is the Key to Unlocking Their Breeding Potential

7

Nunzio D'Agostino, Amalia Barone, and Maria Manuela Rigano

Abstract

Wild tomato species are characterized by wide genetic variability and phenotypic diversity that might be recovered to reintroduce favorable 'wild' genes/alleles back into the 'domesticated' genomes. Within this context, the assessment of the genetic diversity of wild tomato species is gaining momentum and an ever-growing number of genomic resources are now available. The latter provide novel opportunities to broaden the genetic basis of cultivated tomatoes and exploit the overwhelming inter-species genetic variability via 'introgression' breeding and/or new plant breeding techniques to address emerging challenges. This chapter describes the main genomic resources developed so far for wild tomato species; emphasizes that they are an important reservoir of favorable alleles for biotic and abiotic stress tolerance and fruit quality traits and, finally, recalls the concepts of 'introgressomics' and 'rewilding' as founding elements of future breeding strategies.

7.1 Introduction

One of the main requirements in breeding activities is the assessment and exploitation of crop genetic diversity. Major advances in sequencing technologies and bioinformatics have allowed 'sequence space' to be explored in order to (i) investigate the evolutionary history and geographical distribution of a species or a clade, (ii) track wild introgressions, (iii) score single nucleotide polymorphisms (SNPs) at unprecedented level, and (iv) identify useful/favorable alleles for breeding purposes (D'Agostino and Tripodi 2017).

Tomato domestication and artificial selection were mainly focused on increasing yield, fruit size, and firmness as well as on adaptation to different growing systems. The 'domestication syndrome' had a pervasive impact on the genetic diversity of cultivated tomatoes and eroded the genetic variability mainly associated with tolerance to biotic and abiotic stresses (Lin et al. 2014; Sahu and Chattopadhyay 2017).

Wild tomato species are characterized by a wide, often untapped, genetic variability and phenotypic diversity (Bai and Lindhout 2007) that must be recovered and exploited to expedite tomato breeding. Plant breeders have long used wild tomato species as donor of genes/alleles

N. D'Agostino (✉) · A. Barone · M. M. Rigano
Department of Agricultural Sciences, University of Naples Federico II, Via Università 100, 80055 Portici, Italy
e-mail: nunzio.dagostino@unina.it

© Springer Nature Switzerland AG 2021

D. Carputo et al. (eds.), *The Wild Solanums Genomes*, Compendium of Plant Genomes,
https://doi.org/10.1007/978-3-030-30343-3_7

thanks to inter-specific cross compatibility. The *Solanum* section *Lycopersicon* consists of 14 species (Peralta et al. 2005; Spooner et al. 2005) that occupy different habitats along a diversified climatic gradient. The cultivated tomato *Solanum lycopersicum*, which is predominantly self-pollinating and highly inbred, is included in this section. It has a worldwide distribution, and it ranks first among all fruit vegetables for production (Wang et al. 2020).

The genome of *S. lycopersicum* cultivar (cv.) 'Heinz 1706' was published in 2012 (Tomato Genome Consortium 2012) by a consortium of 14 countries. This milestone laid the foundations for several re-sequencing projects of cultivated and wild tomatoes, which have been developed to address different biological questions and enrich trait reservoir (Lin et al. 2014; The 100 Tomato Genome Sequencing Consortium 2014; Tranchida-Lombardo et al. 2018; Ercolano et al. 2014; Causse et al. 2013; Tieman et al. 2017). The sequencing efforts have also further enhanced the use of the available introgression lines (ILs), which consist of lines carrying homozygous chromosome regions from a wild donor in the genetic background of an elite cultivar (i.e., recipient). ILs exist for several wild tomato species (Calafiore et al. 2019; Alseekh et al. 2013; Ballester et al. 2016; Chetelat et al. 2019) and represent a basic resource for genetic and genomic studies as well as a viable option for quantitative trait loci (QTL) detection and breeding purposes (D'Agostino and Tripodi 2017).

This chapter provides an overview of the main genomic resources developed so far for tomato wild species and emphasizes their use for biodiversity-based breeding as they are an important reservoir of favorable alleles for stress tolerance and fruit quality traits, and, finally, recalls the concepts of 'introgressomics' and 'rewilding' as cornerstones at the basis of the breeding strategies of the near future.

7.2 Solanum Pimpinellifolium

S. lycopersicum was domesticated from its wild progenitor *S. pimpinellifolium* (Tomato Genome Consortium 2012). The latter has a bushy appearance, round red fruits weighing 1–2 g, thin pericarp and high seed content, and it is facultative autogamous (Lin et al. 2020). The species is distributed along the coast of Peru, Ecuador, and northern Chile, where plants are exposed to harsh environmental conditions. Precisely, its phenotypic robustness has garnered the attention of breeders.

The Tomato Genome Consortium released into the public domain also the draft sequence of *S. pimpinellifolium* accession 'LA1589'. It was generated from over 55 Gb of 101 bp paired-end Illumina reads. The final assembly resulted in ~627,000 contigs for a total size of 739 Mb.

Despite the fragmented nature of the assembly and the low sequencing depth, this resource allowed to (i) estimate the nucleotide divergence with respect to 'Heinz 1706' (0.6% nucleotide divergence), (ii) investigate the bottlenecks that have narrowed tomato genetic diversity, and (iii) identify signs of recent admixture between the two gene pools ('Heinz 1706' carries introgressions from *S. pimpinellifolium* on chromosomes 4, 9, 11 and 12).

Razali et al. (2018) released the genome assembly and annotation of *S. pimpinellifolium* accession 'LA0480,' which is known for being salt tolerant. The sequencing by Illumina technology (two paired-end short libraries and five mate-pair libraries) produced 160 Gb of raw data, which were assembled into over 160,000 scaffolds for a final assembly size of 811 Mb. The 'LA0480' genome represents a substantial improvement over the 'LA1589' draft sequence, which was assembled to contig level only. A posteriori analysis on the inventory of the 25,970 annotated protein-coding genes revealed a copy number variation (higher number) in genes encoding for inositol-3-phosphate synthase and phosphatase enzymes. These enzymes are responsible for the production of inositol and its derivatives and play a key role in salt stress response (Razali et al. 2018). It is likely that the observed expansion of these two gene families contributed to modifying the inositol phosphate metabolic pathway in *S. pimpinellifolium*. The phenotypic differences observed under field conditions in salt stress tolerance between 'LA0480' and *S. lycopersicum* cv. 'Heinz 1706' could be just ascribed to changes in that pathway.

Wang et al. (2020) used PacBio sequencing and Hi-C scaffolding to generate a chromosome-scale genome of *S. pimpinellifolium* accession 'LA2093.' This high-quality genome represents a robust reference when compared with the fragmented and incomplete drafts of the 'LA1589' and 'LA0480' genomes. It has an estimated genome size of 923 Mb and harbors 35,761 protein-coding genes, 99.4% of which are supported by RNA-Seq data. 'LA2093' was selected for sequencing as it has served as parent of a recombinant inbred line (RIL) population widely used for QTL mapping (Gonda et al. 2019; Ashrafi et al. 2012, 2009). The genome was compared with the reference 'Heinz 1706' for the identification of structural variations (SVs). Although high collinearity was observed, the two genomes differ in 28 inversions and over 90,000 InDels. How these SVs affect important agronomic traits was investigated.

More than 50 *S. pimpinellifolium* accessions, including 'LA1589,' 'LA0480,' and 'LA2093', were subjected to low-pass whole genome sequencing along with over 300 accessions with the goal of providing insights into the history of tomato breeding (Lin et al. 2014). Approximately, 5 Gb of sequence data (100-bp paired-end reads with insert sizes of $\sim$500 bp) were generated for each sample and used for reference-based SNP calling. SNP data points were then used for the genome-wide identification of selective sweeps. Based on QTLs under selection related to domestication and improvement sweeps, it was possible to hypothesize a two-step evolution of tomato fruit size: from *S. pimpinellifolium* to *S. lycopersicum* var. *cerasiforme* (i.e., cherry tomato) and from *S. lycopersicum* var. *cerasiforme* to big-fruited *S. lycopersicum* accessions. In addition, three further accessions of *S. pimpinellifolium*, namely 'LYC2798,' 'LA1584,' and 'LA1578' were selected, along with 27 other accessions of wild species (see also paragraph 6), for shallow whole genome sequencing by The 100 Tomato Genome Sequencing Consortium (2014) in order to investigate nucleotide variation across the tomato clade. Unfortunately, there is no assembly or annotation available for these accessions. However, the consortium has made a variant browser publicly accessible (http://www.tomatogenome.net/VariantBrowser/) where an overview of all SNPs and InDels scored for each of the 84 accessions (cultivated and wild) under study can be found. Altogether the data collected have highlighted the large genetic erosion within the cultivated tomato and the equally large sequence diversity across wild species.

Sixteen *S. pimpinellifolium* accessions were subjected to low coverage whole genome re-sequencing together with 382 additional modern and heirloom *S. lycopersicum* accessions in order to identify loci associated with improved flavor quality (Tieman et al. 2017). The inclusion of these wild accessions in the experimental design was necessary to provide a baseline for fruit chemical composition before human selection pressure. Sequencing was coupled with the analysis of volatiles, sugars, and acids, and allelic and phenotypic data points were combined via genome-wide association studies. Overall, collected data pointed out that modern commercial varieties have lost most of the volatile compounds that are essential for consumers' liking.

7.3 Solanum Pennellii

Solanum pennellii is a wild, green-fruited tomato species native to Peru, where it has adapted to grow in arid habitats (Bolger et al. 2014). In general, individuals of the species exhibit self-incompatibility and are highly heterozygous, with some exceptions (Schmidt et al. 2017). Because of its specific morphology (it has several distinctive features that make it different from other wild tomatoes) and exceptional abiotic stress tolerance, it has long been used as donor in tomato breeding programs. Indeed, *S. pennellii* x

S. lycopersicum introgression lines (see paragraph 8) were used to identify different QTLs affecting traits of agronomic relevance (e.g., yield, stress tolerance, fruit chemical composition, and plant/fruit morphology) (Calafiore et al. 2019; Dariva et al. 2020; Schauer et al. 2006; Muir et al. 2014; Yang et al. 2016).

In 2014, the genome of *S. pennellii* accession 'LA716' was published, precisely to support QTL analysis (Bolger et al. 2014). Genome reconstruction from Illumina reads (three paired-end short libraries and thirteen mate-pair libraries were sequenced) generated an assembly of 942 Mb in size split into 4,579 scaffolds. Genome annotation resulted in the prediction of more than 32,000 high confidence genes and pointed out that long terminal repeat (LTR) retrotransposons have a key role in genome size variation when *S. pennellii* and *S. lycopersicum* are compared. The comparison between orthologous gene pairs of *S. pennellii* and *S. lycopersicum* made it possible to identify genes under positive selection in the wild species. The resulting gene list was enriched in genes encoding for acyl-carrier proteins, known to be involved in lipid biosynthesis. This might help explain the diversity in waxes and cutin accumulation on the fruit cuticle and might reflect *S. pennellii* capability to grow in arid habitats (the thickest cuticle minimizes transpiration and water loss). Also, stress-related gene set was deeply investigated. Half of the *S. pennellii* genes had widespread polymorphisms in the regions upstream the ATG codon, and this has been associated with a difference in gene expression levels between *S. pennellii* and *S. lycopersicum* genes. The co-localization of retrotransposons with stress-responsive genes seems to suggest a potential role of LTR retrotransposons in regulating the expression of these genes by acting as 'conditional' regulatory sequences. Finally, the comparison *S. pennellii* vs *S. lycopersicum* showed divergence in the sequence and in the expression of secondary-metabolism-related genes (i.e., those involved in the biosynthesis of glycoalkaloid, acyl sugar, carotenoid, terpene and volatile molecules). Within the same project, the genome of *S. lycopersicum* cv. M82, which is the recipient parent of the available IL population, was also sequenced and assembled (Bolger et al. 2014).

The accession 'LA716' was also in the collection of wild tomato species subjected to low-pass genome sequencing by The 100 Tomato Genome Sequencing Consortium (2014). Furthermore, its mitochondrial genome (mitogenome) was sequenced, together with those of *S. lycopersicum* accessions 'LA1479' and 'LA1421' (Kim and Lee 2018). They were 97% similar to each other, even though intra- and inter-specific variations were scored. Comparison of these mitogenomes with chloroplast and nuclear genomes allowed also investigating inter-compartmental DNA transfer (i.e., gene transfer from organelles to the nucleus) and identifying numerous nuclear copies of mitochondrial (NUMTs) and plastid (NUPTs) DNA in the corresponding nuclear genomes.

The genome sequence of *S. pennellii* accession 'LYC1722' was recently released (Schmidt et al. 2017). 'LYC1722' is a self-compatible, phenotypically stable biotype of *S. pennellii* that does not exhibit the negative traits typical of the accession 'LA716' (i.e., the presence of the necrotic dwarf gene on chromosome 6 and poor performances in fields).

The 'LYC1722' was de novo assembled from Oxford Nanopore long reads ($\sim$110 Gb of high-quality sequence data) and then polished with short, high-quality Illumina reads. A number of different tools were used for the assembly phase, and the best assembly resulted into 899 contigs covering 1.2 Gb. Based on the assessment of different parameters (e.g., genome contiguity and assembly completeness in terms of gene content), the 'LYC1722' assembly was of better quality than the Illumina-based 'LA716' assembly. The comparison of these two genomes revealed that 'LYC1722' and 'LA716' are relatively diverged accessions harboring different beneficial alleles. In summary, Schmidt et al. (2017) carried out a proof-of-concept experiment to demonstrate that Oxford Nanopore sequencing technology can be successfully used to assemble a gigabase-sized plant genome.

7.4 Solanum Habrochaites

S. habrochaites is a green-fruited wild species whose habitat covers Andean slopes of Peru and Ecuador (Kilambi et al. 2017; Sifres et al. 2011). It is well known for its higher density of trichomes and higher metabolic activity *per* trichome, which are responsible for its wide diversity of natural chemical defenses (Bergau et al. 2015).

A whole-genome reconstruction of the *S. habrochaites* accession 'LYC4' was generated by combining Illumina and 454 reads (total contig size equal to 760 Mb). This efforts was followed by shallow sequencing of additional six accessions, whose sequenced reads were mapped onto the 'LYC4' genome for variant calling (The 100 Tomato Genome Sequencing Consortium 2014). This effort was necessary to support unbiased interpretations of genetic variation among distantly related species (The 100 Tomato Genome Sequencing Consortium 2014). Low-pass whole genome sequencing data were generated and released to the public for the accession 'PI247087' (Lin et al. 2014). Finally, two genome assemblies of *S. habrochaites* accessions 'LA1777' and 'PI127826,' obtained through a combination of 10X Linked-Reads and BioNano Optical Mapping, were also made publicly available (https://zenodo.org/record/3712239#.X4AMci1aab8).

7.5 Solanum Chilense

Solanum chilense is a robust, perennial, outcrossing species characterized by green fruits with a distinctive darker stripe when fully ripe. It grows in a wide area ranging from southern Peru to central Chile and colonizes dry and rocky areas as well as coastal deserts (Chetelat et al. 2009). *S. chilense* has garnered attention due to its resistance to drought, cold and salt, and its adaptation to extreme environments (Stam et al. 2019). Furthermore, as *S. chilense* is an obligate outcrossing species, it has been used as a model to investigate self-incompatibility within the tomato clade (Stam et al. 2019; Igic et al. 2007). Because of all these advantageous features, *S. chilense* has been exploited as donor parent in breeding schemes.

Shallow whole-genome sequencing was performed for the accession 'LA1969' and for two additional accessions (CGN15530, CGN15532) by Lin et al. (2014) and by The 100 Tomato Genome Sequencing Consortium (2014), respectively. However, the lack of a well-annotated reference genome for this species has led Stam et al. (2019) to start a sequencing project with the aim of obtaining the de novo genome assembly for *S. chilense* accession 'LA311,' which originates from a mountain habitat. The Illumina sequencing of four different libraries (i.e., two standard paired-end and two mate-pair libraries) generated ~134 Gb raw data that were assembled into more than 81,300 scaffolds for a total genome size of 914 Mb. Genome annotation resulted into 25,885 high confidence (71% of which supported by RNA-Seq data) and 41,481 low confidence gene loci. Comparative analysis of genes encoding for nucleotide-binding leucine-rich repeat (NLR) proteins pointed out that their number was comparable with that of *S. pennellii* and lower than in cultivated tomatoes. This finding suggests the birth and death of NLR genes in the adaptive evolution of tomato species. Interestingly, several subfamilies of NLR proteins were unique to *S. chilense.*

7.6 Available Sequence Data for Other Wild Species

In this section, we list a series of wild species for which single pass sequence data are available (Table 7.1). To this list, we add the available genome assembly of the *S. arcanum* accession 'LA2157' (The 100 Tomato Genome Sequencing Consortium 2014), which was generated for the same reasons described in paragraph 4.

S. peruvianum is the most widespread species within wild tomatoes, and individuals of the species show a wide phenotypic variability. It is

Table 7.1 List of single pass sequenced accessions from under-exploited wild species

Species	# sequenced accessions* by Lin et al. (2014)	# sequenced accessions§ by The 100 Tomato Genome Sequencing Consortium (2014)
S. arcanum	–	2
S. cheesmaniae	3	2
S. chmielewskii		2
S. corneliomulleri	–	1
S. galapagense	1	1
S. huaylasense		3
S. neorickii	1	–
S. peruvianum	3	2

*Approximately 5 Gb of sequence data for each accession available at SRA under the accession SRP045767
§mean coverage of 36.7 ± 2.3 fold *per* accession.

an obligate outcrossing species, and its habitat is primarily in Peru, where it is widespread along the West coast and in the Andes (Chetelat et al. 2009). *S. arcanum*, *S. huaylasense,* and *S. corneliomulleri* were once classified as subspecies of *S. peruvianum*, but, at present, they are considered distinct species. They are found in mid-elevation Andean valleys and have purplish-green fruits when mature (Peralta et al. 2005). *S. cheesmaniae* and *S. galapagense* are endemic to the Galapagos Islands with fruits from yellow to orange when ripe. Populations of these species are salt tolerant and grow under high intensity light (Darwin et al. 2003). Some populations of *S. cheesmaniae* have leaves with a distinctive citrus-like scent, likely associated with unique natural defensive chemicals (Darwin et al. 2003). *S. neorickii* and its closest relative, *S. chmielewskii*, are found in the mid-elevation valleys of the Andes mountains. Both are self-compatible species that produce green fruits with tiny seeds (Peralta et al. 2005).

Finally, we cannot fail to mention the genomic resource available for *S. lycopersicoides,* which belongs to a different *Solanum* section (i.e., *Lycopersicoides*) compared with the species mentioned above (they are in the section *Lycopersicon*). A chromosome-scale genome assembly for *S. lycopersicoides* accession 'LA2951' has been generated by combining PacBio and Illumina sequencing, and Hi-C scaffolding (Powell et al. 2020). The accession 'LA2951' has been selected for sequencing because (i) it is cold tolerant; (ii) it shows resistance to *Botrytis cinerea* and Pseudomonas; (iii) it is characterized by having fruits with an increased content of anthocyanins; (iv) it is available an IL population generated just using that accession (Canady et al. 2006; Powell et al. 2020). The final assembly is 1.2 Gb in length, over 90% of which is included in 12 chromosomes. This resource was used to define the introgression boundaries of the established ILs, to study gene family expansion/contraction and investigate how this mechanism affects adaptation.

All the species mentioned in this paragraph should be considered as under-exploited genetic resources characterized by large genetic variability and could be used in future breeding programs as a source of potentially favorable/beneficial alleles in biotic and abiotic stress-responsive genes.

7.7 The Tomato Pan-Genome

In order to generate a tomato pan-genome (see Chap. 13), Gao et al. (2019) re-sequenced 725 accessions belonging to the *Lycopersicon* clade, which were collected from different geographical locations and represent phylogenetically distant lineages. More precisely, the collection included 372 big-fruited *S. lycopersicum* var. *lycopersicum*, 267 cherry-sized *S. lycopersicum* var. *cerasiforme*, 78 *S. pimpinellifolium*, 5 *S. galapagense*, and 3 *S. cheesmaniae* accessions. Over 77% of the sequence data were available from previous sequencing efforts. The genome of each accession was assembled de novo, and the resulting contigs were compared with the 'Heinz 1706' reference genome.

The tomato pan-genome has a total size of 1,179 Mb and includes 40,369 protein-coding genes. Genomic sequences marked as 'non-reference' have a total size of 351 Mb, 78% of which is made up of repetitive elements. The 'non-reference' genome includes also 4,873 protein-coding genes, 60% of which were associated with gene ontology terms and Pfam protein domain signatures. There is evidence of the presence of 332 'non-reference' genes in the 'Heinz 1706' genome, even though they were not properly assembled and are missing in the international Tomato Annotation Group (iTAG) official annotation.

The pan-genome was used to evaluate presence/absence variations (PAVs) in protein-coding genes and in promoter regions. PAV calling revealed (i) a 'core' genome with a very high gene content equal to 74.2%; (ii) a 'variable' genome, which comprises genes with lower expression levels most of which involved in defense response, photosynthesis, and biosynthetic processes; (iii) that genomes of wild accessions harbor more genes than early domesticated *S. lycopersicum* var. *cerasiforme* accessions and that the latter, in turn, have a greater number of genes than big-fruited *S. lycopersicum* var. *lycopersicum*. This shows a step-by-step gene loss during tomato breeding history. However, PAVs also revealed partial gene recovery in modern varieties most likely due to introgressions from wild relatives. The main finding that comes from the analysis of PAVs in promoter regions concerns the identification of two alleles in the regulatory region of *TomLoxC* (*Solyc01g006540*). This gene encodes for a 13-lipoxygenas, an enzyme involved in the biosynthesis of apocarotenoids. In addition to the reference allele (4724 bp) present in the 'Heinz 1706' reference genome, a 'non-reference' allele (4151 bp) was captured. PAVs indicate that the latter, which contributes to desirable fruit flavor, was subjected to strong negative selection during domestication and improvement. Indeed, its frequency was highest in *S. pimpinellifolium* and dropped dramatically in early domesticated and heirloom tomatoes, but unexpectedly recovered in modern varieties.

In summary, the construction of the pan-genome allowed capturing the extent and distribution of tomato genetic diversity. This reflects a wide phenotypic variability that can certainly be exploited in future breeding programs.

7.8 Impact on Breeding and Future Directions

Wild species are an important source of favorable genes/alleles especially related to, but not limited to, tolerance to abiotic and biotic stresses. They have the potential to enlarge the genetic basis of cultivated tomato by increasing its genetic diversity. The available genomic resources reviewed in this chapter provide novel opportunities to unlock the breeding potential of wild germplasm and possibly to exploit the overwhelming inter-species genetic variability to address novel challenges imposed by climate change and consumers' need.

The ever-deeper knowledge on the genomes of wild tomato species will expedite the introgression of desirable traits from wild relatives into the elite cultivars, greatly enhancing 'introgressomics,' a strategy underlying the development of pre-breeding materials (Prohens et al. 2017). 'Introgressomics' is about the massive development of introgression materials to provide breeders with enlarged genetic pools, which could be directly incorporated into breeding schemes. This process includes the selection of the most suitable wild species, their hybridization and backcrossing with tomato, and the subsequent development of introgression populations. After being subjected to genomic and phenotypic characterization, these populations are collected

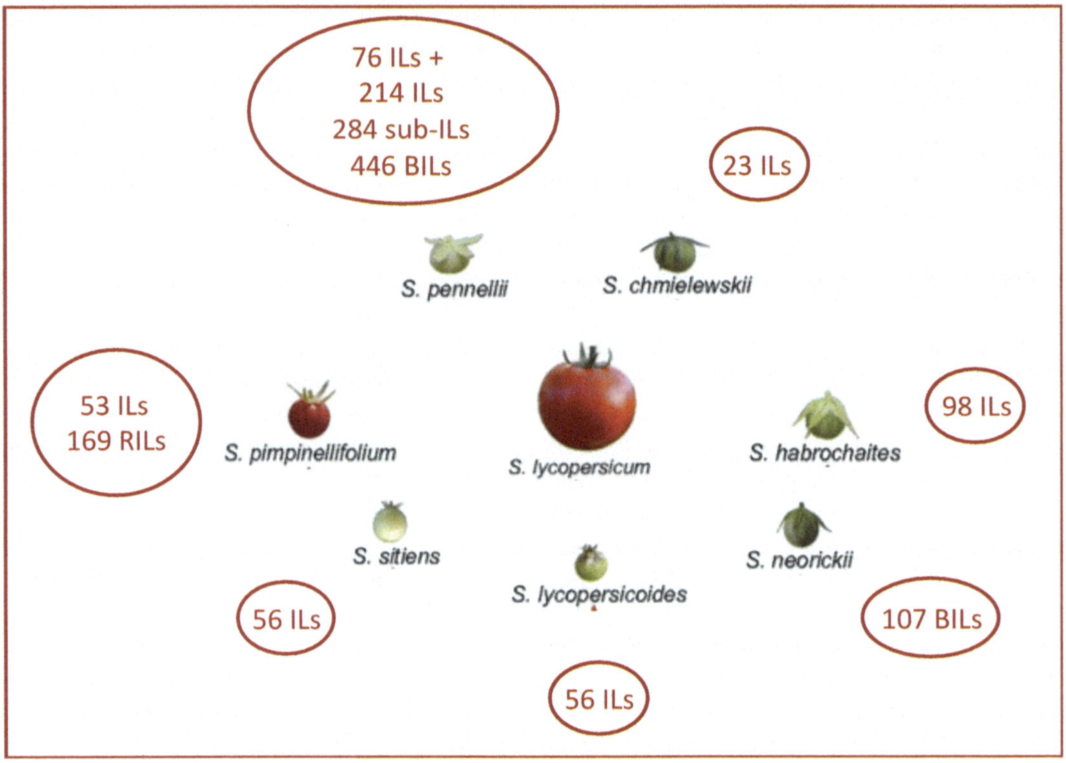

Fig. 7.1 Number of introgression populations currently available for seven wild species related to *S. lycopersicum*. BILs: backcross introgression lines, ILs: introgression lines, RILs: recombinant inbred lines

in suitable repositories, ready to be transferred to breeders. The most common 'introgressomics' populations that have been developed so far for tomato are ILs and RILs. The ever-evolving genomic tools available for this species, such as molecular markers and dense genetic maps, sequenced genomes and transcriptomes, and gene annotations, facilitated the exploitation of these genetic resources.

The development of the first IL population, derived from the wild species *S. pennellii* (Eshed and Zamir 1995), paved the way to the dissection of QTLs controlling the expression of complex traits. Starting from a first group of 50 ILs, which was then enlarged to 76 and 284 ILs and sub-lines, many QTLs have been detected and mapped onto the tomato genome (Lippman et al. 2007) and favorable wild alleles for many traits have been identified. The remarkable results achieved using *S. pennellii* ILs prompted researchers to develop additional tomato segregating populations (namely, ILs, RILs, BILs: backcross inbred lines), which carry introgressions from other wild species using various cultivated tomatoes as recipients. Actually, introgressed populations deriving from seven wild species (Fig. 7.1) are being developed and used with different purposes, such as the identification of QTLs for primary and secondary metabolites in the fruit, flavor volatiles, biomass production, leaf thickness in desert-adapted tomato, leaf and fruit morphology, and yield-associated traits (Alseekh et al. 2015; Ballester et al. 2016; Coneva et al. 2017; Rambla et al. 2017; Caruso et al. 2016; Gur and Zamir 2015; Do et al. 2010).

The QTL (namely QTL Brix9-2–5) responsible for the increased content of total soluble solids (i.e., Brix value) in red ripe tomato fruits was the first to have been identified in a *S. pennellii* introgression sub-line and to be subjected to positional cloning. Within the QTL is the locus encoding for the apoplastic invertase LIN5, which presents a single nucleotide polymorphism that determines the observed phenotype (Fridman et al. 2000, 2002, 2004). In our laboratory, *S. pennellii* sub-lines were obtained from lines IL7-3 and IL12-4 that, in different experiments and under diverse environmental conditions, exhibit high content of ascorbic acid (AsA) in the fruit (Sacco et al. 2013). In particular, some of these sub-lines allowed to dissect the QTLs for high-AsA and to identify candidate wild alleles contributing to a greater content of this metabolite in the fruit (Ruggieri et al. 2015). In case of the sub-line IL7-3-R182, a group of approximately 40 genes were in the wild introgressed region (440 kb in size) some of which are responsible for increased AsA content and higher fruit firmness, without negatively influence other traits, including yield (Aliberti et al. 2020). Therefore, this sub-line, together with sub-line 9–2–5, could be quickly made available to breeders, thus further demonstrating how successful 'introgressomics' is.

As we have been able to point out, the 'rewilding' of crop plants (i.e., the reintroduction of 'wild' genes back into the 'domesticated' genomes of crop plants) can be achieved through traditional 'introgression' plant breeding. A recently developed and much faster approach for the reintroduction of some of the lost properties/traits from the wild relatives of crops is to use the new plant breeding techniques (i.e., genome editing) (Palmgren et al. 2015; Andersen et al. 2015). The latter allow for the precise re-insertion of favorable genes/alleles into elite cultivars after having isolated them from related wild relatives. In conclusion, the 'rewilding' process would allow crops to better tolerate abiotic stress (i.e., adverse environmental conditions), pests, and diseases and have higher nutritional value.

References

Aliberti A, Olivieri F, Graci S, Rigano MM, Barone A, Ruggieri V (2020) Genomic dissection of a wild region in a superior solanum pennellii introgression sub-line with high ascorbic acid accumulation in tomato fruit. Genes (basel) 11(8):847. https://doi.org/10.3390/genes11080847

Alseekh S, Ofner I, Pleban T, Tripodi P, Di Dato F, Cammareri M, Mohammad A, Grandillo S, Fernie AR, Zamir D (2013) Resolution by recombination: breaking up Solanum pennellii introgressions. Trends Plant Sci 18(10):536–538

Alseekh S, Tohge T, Wendenberg R, Scossa F, Omranian N, Li J, Kleessen S, Giavalisco P, Pleban T, Mueller-Roeber B, Zamir D, Nikoloski Z, Fernie AR (2015) Identification and mode of inheritance of quantitative trait loci for secondary metabolite abundance in tomato. Plant Cell 27(3):485. https://doi.org/10.1105/tpc.114.132266

Andersen MM, Landes X, Xiang W, Anyshchenko A, Falhof J, Østerberg JT, Olsen LI, Edenbrandt AK, Vedel SE, Thorsen BJ, Sandøe P, Gamborg C, Kappel K, Palmgren MG (2015) Feasibility of new breeding techniques for organic farming. Trends Plant Sci 20(7):426–434. https://doi.org/10.1016/j.tplants.2015.04.011

Ashrafi H, Kinkade M, Foolad MR (2009) A new genetic linkage map of tomato based on a Solanum lycopersicum× S. pimpinellifolium RIL population displaying locations of candidate pathogen response genes. Genome 52(11):935–956

Ashrafi H, Kinkade MP, Merk HL, Foolad MR (2012) Identification of novel quantitative trait loci for increased lycopene content and other fruit quality traits in a tomato recombinant inbred line population. Mol Breed 30(1):549–567

Bai Y, Lindhout P (2007) Domestication and breeding of tomatoes: what have we gained and what can we gain in the future? Ann Bot 100(5):1085–1094

Ballester A-R, Tikunov Y, Molthoff J, Grandillo S, Viquez-Zamora M, de Vos R, de Maagd RA, van Heusden S, Bovy AG (2016) Identification of loci affecting accumulation of secondary metabolites in tomato fruit of a solanum lycopersicum × solanum chmielewskii introgression line population. Front Plant Sci 7(1428). https://doi.org/10.3389/fpls.2016.01428

Bergau N, Bennewitz S, Syrowatka F, Hause G, Tissier A (2015) The development of type VI glandular trichomes in the cultivated tomato Solanum lycopersicum and a related wild species S. habrochaites. BMC Plant Biol 15(1):289. https://doi.org/10.1186/s12870-015-0678-z

Bolger A, Scossa F, Bolger ME, Lanz C, Maumus F, Tohge T, Quesneville H, Alseekh S, Sørensen I, Lichtenstein G, Fich EA, Conte M, Keller H, Schneeberger K, Schwacke R, Ofner I, Vrebalov J, Xu Y, Osorio S, Aflitos SA, Schijlen E, Jiménez-Goméz JM, Ryngajllo M, Kimura S, Kumar R, Koenig D, Headland LR, Maloof JN, Sinha N, van Ham RCHJ, Lankhorst RK, Mao L, Vogel A, Arsova B, Panstruga R, Fei Z, Rose JKC, Zamir D, Carrari F, Giovannoni JJ, Weigel D, Usadel B, Fernie AR (2014) The genome of the stress-tolerant wild tomato species

Solanum pennellii. Nat Genet 46(9):1034–1038. https://doi.org/10.1038/ng.3046

Calafiore R, Aliberti A, Ruggieri V, Olivieri F, Rigano MM, Barone A (2019) Phenotypic and molecular selection of a superior solanum pennellii introgression sub-line suitable for improving quality traits of cultivated tomatoes. Front Plant Sci 10(190). https://doi.org/10.3389/fpls.2019.00190

Canady MA, Ji Y, Chetelat RT (2006) Homeologous recombination in solanum lycopersicoides introgression lines of cultivated tomato. Genetics 174(4):1775–1788. https://doi.org/10.1534/genetics.106.065144

Caruso G, Gomez LD, Ferriello F, Andolfi A, Borgonuovo C, Evidente A, Simister R, McQueen-Mason SJ, Carputo D, Frusciante L, Ercolano MR (2016) Exploring tomato Solanum pennellii introgression lines for residual biomass and enzymatic digestibility traits. BMC Genet 17(1):56. https://doi.org/10.1186/s12863-016-0362-9

Causse M, Desplat N, Pascual L, Le Paslier M-C, Sauvage C, Bauchet G, Bérard A, Bounon R, Tchoumakov M, Brunel D, Bouchet J-P (2013) Whole genome resequencing in tomato reveals variation associated with introgression and breeding events. BMC Genomics 14(1):791. https://doi.org/10.1186/1471-2164-14-791

Chetelat RT, Pertuzé RA, Faúndez L, Graham EB, Jones CM (2009) Distribution, ecology and reproductive biology of wild tomatoes and related nightshades from the Atacama Desert region of northern Chile. Euphytica 167(1):77–93. https://doi.org/10.1007/s10681-008-9863-6

Chetelat RT, Qin X, Tan M, Burkart-Waco D, Moritama Y, Huo X, Wills T, Pertuzé R (2019) Introgression lines of Solanum sitiens, a wild nightshade of the Atacama Desert, in the genome of cultivated tomato. Plant J 100(4):836–850. https://doi.org/10.1111/tpj.14460

Coneva V, Frank MH, Balaguer MAdL, Li M, Sozzani R, Chitwood DH (2017) Genetic Architecture and molecular networks underlying leaf thickness in desert-adapted tomato *Solanum pennellii*. https://doi.org/10.1104/pp.17.00790

D'Agostino N, Tripodi P (2017) NGS-based genotyping, high-throughput phenotyping and genome-wide association studies laid the foundations for next-generation breeding in horticultural crops. Diversity 9(3):38

Dariva FD, Copati MGF, Pessoa HP, Alves FM, Dias FdO, Picoli EAdT, da Cunha FF, Nick C (2020) Evaluation of anatomical and physiological traits of Solanum pennellii Cor. associated with plant yield in tomato plants under water-limited conditions. Sci Reports 10(1):16052. https://doi.org/10.1038/s41598-020-73004-4

Darwin SC, Knapp S, Peralta IE (2003) Taxonomy of tomatoes in the Galápagos Islands: native and introduced species of Solanum section Lycopersicon (Solanaceae). Syst Biodivers 1(1):29–53

Do PT, Prudent M, Sulpice R, Causse M, Fernie AR (2010) The influence of fruit load on the tomato pericarp metabolome in a <em>Solanum chmielewskii</em> introgression line population. Plant Physiol 154(3):1128. https://doi.org/10.1104/pp.110.163030

Ercolano MR, Sacco A, Ferriello F, D'Alessandro R, Tononi P, Traini A, Barone A, Zago E, Chiusano ML, Buson G (2014) Patchwork sequencing of tomato San Marzano and Vesuviano varieties highlights genome-wide variations. BMC Genomics 15(1):138

Eshed Y, Zamir D (1995) An introgression line population of Lycopersicon pennellii in the cultivated tomato enables the identification and fine mapping of yield-associated QTL. Genetics 141(3):1147–1162

Fridman E, Pleban T, Zamir D (2000) A recombination hotspot delimits a wild-species quantitative trait locus for tomato sugar content to 484 bp within an invertase gene. Proc Natl Acad Sci U S A 97(9):4718–4723. https://doi.org/10.1073/pnas.97.9.4718

Fridman E, Liu Y, Carmel-Goren L, Gur A, Shoresh M, Pleban T, Eshed Y, Zamir D (2002) Two tightly linked QTLs modify tomato sugar content via different physiological pathways. Mol Genet Genomics 266(5):821–826. https://doi.org/10.1007/s00438-001-0599-4

Fridman E, Carrari F, Liu Y-S, Fernie AR, Zamir D (2004) Zooming in on a quantitative trait for tomato yield using interspecific introgressions. Science 305(5691):1786. https://doi.org/10.1126/science.1101666

Gao L, Gonda I, Sun H, Ma Q, Bao K, Tieman DM, Burzynski-Chang EA, Fish TL, Stromberg KA, Sacks GL, Thannhauser TW, Foolad MR, Diez MJ, Blanca J, Canizares J, Xu Y, van der Knaap E, Huang S, Klee HJ, Giovannoni JJ, Fei Z (2019) The tomato pan-genome uncovers new genes and a rare allele regulating fruit flavor. Nat Genet 51(6):1044–1051. https://doi.org/10.1038/s41588-019-0410-2

Gonda I, Ashrafi H, Lyon DA, Strickler SR, Hulse-Kemp AM, Ma Q, Sun H, Stoffel K, Powell AF, Futrell S, Thannhauser TW, Fei Z, Van Deynze AE, Mueller LA, Giovannoni JJ, Foolad MR (2019) Sequencing-based bin map construction of a tomato mapping population, facilitating high-resolution quantitative trait loci detection. Plant Genome 12(1):180010. https://doi.org/10.3835/plantgenome2018.02.0010

Gur A, Zamir D (2015) Mendelizing all components of a pyramid of three yield QTL in tomato. Front Plant Sci 6:1096–1096. https://doi.org/10.3389/fpls.2015.01096

Igic B, Smith W, Robertson K, Schaal B, Kohn J (2007) Studies of self-incompatibility in wild tomatoes: I. S-allele diversity in solanum chilense Dun.(Solanaceae). Heredity 99(5):553–561

Kilambi HV, Manda K, Rai A, Charakana C, Bagri J, Sharma R, Sreelakshmi Y (2017) Green-fruited solanum habrochaites lacks fruit-specific carotenogenesis due to metabolic and structural blocks. J Exp Bot 68(17):4803–4819. https://doi.org/10.1093/jxb/erx288

Kim HT, Lee JM (2018) Organellar genome analysis reveals endosymbiotic gene transfers in tomato. PLoS ONE 13(9):e0202279–e0202279. https://doi.org/10.1371/journal.pone.0202279

Lin T, Zhu G, Zhang J, Xu X, Yu Q, Zheng Z, Zhang Z, Lun Y, Li S, Wang X, Huang Z, Li J, Zhang C, Wang T, Zhang Y, Wang A, Zhang Y, Lin K, Li C, Xiong G, Xue Y, Mazzucato A, Causse M, Fei Z, Giovannoni JJ, Chetelat RT, Zamir D, Städler T, Li J, Ye Z, Du Y, Huang S (2014) Genomic analyses provide insights into the history of tomato breeding. Nat Genet 46(11):1220–1226. https://doi.org/10.1038/ng.3117

Lin Y-P, Lu C-Y, Lee C-R (2020) The climatic association of population divergence and future extinction risk of Solanum pimpinellifolium. AoB Plants 12 (2):plaa012

Lippman ZB, Semel Y, Zamir D (2007) An integrated view of quantitative trait variation using tomato interspecific introgression lines. Curr Opin Genet Dev 17(6):545–552. https://doi.org/10.1016/j.gde.2007.07.007

Muir CD, Pease JB, Moyle LC (2014) Quantitative genetic analysis indicates natural selection on leaf phenotypes across wild tomato species (<em>Solanum</em> sect. <em>Lycopersicon</em>; Solanaceae). Genetics 198 (4):1629. https://doi.org/10.1534/genetics.114.169276

Palmgren MG, Edenbrandt AK, Vedel SE, Andersen MM, Landes X, Østerberg JT, Falhof J, Olsen LI, Christensen SB, Sandøe P, Gamborg C, Kappel K, Thorsen BJ, Pagh P (2015) Are we ready for back-to-nature crop breeding? Trends Plant Sci 20(3):155–164. https://doi.org/10.1016/j.tplants.2014.11.003

Peralta IE, Knapp S, Spooner DM (2005) New species of wild tomatoes (solanum section Lycopersicon: Solanaceae) from Northern Peru. Syst Bot 30(2):424–434

Powell AF, Courtney LE, Schmidt MHW, Feder A, Vogel A, Xu Y, Lyon DA, Dumschott K, McHale M, Sulpice R, Bao K, Duhan A, Hallab A, Denton AK, Mueller LA, Alseekh S, Lie J, Martin C, Fernie AR, Hind SR, Martin GB, Fei Z, Giovannoni JJ, Strickler SR, Usadel B (2020) A <em>Solanum lycopersicoides</em> reference genome facilitates biological discovery in tomato. bioRxiv:2020.2004.2016.039636. https://doi.org/10.1101/2020.04.16.039636

Prohens J, Gramazio P, Plazas M, Dempewolf H, Kilian B, Díez MJ, Fita A, Herraiz FJ, Rodríguez-Burruezo A, Soler S, Knapp S, Vilanova S (2017) Introgressiomics: a new approach for using crop wild relatives in breeding for adaptation to climate change. Euphytica 213(7):158. https://doi.org/10.1007/s10681-017-1938-9

Rambla JL, Medina A, Fernández-del-Carmen A, Barrantes W, Grandillo S, Cammareri M, López-Casado G, Rodrigo G, Alonso A, García-Martínez S, Primo J, Ruiz JJ, Fernández-Muñoz R, Monforte AJ, Granell A (2017) Identification, introgression, and validation of fruit volatile QTLs from a red-fruited wild tomato species. J Exp Bot 68(3):429–442. https://doi.org/10.1093/jxb/erw455

Razali R, Bougouffa S, Morton MJL, Lightfoot DJ, Alam I, Essack M, Arold ST, Kamau AA, Schmöckel SM, Pailles Y, Shahid M, Michell CT, Al-Babili S, Ho YS, Tester M, Bajic VB, Negrão S (2018) The genome sequence of the wild tomato solanum pimpinellifolium provides insights into salinity tolerance. Front Plant Sci 9(1402). https://doi.org/10.3389/fpls.2018.01402

Ruggieri V, Sacco A, Calafiore R, Frusciante L, Barone A (2015) Dissecting a QTL into candidate genes highlighted the key role of pectinesterases in regulating the ascorbic acid content in tomato fruit. Plant Genome 8 (2):plantgenome2014.2008.0038. https://doi.org/10.3835/plantgenome2014.08.0038

Sacco A, Di Matteo A, Lombardi N, Trotta N, Punzo B, Mari A, Barone A (2013) Quantitative trait loci pyramiding for fruit quality traits in tomato. Mol Breed 31(1):217–222. https://doi.org/10.1007/s11032-012-9763-2

Sahu KK, Chattopadhyay D (2017) Genome-wide sequence variations between wild and cultivated tomato species revisited by whole genome sequence mapping. BMC Genomics 18(1):430. https://doi.org/10.1186/s12864-017-3822-3

Schauer N, Semel Y, Roessner U, Gur A, Balbo I, Carrari F, Pleban T, Perez-Melis A, Bruedigam C, Kopka J, Willmitzer L, Zamir D, Fernie AR (2006) Comprehensive metabolic profiling and phenotyping of interspecific introgression lines for tomato improvement. Nat Biotechnol 24(4):447–454. https://doi.org/10.1038/nbt1192

Schmidt MHW, Vogel A, Denton AK, Istace B, Wormit A, van de Geest H, Bolger ME, Alseekh S, Maß J, Pfaff C, Schurr U, Chetelat R, Maumus F, Aury J-M, Koren S, Fernie AR, Zamir D, Bolger AM, Usadel B (2017) De novo assembly of a new *Solanum pennellii* accession using nanopore sequencing. Plant Cell 29 (10):2336. https://doi.org/10.1105/tpc.17.00521

Sifres A, Blanca J, Nuez F (2011) Pattern of genetic variability of Solanum habrochaites in its natural area of distribution. Genet Resour Crop Evol 58(3):347–360. https://doi.org/10.1007/s10722-010-9578-0

Spooner DM, Peralta IE, Knapp S (2005) Comparison of AFLPs with other markers for phylogenetic inference in wild tomatoes [Solanum L. section Lycopersicon (Mill.) Wettst.]. Taxon 54(1):43–61

Stam R, Nosenko T, Hörger AC, Stephan W, Seidel M, Kuhn JMM, Haberer G, Tellier A (2019) The <em>de Novo</em> reference genome and transcriptome assemblies of the wild tomato species <em>Solanum chilense</em> highlights birth and death of NLR genes between tomato species. G3: Genes|Genomes|Genetics 9(12):3933. https://doi.org/10.1534/g3.119.400529

The 100 Tomato Genome Sequencing Consortium (2014) Exploring genetic variation in the tomato (Solanum section Lycopersicon) clade by whole-genome sequencing. Plant J 80(1):136–148

Tieman D, Zhu G, Resende MFR, Lin T, Nguyen C, Bies D, Rambla JL, Beltran KSO, Taylor M, Zhang B, Ikeda H, Liu Z, Fisher J, Zemach I, Monforte A, Zamir D, Granell A, Kirst M, Huang S, Klee H (2017)

A chemical genetic roadmap to improved tomato flavor. Science 355(6323):391. https://doi.org/10.1126/science.aal1556
Tomato Genome Consortium (2012) The tomato genome sequence provides insights into fleshy fruit evolution. Nature 485(7400):635
Tranchida-Lombardo V, Aiese Cigliano R, Anzar I, Landi S, Palombieri S, Colantuono C, Bostan H, Termolino P, Aversano R, Batelli G (2018) Whole-genome re-sequencing of two Italian tomato landraces reveals sequence variations in genes associated with stress tolerance, fruit quality and long shelf-life traits. DNA Res 25(2):149–160
Wang X, Gao L, Jiao C, Stravoravdis S, Hosmani PS, Saha S, Zhang J, Mainiero S, Strickler SR, Catala C, Martin GB, Mueller LA, Vrebalov J, Giovannoni JJ, Wu S, Fei Z (2020) Genome of *Solanum pimpinellifolium* provides insights into structural variants during tomato breeding. bioRxiv:2020.2006.2017.157859. https://doi.org/10.1101/2020.06.17.157859
Yang S, Yu Q, Wang B, Yang T, Li N, Tang Y, Aisimutuola P, Wang Q, Xu J, Gao J (2016) Identification of QTLs for red fruit firmness using the wild tomato species Solanum pennellii LA716 introgression lines. Plant Breed 135(6):728–734. https://doi.org/10.1111/pbr.12423

The *Solanum Commersonii* Genome Sequence

8

Riccardo Aversano, Vincenzo D'Amelia, Salvatore Esposito, Raffaele Garramone, Clizia Villano, and Domenico Carputo

Abstract

Solanum commersonii ($2n = 2x = 24$, 1EBN) is an important wild potato relative that has garnered attention for its high tolerance to cold and bacterial wilt. Although several efforts have been devoted to traits introgression from this wild species into the cultivated genepool, its genetic potential remains largely untapped. Now, the development of genomic resources for *S. commersonii* is gaining momentum, and we expect that they will soon impact the harnessing of this promising wild species in the release of new potato verities displaying introgression of traits from it. After illustrating the whole-genome structure and organization of *S. commersonii*, this chapter describes the main genomic resources developed so far and how they have been used to distill the diversity of several gene families playing key biological roles, such as RNA silencing mechanisms, secondary metabolites biosynthesis, and transcriptional factors. Finally, we provide a general overview of the breeding strategies used to exploit S. *commersonii* genetic potential and provide perspectives to develop superior stress-tolerant cultivars.

R. Aversano · V. D'Amelia · S. Esposito ·
R. Garramone · C. Villano · D. Carputo (✉)
Department of Agricultural Sciences, University of Naples Federico II, Via Università 100, 80055 Portici, Italy
e-mail: carputo@unina.it

8.1 Introduction

Solanum commersonii is a wild potato species endemic to the coastal belt of Argentina and Uruguay and coastal regions of South Brazil. It grows in a wide variety of habitats but very frequently in marshy places, fields, riverbanks, woods, and sandy shores, from sea level to about 400 m. Taxonomically, it is classified within the primitive superseries Stellata, which includes mostly 1EBN potatoes lacking a fused corolla and displaying a typically star-shaped flower (Hawkes 1990). It belongs to the tertiary genepool because it can be sexually hybridized to potato through unreduced gametes and bridge crosses (Jansky 2006). *S. commersonii* is phylogenetically separated from *S. tuberosum* as revealed by analyses of plastome restriction sites and nitrate reductase gene sequence (Rodriguez and Spooner 2009). *S. commersonii* is an important donor of germplasm. Indeed, several resistance traits lacking in the cultivated potato have been reported for this species, including resistance to late blight (the most critical potato disease) (Micheletto et al. 2000), bacterial wilt (the second most crucial potato disease) (Gonzalez et al. 2013), root-knot nematodes, soft rot and blackleg, verticillium wilt, Potato Virus X, Tobacco Etch Virus, and common scab

© Springer Nature Switzerland AG 2021

D. Carputo et al. (eds.), *The Wild Solanums Genomes*, Compendium of Plant Genomes,
https://doi.org/10.1007/978-3-030-30343-3_8

(Hanneman and Bamberg 1986; Hawkes 1990; Lafierrere et al. 1999; Micheletto et al. 2000; Carputo et al. 2000).

Additionally, *S. commersonii* has long received attention because of its frost tolerance and capacity to cold acclimate (increase in freezing tolerance upon exposure to chilling temperatures) (Vega et al. 2000). By contrast, the cultivated potato is sensitive to low temperatures and lacks the capability to cold acclimate (Palta and Simon 1993). Both traits are fundamental since freezing temperatures cause significant yield losses in potato production worldwide. Unfortunately, only limited success has been achieved by using traditional plant breeding methods to improve freezing stress resistance in potato. Probably because of the complex nature of the frost and winter survival trait, in which many factors may interact (Chinnusamy et al. 2007). Considering the value of *S. commersonii* as a source of useful genes, in the last decade, several genomic resources have been developed to pave a path toward further potato improvement and decipher the mechanisms underlying agronomic traits that can be improved through the exploitation of this valuable species. In the following paragraphs, we overview the leading genomic resources developed for this wild species, such as genome and transcriptomic sequences, and how they have been exploited to distill the diversity of several important gene families. Finally, we provide a general overview of the breeding strategies used so far to introgress *S. commersonii* traits into the cultivated potato and provide perspectives for its efficient exploitation.

8.2 Genomic Resources

Since 2015, the genome sequences of *S. commersonii* have been released (Aversano et al. 2015; Cho et al. 2016, 2018), allowing the understanding of the genome structure and organization of this species and its intriguing evolutionary roots. Among the main goals achieved by these efforts, several protein sequences predicted to be responsive to biotic and abiotic stresses and genes lacking in *S. tuberosum* have been identified. Differentiation of major repetitive sequence classes between *S. commersonii* and *S. tuberosum* in terms of dynamics and evolutionary processes have been elucidated. Moreover, the transcript's analysis through the massive sequencing of *S. commersonii* mRNAs revealed molecular changes and genome regulation conferring this species adaptive characteristics such as tolerance to cold and bacterial wilt. Below we provide details on the main genomic resources developed over the last years for *S. commersonii* and how they have been exploited to gain new insights.

8.2.1 Whole-Genome Sequence

The genome of *S. commersonii* (clone cmm1t of PI243503) was sequenced using the Illumina HiSeq1000 platform (Aversano et al. 2015). A 105-fold base pair coverage of the estimated genome size was achieved starting from five libraries (400, 550, and 700 bp paired-end libraries and 3, 5, and 10 kb mate-pair libraries). The whole-genome shotgun sequencing approach generated 146 Gb of the raw sequence, and the final assembled sequence was 830 Mb, with a scaffold N50 size of 44 Mb. Scaffold order along each chromosome was accomplished by an interactive mapping approach using as reference the genome sequence of the doubled monoploid clone DM1-3 516 R44 of *S. tubersoum* Group Phureja. This approach enabled to obtain 12 pseudomolecules representing the 12 chromosomes of *S. commersonii*. Gene model annotation profited considerably from a high-throughput RNA sequencing (RNA-seq) analysis performed on four tissues (leaf, flower, stolon, and tuber). Thus, the identification of genuine transcription starts and termination sites and the annotation of more than 37,600 protein-coding genes was accomplished. A functional categorization through Gene Onthology (GO) enrichment analysis revealed that "mRNA metabolism", "regulation of signal transduction", "organelle organization" and "programmed cell death" were among the most enriched terms related to biological process, whereas "methyltransferase activity", "motor activity" "cobalt ion binding" and "passive transmembrane transporter activity" were enriched within molecular function terms. Finally, as far as cellular component is concerned, "endoplasmatic reticulum" and "plasma membrane part" were the most abundant. The coverage of the gene space was further assessed using ESTs and using the core eukaryotic gene (CEG) mapping strategy. The assembly described a vast majority of the gene space since 98% of the *S. commersonii* CEG homologs mapped to its respective assemblies. Sequence variation analyses have been carried out through the detection of single-nucleotide polymorphisms (SNPs). Overall, 9,894,571 SNPs among 662 Mb reliable genome bases were identified, yielding a SNP frequency of 1.49% across the whole genome. The SNPs frequency found in *S. commersonii* was slightly higher than that reported in the wild potato *S. chacoense* (0.68%), where 1,414,890 biallelic SNPs were identified from a total of 208 Mb of assayable nucleotides. This is probably due to the inbred nature of *S. chacoense* M6 clone enabled by the presence of the *S-locus inhibitor* (*Sli*) dominant allele (Hosaka and Hanneman 1998), which allowed selfing and the development of the inbred (Jansky et al. 2014). Recently, Hardigan et al. (2017) detected mean heterozygous nucleotide frequencies of 1.05% in diploid potato landraces and 2.73% in tetraploid cultivars, using a panel of 63 accessions, including wild species. When SNPs were analyzed in respect to their location, *S. commersonii* displayed 12,412 genes encompassing SNPs, which overrepresented some major functional categories, such as "macromolecule metabolic processes", "response to stimuli", "carbohydrate derivative binding", "localization", and "ion binding". Furthermore, microsynteny analyses between *S. commersonii* and *S. tuberosum* pointed out an enrichment of SNPs and, particularly, of insertion-deletions (indels) within the intergenic regions. Such latter differences are consistent with dissimilarities in their genome sizes.

8.2.2 Organelle Genomes

Plastids and mitochondria possess their genomes, the plastome, and chondrome, respectively, and specific machineries to decode important cellular functions, e.g., photosynthesis and male sterility (Ruiz and Daniell 2005; Jheng et al. 2012). The critical importance of these organelles and the genes they retain has been confirmed repeatedly by observations that their malfunction and minute changes at the DNA level can have severely debilitating consequences and may even culminate in lethality. Plant cytoplasmic genome sequences are being collected at an unprecedented pace, and in the last few years, both *S. commersonii* plastome and chondrome have been sequenced. In 2016, Cho and collaborators presented the first report on the complete plastidial genome sequence of *S. commersonii*, clone Lz3.2. Its chloroplast genome was completed using an Illumina HiSeq 2000 analyzer. It was 155,525 bp in length, including a large single copy (LSC) region of 85,973, two copies of inverted repeats (IR) of 25,593 bp, and a small single copy (SSC) region of 18,366 bp. The genome is a circular DNA molecule containing 87 protein-coding genes that include synthesis-related genes and transcription and translation-related genes. Gene order, orientation and content of the *S. commersonii* chondriome were conserved with those of other *Solanaceae* species, and the phylogenetic tree provided evidence that *S. commersonii* colocalize with *S. tuberosum*. However, nine indels between *S. commersonii* and *S. tuberosum* were found in their chloroplast genomes, letting two Indel markers be developed. These markers empowered the two species to be distinguished and were successfully applied to chloroplast genotyping (chlorotype) in somatic hybrids and their progenies (Cho et al. 2016).

Regarding *S. commersonii* chondrome, the completion of its genome sequence has been very recently reported by a South Korean group (Cho et al. 2018). The team sequenced two mitochondrial genomes of the *S. commersonii* clone

Lz3.3. The circular DNA molecules displayed an average length of 276,051 bp and owned 48.3% of sequence similarity. Comparative analysis highlighted two collinear regions with lengths of 126 and 12 kb between each other. The predicted genes were 65 and 51 in the two genomes, constituting approximately 12.8% and 13.7% of the two genome sequences, respectively. In particular, 80 unique genes were identified, such as 34 protein-coding sequences, 25 ORFs, 18 tRNA, and 3 rRNA genes. The exons of the trans-spliced genes nad1, nad2, and nad5 exist entirely or partially in each genome.

8.2.3 Repetitive Sequences

Introgression of useful traits from *S. commersonii* into potato cultivars requires effective recombination between hom(e)olog chromosomes, which is partially determined by the degree of genome differentiation between the crop and its wild relatives. This means that the higher the differentiation, the harder it is to introgress genes of interest. The divergence between the two genomes can be explained in terms of both large-scale structural differences and nucleotide-level variations, particularly of repetitive DNA sequences (Belyayev 2014; Bennetzen and Wang 2014). Therefore, the repeat profiles can give information on the phylogenetic relationships within and between species. Despite the *S. commersonii* and *S. tuberosum* genome sequences are available, only a few comparative studies on the characterization of LTR-RTs between the potato and its wild relatives have been reported. Gaiero et al. (2018) carried out a comparative analysis of repetitive sequences in several *Solanaceae,* including *S. commersonii* and *S. tubersoum*. The authors found that wild potatoes displayed 3–6 times lower proportions of tandem repeats (including satellites, rDNA, and telomeric repeats) than the cultivated potato. It appeared that this discrepancy was mostly caused by one satellite repeat that showed high homology with the satellite CL14 (Torres et al. 2011). Some lineage-specific tandem repeats have been described, such as a 90 bp satellite that is more prevalent in wild potatoes than in *S. tuberosum*. Analyzing the *S. commersonii* genome, a high proportion of annotated repeats were identified as LRT-*Gypsy* elements, with Chromovirus being the most abundant lineage (Aversano et al. 2015; Gaiero et al. 2018). As described in Chaps. 7 and 9, this higher abundance has already been reported using very different approaches for potato and tomato (Datema et al. 2008; Peters et al. 2009; Xu et al. 2011; The Tomato Genome Consortium 2012), *S. pennellii* (Bolger et al. 2014), and *S. chacoense* (Leisner et al. 2018). However, the idea of elevated *Gypsy* abundance in plant genomes is not universal. Indeed, Argout and colleagues (2011) and Jaillon et al. (2007) revealed a higher abundance of *Copia* rather than *Gypsy* in cacao and grapevine, respectively. Using the LTRs classification by Gaiero et al. (2018), our laboratory analyzed their insertion time and activity to understand the evolutionary dynamics of LTR retrotransposon families in *S. commersonii*. The data revealed that in *S. commersonii, Copia* elements were younger than *Gypsy* ones, suggesting a recent proliferation of the former elements in potato (Esposito et al. 2019). Confirming this hypothesis, *Copia* elements with insertion time equal to 0 were also predicted since they showed 100% of identity between the two LTRs. These represent the youngest LTR-RTs originating from very recent activation and mobilization.

8.2.4 Transcriptome and Non-coding RNAs

Plant researchers working with *S. commersonii* have utilized a variety of approaches to understand gene expression to evaluate how genes work together and how their altered expression contributes to complex plant phenotypes. In some cases, RNA-seq has been preferred to the conventional array-based approaches to improve qualitatively and quantitatively analyses of transcriptomes, quantify alternative spliced isoforms, and identify sequence variants and novel transcripts. For example, as aforementioned, a global

RNA profiling on different *S. commersonii* tissues was carried out by Aversano et al. (2015). The comparative analysis between the wild species and *S. tuberosum* indicated that the number of transcripts was higher in the latter species, even though the number of their genes was similar. This might indicate the presence of more prominent alternative splicing activities in potato than in *S. commersonii* and confirms previous observations in potato, where ~25% of genes encoded two or more isoforms (Potato Genome Sequencing Consortium 2011). To identify *S. commersonii* genetic determinants related to its tolerance to cold and bacterial wilt, whole-transcriptome analyses were performed. In the next sections, we summarize the main findings obtained so far and how transcriptome and non-coding RNA resources have facilitated the understanding of both traits.

8.2.4.1 Cold

The analysis of gene expression in *S. commersonii* showed that an extensive reorganization of the transcriptome appears in plants challenged for 30 min at −2 °C (non-acclimated condition, NAC) and those acclimated at 4 °C for 2 weeks and then transferred for 30 min at −2 °C (acclimated condition, AC) (Aversano et al. 2015). In particular, we observed augmented activity of genes affecting ROS scavenging enzymes (e.g., *superoxide dismutase*, *SOD*; *catalase*, *CAT*; *ascorbate peroxidase*, *APX*), those encoding proteins that may function as osmoprotectants, and those involved in cell repair (such as *heat shock proteins,* HSPs, and *dehydrins*, *DHNs*). One notable observation was that in *S. commersonii,* several genes were responsive to cold but with contrasting kinetics under AC vs. NAC conditions. For instance, *brassinosteroid-signaling kinase 1* (*BSK1*) was activated under AC and suppressed under NAC. Conversely, one MYB and one bHLH transcription factor (TF) were repressed under AC and activated under NAC. Since MYB and bHLH proteins often interact with each other to control transcription (Stracke et al. 2001; Heim et al. 2003; Ramsay and Glover 2005), this differential expression of MYB and bHLH TF suggests that the regulation of some cold-responsive genes may be achieved by modulating the ratio of these partners. TFs were mostly up-regulated under both conditions, as observed in *Arabidopsis* (Lee et al. 2005); only a few were repressed. Among the negatively correlating TFs, there was a Cys-2/His-2-type (C2H2) zinc-finger protein, which has been found to work downstream of DREB1/CBF and to be responsible for stress tolerance in plants (Sakamoto et al. 2004). The comparison of cold-responsive gene expression profiles between AC and NAC stressed plants also highlighted remarkable features of *S. commersonii* genes known to be critical in cold sensing and signaling pathways. For instance, two calcium-dependent protein kinases, *CDPK17* and *CDPK 19 (CPK8)*, were differentially expressed. Interestingly, *CDPK19* was up-regulated only under NAC conditions, whereas *CDPK17* expression required acclimation. Neither of the two genes has been previously reported to be involved in cold stress response. Hence, the activation of *CDPK19 (CPK8)* and *CDPK17* in *S. commersonii* hints at potential independent roles in freezing and cold acclimation responses, respectively. Regarding the CBFs (C-repeat binding factors), data available showed that all *S. commersonii* CBFs were up-regulated both under NAC and AC conditions relative to controls. Such a high expression of CBFs and genes regulated by CBF proteins (e.g., COR genes) may be directly responsible for enhanced cold tolerance and acclimation ability in this species.

Emerging evidence has revealed that non-coding (ncRNAs) are major products of the plant transcriptome (Rymarquis et al. 2008) and that they may have significant regulatory importance, especially during stress situations (Matsui et al. 2013). Recently, *S. commersonii* clone cmm1T has been the object of a small ncRNAome analysis to decipher the root cause behind its cold tolerance (Esposito et al. 2018a). Twenty-one thousand ncRNAs were identified, and 273 different miRNAs (of which 24-nt RNAs were the most abundant) were annotated. Overall, out of 273 *S. commersonii* miRNAs, 229 were considered novel and *S. commersonii*-specific, whereas

44 were conserved. Among these latter, miR166 and miR169 were the most abundant. Indeed, we identified six members belonging to the former family and five to the latter in concordance with results in *S. tuberosum* (Lakhotia et al. 2014). Gene expression analysis provided evidence that the transcription of several miRNAs was differentially regulated in response to cold stress conditions (Fig. 8.1a). Most of miRNAs targeted more than one transcript, and analysis of gene ontology (GO) annotation enabled the identification of enriched terms both in the molecular functions category (e.g., nucleotide binding, transferase activity) and biological processes (defense response, regulation of transcription, regulation of cellular metabolic process) (Fig. 8.1b). Conserved miR408a and miR408b changed their expression under NAC stress. MiR408 has been reported to target transcripts belonging to the laccase and plantacyanin families, which catalyze the oxidation of phenols and arylamines and regulate cell wall function (Jeon et al. 2011). Interestingly, the down-regulation of miR408 affects the cell wall composition, and, in agreement with this data, its overexpression results in enhanced biomass and seed yield (Song et al. 2017). Given that, it can be speculated that in *S. commersonii,* miR408 could negatively influence cell wall stratification and improve membrane permeability, an important step to avoid breakages caused by ice formation. A simplified model in Fig. 8.1c describes the molecular mechanism of this microRNA in cold-stressed *S. commersonii.* Another interesting miRNA is miR4376. It was down-regulated following stress in AC conditions. In *S. lycopersicum,* Wang et al. (2011) demonstrated that miR4376 regulates the expression of an auto-inhibited Ca^{2+}-ATPase (*ACA10*). This has likely broad implications in light of the role of Ca^{2+} signaling under stress conditions (McAinsh and Pittman 2009; Dodd et al. 2010; Kudla et al.

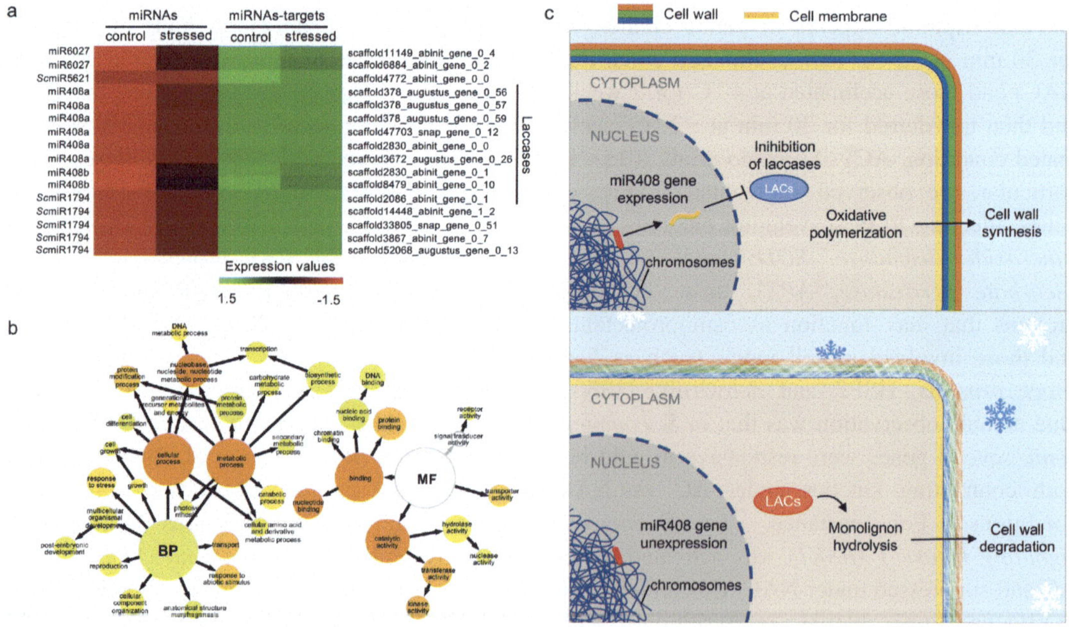

Fig. 8.1 **a** Heat map of microRNA negatively correlated with their predicted target genes (Pearson correlation > −0.75). **b** Semantic diagram summarizing enriched GO terms. **c** Simplify model of genes targeted by miR408 **c**. Under normal conditions, miR408 is expressed in *S. commersonii* leaves, inhibiting its targets (e.g., laccases and plantacyanins). By contrast, under cold stress conditions, miR408 is repressed, enabling the degradation of cell wall through the monolignon hydrolysis. Thus, miR408 might be involved in plant cold tolerance in *S. commersonii* via reduction and degradation of lignin content in the cell wall, thereby increasing cell wall permeability

2010). Indeed, growing evidence indicates that stress alert begins through Ca^{2+} signals that can be propagated as waves of Ca^{2+} (Evans et al. 2001; Sanders et al. 2002). In this scenario, Ca^{2+}-ATPase pumps and Ca^{2+}exchangers, which remove cytosolic Ca^{2+} from storage sites within and outside of the cell and release it for signal transduction mechanisms, deserve future studies (Table 8.1).

8.2.4.2 Bacterial Wilt

In 2015, deep sequencing of *S. commersonii* RNA (RNA-seq) was carried out to determine its below-ground transcriptional responses to *R. solanacearum* (Zuluaga et al. 2015). This is one of the most aggressive bacterial pathogens infecting the cultivated potato. Disease control of bacterial wilt is very challenging because of the bacterium aggressiveness, its persistence in the field, and the lack of resistant varieties. Potato breeding programs have used *S. commersonii* as a source of resistance against bacterial wilt (Kim-Lee et al. 2005 2005; Siri et al. 2009). Zuluaga and colleagues (2015), through RNA-seq, profiled two *S. commersonii* clones contrasting in bacterial wilt resistance. RNA was extracted from inoculated and control *S. commersonii* roots or from a mixture of tissues containing flowers, roots, stolons, shoots, and leaves, and subjected to Illumina sequencing. Data analysis highlighted that *R. solanacearum* infection preferentially impacted the *S. commersonii*-specific genes and triggered stress responses. In particular, both clones showed a similar proportion of up- and down-regulated genes after bacterial inoculation (118 up vs. 103 down in the resistant clone and 339 up vs. 305 down in the susceptible one). The authors found that 2% of *S. commersonii* genes were differentially expressed in any of the infected asymptomatic plants compared to non-inoculated controls, probably due to the quantitative nature of resistance of this wild species. Differences in the transcriptome of the susceptible and resistant *S. commersonii* clones upon inoculation with *R. solanacearum* allowed the identification of four up-regulated genes in the susceptible clone and down-regulated in the resistant ones. In particular, the *3-ketoacyl-CoA thiolase 5* (*KAT5*) and lipid-binding protein are *S. commersonii*-specific-genes not present in the *S. tuberosum* Group Phureja DM genome. They belong to the GO category fatty acid metabolism/beta-oxidation, involved in the metabolism of the phytohormone jasmonic acid (JA) and might correspond to bacterial wilt susceptibility genes. Therefore, these genes could represent good candidates to engineer potato varieties resistant to such disease.

8.3 Gene Discovery

The development of *S. commersonii* genomic resources has ushered in a new era for wild potato analyses. Genome sequences and expression data are allowing specific plant comparative

Table 8.1 Statistics on de novo *S. commesonii* genome assembly

Genome assembly statistics	
Paired-end libraries size, bp	400, 550, 700
Mate-pair libraries, kb	3, 5, 10
Draft genome size, Mb	830
Coverage (filtered reads)	105X
N50 index (scaffolds), number	4833
N50 length (scaffolds), bp	44,298
Longest scaffold, bp	458,668
Average scaffold length, bp	13,543
Genome SNP frequency, %	1.49
CDS SNP frequency, %	0.28

studies to understand the pattern of gene family size variation in potato, an important mechanism shaping the natural variation for adaptation in various species. To date, few studies have been conducted. Those available refer mainly to the identification of orthologous gene pairs, which play essential biological roles, such as RNA silencing mechanisms, secondary metabolites biosynthesis, and transcriptional factors.

8.3.1 Riboregulators

Plants require elaborate mechanisms to produce small regulatory RNAs. Several proteins are involved in their biogenesis. Among them, *DICER-like* (*DCL*) and *RNA-dependent RNA polymerase* (*RDR*) are key components. The *S. commersonii* genome has been recently exploited to analyze *DCL* and *RDR* genes to enhance understanding of their molecular diversification (Esposito et al. 2018b). Seven *DCLs* were identified in this species and *S. tuberosum*, suggesting that *DCL* loci were substantially conserved in the two genomes after their evolutionary divergence. A significant expansion was found in the *DCL2* clade, where four paralogous copies of DCL2 were identified. The ratios of non-synonymous versus synonymous substitution rate (Ka/Ks) for the orthologous gene pairs suggested that *DCL2* duplications appeared before the recent divergence between the two species, dated 2,3 Mya (Aversano et al. 2015) since all values found were <1. Six and seven *RDR* genes were identified in *S. commersonii* and *S. tuberosum*, respectively. Compared with tomato species (*S. lycopersicum* and *S. pennellii*), both potatoes harbor two copies of *RDR1*, hinting at a potato-specific duplication that derived after their divergence, as also confirmed by Ks values. The variability in the number of *RDR1* genes in different *Solanum* species is intriguing since potato and tomato experienced different life histories after their divergence (Tomato Genome Consortium 2012). In particular, it might be speculated that the nature of selection pressure imposed by their environmental conditions might cause the variability in the *RDR1* clade, as already proposed for gene families involved in plant stresses (Hanada et al. 2008).

8.3.2 Natural Compounds

The availability of the *S. commersonii* and *S. tuberosum* genome sequences, enabled to verify whether differences in their metabolite levels were caused by variations in the abundance of genes involved in various metabolic processes. In particular, the genes involved in the ascorbic acid pathway in the biosynthesis of aromatic amino acids and the phenylpropanoid and glycoalkaloid pathways were investigated by Aversano et al. (2017) and Villano et al. (2020a). Genome analysis enabled the identification of protein-coding genes implicated in the metabolism of ascorbic acid, tryptophan, and tyrosine both in *S. commersonii* (51, 8, and 7, respectively) and *S. tuberosum* (37, 16, and 12, respectively). Among all the gene families taken into consideration, the phosphomannomutase (ascorbic acid pathway) and N-5 phosphoribosyl-anthranilate isomerase (tryptophan pathway) showed the most considerable differences in terms of the number of predicted genes between *S. commersonii* and *S. tuberosum* (3 vs. 8 and 1 vs. 8, respectively). The concentration of ascorbic acid, tryptophan, and tyrosine was consistent with the number of corresponding protein-coding genes. In the phenylpropanoid pathway, the number of annotated genes was greater in *S. commersonii* than in *S. tuberosum* (78 vs. 61). The families cytochrome P450 cinnamate 4hydroxylase, anthocyanidin 3-O-glucosyltransferase/hydroxycinnamoyl transferase-like protein showed considerable differences in terms of number of genes between *S. tuberosum* and *S. commersonii* (3 vs. 16 and 1 vs. 5, respectively). As for the glycoalkaloid pathway, an opposite scenario was observed: a more significant number of genes was found in *S. tuberosum* than *S. commersonii*. In this case, the UDP-glucosyltransferase and aminotransferase family showed the most considerable differences in gene copy number (333 vs. 219; 22 vs. 12, respectively). Phylogenetic analysis showed that *S. commersonii* UDP-glucose pyrophosphorylase

(ScUGPase) was unrelated to the *S. tuberosum* UDP-glucose pyrophosphorylase (StUGPase) clade, probably because the low homology between their P_GLUCOSE_I-SOMERASE_1 domain sequence. UGPase amino acid sequences are highly conserved among eukaryotes, with more than 80% identity for plant UGPases (Eimert et al. 1996; Geisler et al. 2004). Therefore, slight differences in amino acid composition could be the base of a considerable variation in the enzyme's catalytic properties in *S. commersonii* and *S. tuberosum*.

Recently, we studied the 3-hydroxy-3-methylglutaryl coenzyme A reductase (HMG-CoA) in *S. commersonii* (Villano et al. 2020a). It plays an important role into sterol precursors biosynthesis, and its expression has been reported to be affected by biotic and abiotic stressors and in a tissue-specific manner. Phylogenetic analyses showed that both *S. commersonii* and its cultivated counterpart harbor four HMG homologs, which arose from HMG1 through segmental duplication events rather than other mechanisms of gene duplication. ScHMG1 is also the direct ortholog of AtHMG1, a key gene in steroidal glycoalkaloid regulation in *Arabidopsis*. Using transcriptional and functional approaches, ScHMG1 was overexpressed in *S. commersonii*, allowing the identification of a correlation between this gene and dehydrocommersonine accumulation, known to be among the most poisonous compounds against different potato pests (Manrique-Carpintero et al. 2013).

8.3.3 Transcription Factors

The variations that differentiate wild and cultivated plants are often attributed to a small number of genetic loci. For instance, changes within a single TF sequence can have an overwhelming impact, leading to alterations in plant architecture, developmental patterning, and secondary metabolite accumulation. These modifications are due to the capacity of TFs to regulate the expression of many downstream genes simultaneously. In a very recent work, the WRKY transcriptional factor family has been studied in S. *commersonii* due to their ability to coordinate signals in plant immunity response to biotic and abiotic stresses (Villano et al. 2020b). Compared to the cultivated potato, the authors identified interspecific differentiation in terms of the number of genes encoding putative WRKY TFs (79 in *S. commersonii* and 81 in *S. tubersoum*) and sequence variability. In particular, an expansion of the number of binding domain (WD) were found in two WRKYs (ScWRKY010 and ScWRKY002) probably occurred during evolution, as previously hypothesized for other genes (Aversano et al. 2015; Esposito et al. 2018b). Analysis of gene expression profiles in different tissues and under various stresses allowed to identify expression changes of selected ScWRKYs. In particular, most of them were up-regulated under wounding and biotic (Potato Virus Y and *P. carotovorum*) stress conditions and across different tissues (flowers, leaves, and tubers), hinting at a possible role in the crosstalk between plant and environmental cues in potato species.

S. commersonii TFs were used in a different work to highlight the possible role of gene duplication in determining variation between landraces/cultivated genotypes and wild species (Hardigan et al. 2017). Generally, TFs are among genes mostly affected by gene duplication and tend to have many paralogs into a plant genome (Panchy et al. 2016). Considering their role in altering gene expression, the duplication of a TF or, in general, of regulatory genes may strongly influence the phenotype and have evolutionary implications. A compelling case observed in *S. commersonii* is the tandem duplication of R2R3-MYB TFs placed in the locus Developer (D), which controls phenylpropanoids and, specifically, anthocyanin biosynthesis activation (Jung et al. 2009; D'Amelia et al. 2018). The presence of duplicate copies is often beneficial: the increasing gene dosage confers an advantage by meeting metabolic demands in situations where specific compounds are requested (Panchy et al. 2016). The double amount of total phenylpropanoids detected in *S. commersonii,* compared to *S. tuberosum*, may be in part the result of an additional TF copy (*ScAN1*-like) found in the locus *D*, whose presence confers an advantage to face more environmental stresses (Fig. 8.2a; Aversano et al. 2017; D'Amelia et al. 2018). Another possibility to retain duplicates is a significant change in function (Panchy et al. 2016). The retention of *AN2* and *AN1* in the two species supports this hypothesis (Fig. 8.2a). In D'Amelia et al. (2018), it was observed that *AN1* was sub-functionalized: *AN1* increased the ability to induce anthocyanin accumulation compared to *AN2* that, conversely, retained a more reliable capacity to activate biosynthesis of hydroxycinnamic acid (HA) derivatives (other types of phenylpropanoid compounds) and relatively anthocyanins. *AN2* coding sequence resulted highly conserved between *S. commersonii* and *S. tuberosum*, and it is likely the ancestral gene. Conversely, *AN1* is highly variable also between cultivated varieties and has been probably subjected to an evolutive diversification over time (D'Amelia et al. 2014). Another interesting aspect is that the duplications of the two genes predated the divergence of potato lineages and the different environmental conditions faced by the two species also exerted different selection pressure. Indeed, it has been observed that *AN2* lost its ability to be cold stress-activated in *S. tuberosum* as a consequence, hypothetically, of domestication and breeding processes (D'Amelia et al. 2018). Among HAs with a known capacity to face cold stress, some types of ferulic acids were detected in CMM1T, whereas they were not detected in the cold-sensitive potato (Fig. 8.2b). The cold induction of *AN2* in *S. commersonii* and the consequent accumulation of stress-adaptive compounds can contribute to frost tolerance and cold acclimation capacity of this wild species.

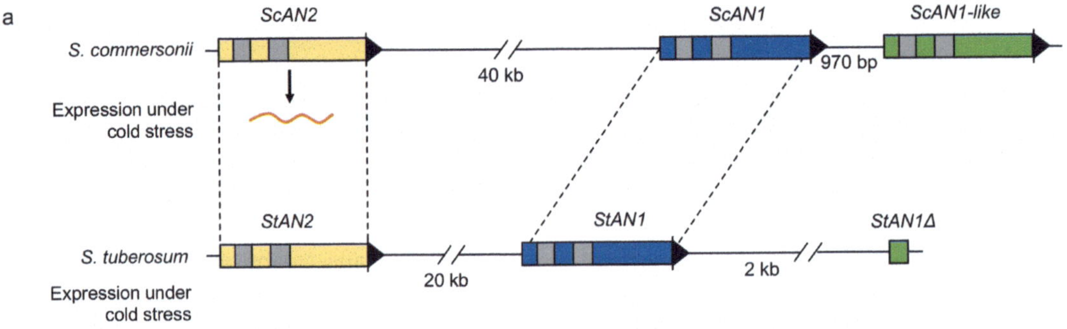

b

	S. commersonii		*S. tuberosum*	
	Control	Cold-Stressed	Control	Cold-Stressed
Metabolites	Mean Peak Area	Mean Peak Area	Mean Peak Area	Mean Peak Area
CGA_1	7952757 d*	21626518 c	45055839.33 b	65342066 a
CGA_2	ND b	ND b	297252.67 a	249313 a
CGA_3	ND c	ND c	11338220 b	20019170 a
CGA_4	ND c	131590**c** c	891588 b	3226120 a
MeCGA_1	10655405 d	21415499 c	36619251.33 b	54257901 a
FerQuin	3924380.68 a	7033435 a	b	ND b
MeCGA_3	ND c	10911037 c	11631043.33 b	20330824 a
FerPutr_1	22533976.33 a	17908420 c	ND c	ND c
FerPutr_2	9703852 a	10743795 a	ND b	ND d

* Means denoted by the same letter did not differ significantly at P ≤0.05 according to Duncan's multiple range test.

Fig. 8.2 **a** Locus D in *S. commersonii* and *S. tuberosum*. Tandem R2R3Myb repeats are colored differently. **b** Peak areas of identified hydroxycinnamic acid derivatives in *S. commersonii* and *S. tuberosum* in control and cold-stressed condition. Feruloyl putrescine isomer 2 (FerPutr_2), feruloyl putrescine isomer 1 (FerPutr_1), methyl-chlorogenic acid (CGA) isomer 3 (MeCGA_3), feruloyl quinic acid (FerQuin), methyl-CGA isomer 1 (MeCGA_1), CGA isomer 4 (CGA_4), CGA isomer 3 (CGA_3), CGA isomer 2 (CGA_2), CGA isomer 2 (CGA_1)

8.4 Impact on Breeding and Future Directions

Due to its EBN, *S. commersonii* is sexually isolated from 2EBN haploids of the cultivated potato. Thus, considerable efforts have been made to bypass its incongruity. Among them, Carputo et al. (1997) first reported the possibility to produce bridge ploidies that balance the maternal to paternal EBN ratio in the *S. tuberosum-S. commersonii* hybrid endosperm (Fig. 8.3). In this breeding approach, the key step is the production of triploid hybrids from crosses between tetraploid (and thus 2EBN) *S. commersonii* and 2x(2EBN) *S. tuberosum*. Triploids have also been produced through 2x(1EBN) *S. commersonii*—2x(2EBN) *S. tuberosum* crosses. In this case, the *S. commersonii* clone produced 2n gametes and was thus functionally tetraploid (Gaiero et al. 2017). Following triploid production, the breeding scheme is quite straightforward. Indeed, selected triploid hybrids that produce 2n gametes are then used in backcross programs (Fig. 8.3). Taking advantage from this strategy, Gaiero et al. (2017) proposed a breeding approach that combines sexual hybridization with genetic engineering. They selected a BC2 *S. tuberosum-S. commersonii* sexual hybrid showing partial resistance to bacterial wilt to transgenically express the *Elongation Factor Tu Receptor* (*AtEFR*) gene that conferred an enhanced and durable resistance.

Breeding efforts have also employed somatic fusion to generate *S. tuberosum* (+) *S. commersonii* hybrids, subsequently included in backcross programs. As known, this breeding method allows the combination of intact genomes from the two parents, and its advantages in potato breeding have been recently reviewed (Tiwari et al. 2018). Through this approach, frost tolerance and capacity to cold acclimate (Cardi et al. 1993; Chen et al. 1999), resistance to bacterial wilt (Laferriere et al. 1999; Kim-Lee et al. 2005), and tuber soft rot (Carputo et al. 2000) have been transferred into the cultivated background. It should be pointed out that, to our knowledge, up to now, these breeding efforts have not been successful in the release of new potato varieties displaying introgression of traits from *S. commersonii*. Among the main reasons for this are

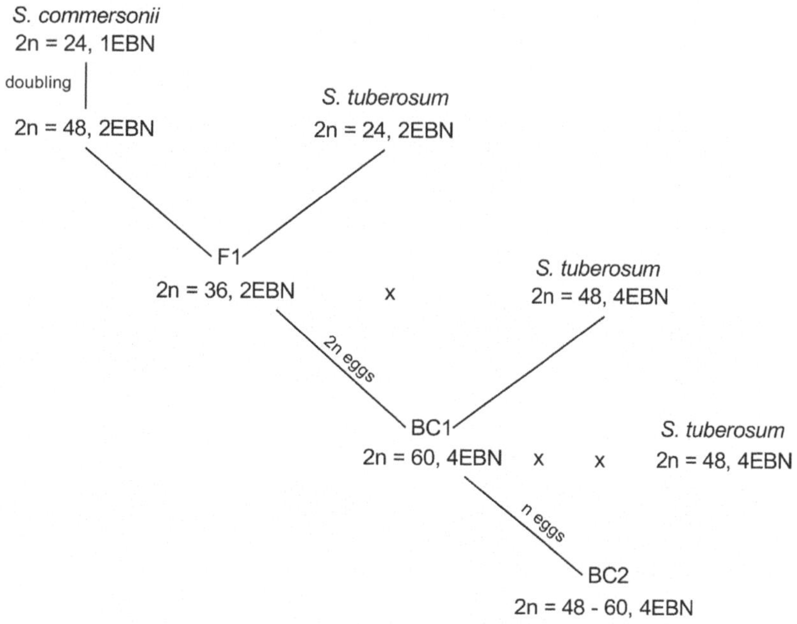

Fig. 8.3 Introgressive hybridization breeding scheme between *S. commersonii* and *S. tuberosum* (Carputo et al. 1997)

linkage drag and the genetic complexity of traits of interest, in which many factors may interact. Fortunately, as described in this chapter, over the last decade, the impressive advancements in 'omics' techniques coupled with developments in bioinformatic methods are hastening the access to complex traits in *S. commersonii*, but much more need to be done. We believe that to accelerate the exploration of the genotype–phenotype relationships at the whole-genome level, we need to move toward through (1) high-density genotyping of large populations to discover new genes and QTLs and provide the basis for modeling *S. commersonii* complex traits, (2) the development of an integrated strategy based on molecular breeding and modern targeted gene delivery utilizing *S. commersonii*. This will fast-forward the development of climate-resilient elite potato genotypes and, more broadly, provide an unprecedented opportunity to face the critical challenge to meet the demands for a growing human population living in a changing and unstable climate.

References

Argout X, Salse J, Aury JM et al (2011) The genome of *Theobroma cacao*. Nat Genet 43(2):101–108

Aversano R, Contaldi F, Ercolano MR et al (2015) The *Solanum commersonii* genome sequence provides insights into adaptation to stress conditions and genome evolution of wild potato relatives. Plant Cell 27:954–968

Aversano R, Contaldi F, Adelfi MG et al (2017) Comparative metabolite and genome analysis of tuber-bearing potato species. Phytochemistry 137:42–51

Belyayev A (2014) Bursts of transposable elements as an evolutionary driving force. J Evolution Biol 27:2573–2584

Bennetzen JL, Wang H (2014) The contributions of transposable elements to the structure, function, and evolution of plant genomes. Ann Rev Plant Biol 65:505–530

Bolger A, Scossa F, Bolger ME et al (2014) The genome of the stress-tolerant wild tomato species *Solanum pennellii*. Nat Genet 46:1034–1038

Cardi T, D'Ambrosio E, Consoli D, Puite KJ, Ramulu KS (1993) Production of somatic hybrids between frost-tolerant *Solanum commersonii* and *S. tuberosum*: characterization of hybrid plants. Theor Appl Genet 87:193–200

Carputo D, Barone A, Cardi T et al (1997) Endosperm Balance Number manipulation for direct in vivo germplasm introgression to potato from a sexually isolated relative (*Solanum commersonii* Dun.). Proc Natl Acad Sci USA 94:12013–12017

Carputo D, Basile B, Cardi T, Frusciante L (2000) Erwinia resistance in backcross progenies of *Solanum tuberosum* X *S. tarijense* and *S. tuberosum* (+) *S. commersonii* hybrids. Potato Res 43:135–142

Chen Y-KH, Palta JP, Bamberg JB et al (1999) Expression of non-acclimated freezing tolerance and cold acclimation capacity in somatic hybrids between hardy wild *Solanum* species and cultivated potatoes. Euphytica 107:1–8

Chinnusamy V, Zhu J, Zhu J-K (2007) Cold stress regulation of gene expression in plants. Trends Plant Sci 12(10):444–451

Cho K-S, Cheon K-S, Hong S-Y et al (2016) Complete chloroplast genome sequences of *Solanum commersonii* and its application to chloroplast genotype in somatic hybrids with *Solanum tuberosum*. Plant CEll Rep 35:2113–2123

Cho KS, Cho JH, Im JS et al (2018) Mitochondrial genome sequence of tuber-bearing wild potato, *Solanum commersonii* Dunal. Mitochondr DNA B 3 (1):198–199

D'Amelia V, Aversano R, Batelli G et al (2014) High AN1 variability and interaction with basic helix-loop-helix co-factors related to anthocyanin biosynthesis in potato leaves. Plant J 80:527–540

D'Amelia V, Aversano R, Ruggiero A et al (2018) Subfunctionalization of duplicate MYB genes in *Solanum commersonii* generated the cold-induced ScAN2 and the anthocyanin regulator ScAN1. Plant Cell Environ 41(5):1038–1051

Datema E, Mueller LA, Buels R et al (2008) Comparative BAC end sequence analysis of tomato and potato reveals overrepresentation of specific gene families in potato. BMC Plant Biol 8(1):34

Dodd AN, Kudla J, Sanders D (2010) The language of calcium signaling. Ann Rev Plant Biol 61:593–620

Eimert K, Villand P, Kilian A, Kleczkowski LA (1996) Cloning and characterization of several cDNAs for UDP-glucose pyrophosphorylase from barley (*Hordeum vulgare*) tissues. Gene 170(2):227–232

Esposito S, Aversano R, Bradeen JM et al. (2018a) Deep-sequencing of *Solanum commersonii* small RNA libraries reveals riboregulators involved in cold stress response. Plant Biol.https://doi.org/10.1111/plb.12955

Esposito S, Aversano R, D'Amelia V et al. (2018b) Dicer-like and RNA-dependent RNA polymerase gene family identification and annotation in the cultivated Solanum tuberosum and its wild relative *S. commersonii*. Planta 248(3):729–743

Esposito S, Barteri F, Casacuberta J, Mirouze M, Carputo D, Aversano R (2019). LTR-TEs abundance, timing and mobility in *Solanum commersonii* and *S. tuberosum* genomes following cold-stress conditions. Planta 250(5), 1781–1787

Evans NH, McAinsh MR, Hetherington AM (2001) Calcium oscillations in higher plants. Curr Opin Plant Biol 4(5):415–420

Gaiero P, Mazzella C, Vilaró F et al. (2017) Pairing analysis and in situ hybridisation reveal autopolyploid-like behaviour in *Solanum commersonii* × *S. tuberosum* (potato) interspecific hybrids. Euphytica 213:137

Gaiero P, Vaio M, Peters SA et al (2018) Comparative analysis of repetitive sequences among species from the potato and the tomato clades. Ann Bot. https://doi.org/10.1093/aob/mcy186

Geisler M, Wilczynska M, Karpinski S, Kleczkowski LA (2004) Toward a blueprint for UDP-glucose pyrophosphorylase structure/function properties: homology-modeling analyses. Plant Mol Biol 56(5):783–794

González M, Galván G, Siri MI et al (2013) Resistencia a la marchitez bacteriana de la papa en *Solanum commersonii*. Agrociencia Uruguay 7:45–54

Hanada K, Zou C, Lehti-Shiu MD et al (2008) Importance of lineage-specific expansion of plant tandem duplicates in the adaptive response to environmental stimuli. Plant Physiol 148:993–1003

Hanneman RE, Bamberg JB (1986) Inventory of tuber bearing solanum species. University of Wisconsin, Madison USA

Hardigan MA, Laimbeer FPE, Newton L et al (2017) Genome diversity of tuber-bearing Solanum uncovers complex evolutionary history and targets of domestication in the cultivated potato. P Natl A Sci 114(46): E9999–E10008

Hawkes JG (1990) The potato: evolution, biodiversity and genetic resources. Belhaven Press, London, p 259

Heim MA, Jakoby M, Werber M et al (2003) The basic helix–loop–helix transcription factor family in plants: a genome-wide study of protein structure and functional diversity. Mol Biol Evol 20(5):735–747

Hosaka K, Hanneman RE (1998) Genetics of self-compatibility in a self-incompatible wild diploid potato species *Solanum chacoense*: 1. Detection of an S locus inhibitor (Sli) gene. Euphytica 99(3):191–197

Jaillon O, Aury JM, Noel B et al (2007) The grapevine genome sequence suggests ancestral hexaploidization in major angiosperm phyla. Nature 449(7161):463–467

Jansky SH (2006) Overcoming hybridization barriers in potato. Plant Breed 125:1–12

Jansky SH, Chung YS, Kittipadukal P (2014) M6: a diploid potato inbred line for use in breeding and genetics research. J Plant Regist 8(2):195–199

Jeon J, Baldrian P, Murugesan K, Chang YS (2011) Laccase-catalysed oxidations of naturally occurring phenols: from in vivo biosynthetic pathways to green synthetic applications. J Microbiol Biotechn 5:318–332

Jheng CF, Chen TC, Lin JY et al (2012) The comparative chloroplast genomic analysis of photosynthetic orchids and developing DNA markers to distinguish *Phalaenopsis orchids*. Plant Sci 190:62–73

Jung CS, Griffiths HM, De Jong DM et al (2009) The potato developer (D) locus encodes an R2R3 MYB transcription factor that regulates expression of multiple anthocyanin structural genes in tuber skin. Theor Appl Genet 120(1):45–57

Kim-Lee H, Moon JS, Hong YJ et al (2005) Bacterial wilt resistance in the progenies of the fusion hybrids between haploid of potato and *Solanum commersonii*. Am J Potato Res 82(2):129–137

Kudla J, Batistic O, Hashimoto K (2010) Calcium signals: the lead currency of plant information processing. Plant Cell 22:541–563

Laferriere LT, Helgeson JP, Allen C (1999) Fertile *Solanum tuberosum* + *S. commersonii* somatic hybrids as sources of resistance to bacterial wilt caused by *Ralstonia solanacearum*. Theor Appl Genet 98:1272–1278

Lakhotia N, Joshi G, Bhardwaj AR et al (2014) Identification and characterization of miRNAome in root, stem, leaf and tuber developmental stages of potato (*Solanum tuberosum* L.) by high-throughput sequencing. BMC Plant Biol 14:1–6

Lee BH, Henderson DA, Zhu JK (2005) The *Arabidopsis* cold-responsive transcriptome and its regulation by ICE1. Plant Cell 17(11):3155–3175

Leisner CP, Hamilton JP, Crisovan E et al (2018) Genome sequence of M6, a diploid inbred clone of the high glycoalkaloid-producing tuber-bearing potato species *Solanum chacoense*, reveals residual heterozygosity. Plant J 94:562–570

Manrique-Carpintero NC, Tokuhisa JG, Ginzberg I, Holliday JA, Veilleux RE (2013) Sequence diversity in coding regions of candidate genes in the glycoalkaloid biosynthetic pathway of wild potato species. G3 Genes Genomes Genet 3:1467–1479.

Matsui A, Nguyen A, Nakaminami K, Seki M (2013) *Arabidopsis* non-coding RNA regulation in abiotic stress responses. Int J Mol Sci 14(11):22642–22654

McAinsh MR, Pittman JK (2009) Shaping the calcium signature. New Phytol 181:275–294

Micheletto S, Boland R, Huarte M (2000) Argentinian wild diploid Solanum species as sources of quantitative late blight resistance. Theor Appl Genet 101:902–906

Palta JP, Simon G (1993) Breeding potential for improvement of freezing stress resistance: genetic separation of freezing tolerance, freezing avoidance, and capacity to cold acclimate. In: Li PH, Christersson L (eds) Advances in plant cold hardiness. CRC Press, Boca Raton, FL, pp 299–310

Panchy N, Lehti-Shiu MD, Shiu SH (2016) Evolution of gene duplication in plants. Plant Physiol 171:2294–2316

Peters SA, Datema E, Szinay D et al (2009) *Solanum lycopersicum* cv. Heinz 1706 chromosome 6: distribution and abundance of genes and retrotransposable elements. Plant J 58:857–869

Potato Genome Sequencing Consortium (2011) Genome sequence and analysis of the tuber crop potato. Nature 475:189–195

Ramsay NA, Glover BJ (2005) MYB–bHLH–WD40 protein complex and the evolution of cellular diversity. Trends Plant Sci 10(2):63–70

Rodríguez F, Spooner DM (2009) Nitrate reductase phylogeny of potato (Solanum sect. Petota) genomes with emphasis on the origins of the polyploid species. Syst Bot 34:207–219

Ruiz ON, Daniell H (2005) Engineering cytoplasmic male sterility via the chloroplast genome by expression of β-ketothiolase. Plant Physiol 138(3):1232–1246

Rymarquis LA, Kastenmayer JP, Hüttenhofer AG, Green PJ (2008) Diamonds in the rough: mRNA-like non-coding RNAs. Trends Plant Sci 13(7):329–334

Sakamoto H, Maruyama K, Sakuma Y et al (2004) *Arabidopsis* Cys2/His2-type zinc-finger proteins function as transcription repressors under drought, cold, and high-salinity stress conditions. Plant Physiol 136 (1):2734–2746

Sanders D, Pelloux J, Brownlee C, Harper JF (2002) Calcium at the crossroads of signaling. Plant Cell 14:401–417

Siri MI, Galván GA, Quirici L et al (2009) Molecular marker diversity and bacterial wilt resistance in wild *Solanum commersonii* accessions from Uruguay. Euphytica 165(2):371–382

Song Z, Zhang L, Wang Y, Li H, Li S, Zhao H, Zhang H (2017) Constitutive expression of miR408 improves biomass and seed yield in *Arabidopsis*. Front Plant Sci 8:2114

Stracke R, Werber M, Weisshaar B (2001) The R2R3-MYB gene family in *Arabidopsis thaliana*. Curr Opin Plant Biol 4(5):447–456

The Tomato Genome Consortium (2012) The tomato genome sequence provides insights into fleshy fruit evolution. Nature 485:635–641

Tiwari JK, Devi S, Ali N et al (2018) Progress in somatic hybridization research in potato during the past 40 years. Plant Cell Tiss Org 132(2):225–238

Torres GA, Gong Z, Iovene M et al. (2011) Organization and evolution of subtelomeric satellite repeats in the potato genome. G3-Genes Genom Genet 1:85–92

Vega SE, Palta JP, Bamberg JB (2000) Variability in the rate of cold acclimation and deacclimation among tuber-bearing Solanum (potato) species. J Am Soc Hort Sci 125(2):205–211

Villano C, D'Amelia V, Esposito S, Adelfi MR, Contaldi F, Ferracane R, Vitaglione P, Aversano R, Carputo D (2020a) Genome-wide HMG family investigation and its role in glycoalkaloid accumulation in wild tuber-bearing *Solanum commersonii. Life* 10 (4):37

Villano C, Esposito S, D'Amelia V, Garramone R, Alioto D, Zoina A, Aversano R, Carputo D (2020b) WRKY genes family study reveals tissue-specific and stress-responsive TFs in wild potato species. Sci Rep 10:7196

Wang Y, Itaya A, Zhong X et al (2011) Function and evolution of a microRNA that regulates a Ca2+-ATPase and triggers the formation of phased small interfering RNAs in tomato reproductive growth. Plant Cell 23:3185–3203

Xu X, Pan S, Cheng S et al (2011) Genome sequence and analysis of the tuber crop potato. Nature 475:189–195

Zuluaga AP, Solé M, Lu H et al (2015) Transcriptome responses to *Ralstonia solanacearum* infection in the roots of the wild potato *Solanum commersonii*. BMC Genomics 16(1):24

Solanum Chacoense Genome Sequence

9

Courtney P. Leisner

Abstract

Cultivated potato (*Solanum tuberosum* L.) was domesticated nearly 8,000 years ago and stands as the world's most important tuber crop. Potato breeding has been hampered by the polyploid nature of the potato genome, inherent self-incompatibility, and high degree of heterozygosity, preventing breeders from utilizing inbred lines or recombinant inbred line populations for more traditional mapping designs that are available for other major crops. *S. chacoense* is a wild potato relative that is an excellent germplasm source for desirable traits in cultivated potato. As a self-compatible diploid relative of potato with key agronomic traits, *S. chacoense* represents a key advancement in the generation of diploid inbred lines in potato. *S. chacoense* M6 (Reg. No. GP-1, BS 228) is a diploid inbred clone, which is a vigorous homozygous breeding line generated from seven generations of self-pollination. This chapter will describe the genome sequence of M6 and the attributes that make it an important resource in diploid potato breeding.

C. P. Leisner (✉)
Department of Biological Sciences, Auburn University, Auburn, AL 36849, USA
e-mail: cpl0013@auburn.edu

9.1 Challenges of Current Potato Breeding

Cultivated potato (*Solanum tuberosum* L.) is the fifth most important food crop globally in terms of production quantity (FAOSTAT 2017) and the world's most important tuber crop (Zhou et al. 2020). Domesticated nearly 8,000 years ago, potatoes are grown from below-ground storage tubers. Potatoes are an excellent source of certain vitamins and minerals, meeting 4% of the daily recommended intake of vitamin C, and 12% for potassium (Navarre et al. 2009). Potatoes also contain phytonutrients such as flavanols, anthocyanins, carotenoids, and polyphenols (Navarre et al. 2009). In spite of its importance, over 100 years of breeding in potato has led to few genetic gains (Douches et al. 1996). The lack of improvement in yield is due partly from the narrow genetic base of cultivated potato and its biological attributes (Douches et al. 1996; Bethke et al. 2019). A high degree of relatedness is observed among potato cultivars based on pedigree analysis, causing some contribution to yield stasis (Mendoza and Haynes 1974; Douches et al. 1996). Furthermore, the need for clonal propagation from tubers causes longer selection cycles and time to homozygosity (Howard 1970; Douches et al. 1996). This fact is starkly highlighted by the cultivars Russet Burbank and Bintje; Russet Burbank was released in 1902 (Bethke et al. 2014) and Bintje was bred in 1904 (Ramulu

© Springer Nature Switzerland AG 2021

D. Carputo et al. (eds.), *The Wild Solanums Genomes*, Compendium of Plant Genomes,
https://doi.org/10.1007/978-3-030-30343-3_9

et al. 1983), yet these cultivars are still highly cultivated and used for commercial production.

The main limitations to breeding in potato center on the polyploid nature of potato, its self-incompatibility, and heterozygosity (Endelman and Jansky 2016; Jansky et al. 2014; Kyriakidou et al. 2020). These features lead to (or are compounded by) low recombination, long regeneration cycle, requirement for vegetative propagation, inbreeding depression, and poor adaptation of wild germplasm (Visser et al. 2009; Lindhout et al. 2011; The Potato Sequencing Consortium 2011; Hardigan et al. 2017; Zhang et al. 2019; Zhou et al. 2020). Cultivated potato is an autotetraploid ($2n = 4x = 48$), which creates a challenge to breeders due to complex segregation ratios in autotetraploid crosses (Endelmen and Jansky 2016). Genetic mapping studies therefore have been done at the diploid level through utilization of cultivated Andean diploids from *S. tuberosum* Group Phuereja or Stenotomum, haploids derived from *S. tuberosum* Group Chilotanum tetraploids, or wild *Solanum* relatives (Gebhardt et al. 1989; van Eck et al. 1994; Li et al. 2005; Prashar et al. 2014; Endelman and Jansky 2016). Therefore, potato breeders and researchers have not been able to utilize inbred lines or recombinant inbred line populations for more traditional mapping designs that are available to other major crops such as maize, rice, wheat, soybean, and tomato (however, see Endelman and Jansky 2016; Marand et al. 2019).

One current strategy for overcoming these limitations in potato breeding is the creation of inbred diploid lines that can be propagated by true seed (Birham and Hosaka 2000; Phumichai et al. 2005; Lindhout et al. 2011; Li et al. 2013; Jansky et al. 2014, 2016). In fact, approximately 70% of the potato germplasm is composed of diploid wild relatives or landraces of potato, whose diversity and taxonomy are described in detail in Chap. 4. The key hurdle for domestication of these diploid relatives is the high prevalence of gametophytic self-incompatibility system present in most diploid potatoes (Pushkarnath 1942; Hawkes 1958; Pandey 1962; Cipar et al. 1964). A major breakthrough has been the generation of diploid self-compatible inbred lines (Ye et al. 2018). Clustered regularly interspaced short palindromic repeats (CRISPR)-associated protein 9 (Cas 9) was used to create self-compatible diploid lines of *S. tuberosum* group Phureja without the introduction of any wild potato DNA fragments (Ye et al. 2018). Another key step in diploid potato breeding is the development of M6, a self-compatible inbred line of the wild potato relative *S. chacoense* without the use of gene editing (Jansky et al. 2014).

9.2 *S. Chacoense*—A Wild Potato Relative

S. chacoense is a wild potato relative that is an excellent source for desirable traits in cultivated potato (Fig. 9.1). These include resistance to late blight, viruses, and cyst nematodes (Spooner et al. 2014; Kaiser et al. 2020) and important processing traits such as high specific gravity and fry color after cold storage (Jansky et al. 2014). One issue with the *S. chacoense* accession is a high glycoalkaloid content in the tubers. Steroidal glycoalkaloids are compounds that are toxic upon consumption and therefore represent an issue to breeders as highlighted by Jansky and colleagues in the Chap. 12. As a self-compatible diploid relative of potato with key agronomic traits, *S. chacoense* represents a key advancement in the generation of diploid inbred lines in potato. The genome sequence of *S. chacoense* is of the diploid inbred clone M6 (Reg. No. GP-1, BS 228), which is a vigorous homozygous breeding line generated from seven generations of self-pollination (Jansky et al. 2014; Leisner et al. 2018). This chapter will describe the genome sequence of M6 *S. chacoense* (Leisner et al. 2018) and the attributes that make it an important resource in diploid potato breeding.

Fig. 9.1 Morphology of the plant, leaves, and flowers of M6 Solanum chacoense. Whole plant (left), leaves (top right), and flowers (bottom right). Adapted from Leisner et al. (2018)

9.2.1 Assembly Quality and Landscape of the M6 *S. Chacoense* Genome

Sequencing and assembly of the M6 *S. chacoense* genome yielded a draft genome assembly of 825,767, 562 bp in 8,260 scaffolds (Leisner et al. 2018). The overall assembly had an N50 scaffold size of 713,602 bp. The estimated genome size of M6 *S. chacoense* is 882 Mb, which is similar in size to the estimated genomes of *S. tuberosum* (DM) (The Potato Sequencing Consortium; Pham et al. 2020) and *S. commersonii* (Aversano et al. 2015) and slightly larger than *S. tuberosum* "Solyntus" (van Lieshout et al. 2020). Scaffolds of the draft genome assembly were anchored to 12 chromosomes using data from two genetic maps (Leisner et al. 2018). Overall, 62% of the genome sequence was anchored to the 12 chromosomes, representing 748 scaffolds and 508,150,181 Mb of sequence. Sequence comparison of the 12 M6 *S. chacoense* pseudomolecules with the published *S. tuberosum* group Group Phureja DM1-3 R44 genome (Hardigan et al. 2016) showed concordance across all 12 chromosomes (Leisner et al. 2018).

The M6 *S. chacoense* genome annotation resulted in 49,124 high-confidence gene models that represent 37,704 loci. Construction of the M6 *S. chacoense* pseudomolecules resulted in anchoring of 29,989 high-confidence genes across the 12 chromosomes (Leisner et al. 2018). Analysis of the representation of genic sequences in the annotated gene set with Benchmarking Universal Single-Copy Orthologs (BUSCO) (Simão et al. 2015) found 95.4% of the core plant ortholog genes were represented in the genome annotation and 2.6% were fragmented (Leisner et al. 2018).

Gene density, repeat coverage, single nucleotide polymorphism (SNP) density, and recombination rate were analyzed across the M6 *S. chacoense* genome (Leisner et al. 2018). As described in Jansky et al. (2014), M6 *S. chacoense* is a breeding line derived from self-pollinating diploid *S. chacoense* for seven generations. Therefore, we would expect only 0.8% of SNPs to be heterozygous after self-pollination for seven generations if we assume all SNPs were heterozygous in the clone originally self-pollinated (Jansky et al. 2014). Previous analysis of heterozygosity in M6 *S. chacoense* found that of the 7,845 high-quality SNPs analyzed using the SolCAP Illumina SNP array (Felcher et al. 2012), 7,060 (90%) were homozygous. Analysis of SNPs at the whole-genome scale in M6 *S. chacoense* identified 1,414,890 biallelic SNPs from a total of 208 Mb of assayable nucleotides, or a genome-wide SNP frequency of 0.68% (Leisner et al. 2018). This indicates that there is residual heterozygosity present in the M6 *S. chacoense* genome.

Analysis of heterozygosity across the M6 *S. chacoense* genome revealed an enrichment of biallelic loci on chromosomes 4, 8, and 9 (Leisner et al. 2018). There were 1.73%, 2.37%, and 2.10% heterozygous positions on chromosome 4, 8, and 9, respectively, compared to a heterozygous SNP frequency range of 0.29–0.69% across the remaining nine chromosomes. These enriched areas of heterozygous SNPs corresponded to regions of low gene density (per Mb), high repeat coverage (% per Mb), and low recombination (Leisner et al. 2018). It is possible that the retained regions of heterozygosity in the M6 *S. chacoense* genome are due to linkage of beneficial alleles with deleterious alleles that are maintained in repulsion and would require increased recombination through sexual propagation to be purged.

The overall limited heterozygosity in M6 *S. chacoense* contrasts with other potato genomes, both diploid and tetraploid. Analysis of heterozygosity using SNP genotyping arrays found heterozygosity rates of 56% at the assayed loci in tetraploid cultivars (Hirsch et al. 2013) and 0.67–37.2% at the assayed loci in a diversity panel of *Solanum* that included wild relatives and landraces (Hardigan et al. 2014). These SNP array heterozygosity analyses may be slightly biased as these loci were selected to be polymorphic across potato accessions (Hamilton et al. 2011). SNP frequency in *S. commersonii* assessed using whole-genome assembly analysis identified 1.49% heterozygous SNPs (Aversano et al. 2015), which is slightly higher than M6 *S. chacoense*. An additional study analyzing 63 diverse accessions of potato using whole-genome heterozygosity analysis found a range of 1.05–2.73% between diploid landraces and tetraploid cultivars in heterozygous nucleotide frequencies (Hardigan et al. 2017). Therefore, while localized regions of heterozygosity in M6 *S. chacoense* exist, the overall rate of heterozygous nucleotide frequencies in the genome is low, making it an excellent candidate for development of diploid inbred potato lines (for review of heterogeneity among wild potato species populations see Bamberg and del Rio 2020).

9.2.2 Self-Compatibility in M6 *S. Chacoense*

The *S. chacoense* germplasm is unique in that it is self-compatible, unlike most diploid wild potato relatives (Pushkarnath 1942; Pandey 1962). Self-incompatibility is controlled by the self-incompatibility (S) locus, which is highly polymorphic (Kao and McCubbin 1996). In some *S. chacoense* germplasm, self-incompatibility is overcome due to inactivation by the dominant allele of the S-locus inhibitor gene *Sli* (Hosaka and Hanneman 1998a; b). Previous analysis has shown plants carrying the *Sli* gene exhibit high levels of inbreeding depression (flower bud abortion, poor vigor, sterility) after the first self-pollination (Birhman and Hosaka 2000), but regain their vigor and fertility after subsequent generations of self-pollination (Jansky et al. 2014; also see Hosaka and Sanetomo 2020).

The single dominant *Sli* gene prevents gametophytic self-incompatibility by the stylar S gene (s) (Hosaka and Hanneman 1998a; b). The *Sli* gene has been mapped to the end of chromosome 12 (Hosaka and Hanneman 1998b; Clot et al. 2020), and the S locus has been mapped to chromosome 1 (Gebhard et al. 1991; Jacobs et al. 1995; Rivard et al. 1996; Hosaka and Hanneman 1998b). Analysis of self-pollinated M6 *S. chacoense* lines demonstrated that M6 *S. chacoense* is homozygous for *Sli* (Jansky et al. 2014). This is unique as it was previously believed *Sli* could only be present in the heterozygous state (Hosaka and Hannemena 1998a). The M6 *S. chacoense* genome provides another important resource for characterizing self-incompatibility in potato, opening the door for cultivation of other wild diploid *Solanum* relatives that might further facilitating diploid potato breeding.

9.2.3 Glycoalkaloid Content in M6 *S. Chacoense*

One limitation of the use of M6 *S. chacoense* as a diploid breeding line is the high level of glycoalkaloids present in the tubers. The production of glycoalkaloids is prevalent in tuber-bearing species and results in production of toxic compounds in the fruits and tubers (Friedman 2006). Current acceptable levels of glycoalkaloids are 20 mg 100 g^{-1} fresh weight (Sinden and Webb 1974). M6 *S. chacoense* tubers contain 28.3 mg glycoalkaloids 100 g^{-1} fresh weight (Jansky et al. 2014). Reduction in glycoalkaloid content can be achieved through crosses of M6 *S. chacoense* with low glycoalkaloid breeding lines (Sanford et al. 1996; Jansky et al. 2014). Analysis of the M6 *S. chacoense* genome identified a cluster of sesquiterpene synthase genes on the bottom arm of chromosome 6 and 9, which is syntenic to the location of these genes in DM1-3 and tomato (Leisner et al. 2018). Further exploration of the M6 *S. chacoense* genome can be done in order to understand glycoalkaloid production and reduce its content in tubers.

9.2.4 Disease Resistance in M6 *S. Chacoense*

M6 *S. chacoense* has demonstrated resistance to several important potato diseases. This includes decreased damage to soft rot disease (*Pectobacterium carotovorum*) compared to Atlantic, which is a susceptible cultivar (Jansky et al. 2014; Chung et al. 2017). It also expresses resistance to Verticillium wilt compared to the resistant cultivar Ranger Russet (Jansky et al. 2014). Marker analysis has revealed M6 is homozygous for a homolog of the *Ve* gene which likely confers resistance to Verticillium wilt (Uribe et al. 2014; Jansky et al. 2014). Utilizing a diploid F_2 mapping population with M6 and USDA8380-1, a major quantitative trait loci (QTL) for Colorado potato beetle resistance has also been identified for *S. chacoense* (Kaiser et al. 2020). The M6 *S. chacoense* genome sequence provides a key resource for exploring other modes of disease resistance in this diploid inbred line.

9.3 Diversity of *S. chacoense*

In addition to M6 *S. chacoense*, there are other *S. chacoense* resources currently available. Recent work by Li et al. (2018) performed whole-genome sequencing on 201 accessions of *Solanum* section *Petota*. This work identified several markers for important agronomic traits useful for potato breeding. In this study, they sequence five accessions of *S. chacoense*, all originating from Argentina. *S. chacoense* grows from central Bolivia south toward Argentina, Brazil, and Paraguay and is one of the most widely distributed wild potato species (Miller & Spooner, 1996). Additional germplasm resources for *S. chacoense* include work done by Hardigan et al. (2016), who analyzed structural variation in 12 monoploid/doubled monoploid clones of potato using next-generation sequencing analysis. This work included four monoploid/doubled monoploid clones of potato that contained from 3 to 50% genetic input from *S. chacoense*. Finally, Hardigan et al. (2017) examined genome diversity and genetic variation across *Solanum* through resequencing across a panel of 67 genotypes. This included sequencing of a separate *S. chacoense* accession different than M6 *S. chacoense* and Li et al. (2018). Structural genome analysis was also completed for *S. chacoense* M6 by Kyriakidou et al. (2019), along with 11 other potato taxa to describe the importance of structural variation and copy number variants to the potato genome sequence. Taken together, there is considerable diversity of *S. chacoense* in the potato germplasm and several additional genomics resources present for this diploid wild potato relative.

9.4 Conclusions

M6 *S. chacoense* represents a key step toward diploid breeding in potato. It is homozygous, self-compatible, possesses key market quality traits, shows level of disease resistance, and produces tubers under both short and long photoperiods, unlike other wild potato relatives (Jansky et al. 2014). Additionally, M6 *S. chacoense* has been used to create the first diploid inbred lines derived from an F_2 population in potato (Endelman and Jansky 2016) and was used to develop an interspecific pseudotestcross F_1 population with another diploid potato clone to investigate QTL associated with yield traits (Marand et al. 2019). The genome sequence of M6 *S. chacoense* is a high-quality representation of the genome, with large portion of the scaffolds and gene spaced anchored into 12 pseudomolecules (Leisner et al. 2018). The gene annotation of M6 *S. chacoense* produced 37,740 functionally annotated genes and identified core sets of Solanaceae genes (Leisner et al. 2018). This genome resource will no doubt accelerate potato breeding through the development of diploid inbred potatoes.

References

Aversano R, Contaldi F, Ercolano MR et al (2015) The *Solanum commersonii* genome sequence provides insights into adaptation to stress conditions and genome evolution of wild potato relatives. Plant Cell 27:954–968

Bamberg J, del Rio A (2020) Assessing under-estimation of genetic diversity within wild potato (*Solanum*) species populations. Am J Pot Res. https://doi.org/10.1007/s12230-020-09802-3

Bethke PC, Halterman DA, Jasnky SH (2019) Potato germplasm enhancement enters the genomics era. Agronomy 9:575

Bethke PC, Nassar AMK, Kubo S, Leclerc YN, Li X-Q, Haroon M, Molen T, Bamber J, Martin M, Donnelly DJ (2014) History and origin of Russet Burbank (Netted Gem) a sport of Burank. Am J Potato Res 91:594

Birham RK, Hosaka K (2000) Production of inbred progenies of diploid potatoes using an S-locus inhibitor (*Sli*) gene, and their characterization. Genome 43:395–502

Chung YS, Kim C, Jansky SH (2017) New source of bacterial soft rot resistance in wild potato (*Solanum chacoense*) tubers. Genet Resour Crop Evol 64:1963–1969

Cipar MS, Peloquin SJ, Hougas RW (1964) Variability in the expression of self-incompatibility in tuber-bearing diploid *Solanum* species. Am Potato J 41:155–162

Clot CR, Polzer C, Prodhomme C, Schuit C, Engelen CJM, Hutten RCB, van Eck HJ (2020) The origin and widespread occurrence of *Sli*-based self-compatibility in potato. Theor Appl Genet 133:2713–2728

Douches DS, Maas D, Jastrzebski K, Chase RW (1996) Assessment of potato breeding progess in the USA of the last century. Crop Sci 36:1544–1552

Endelman JB, Jansky SH (2016) Genetic mapping with an inbred line-derived F2 population in potato. Theor Appl Genet 129:935–943

FAOSTAT (2017) Food and agriculture organization of the United Nations statistics division. Accessed 22 March 2019 http://faostat3.fao.org/

Felcher KJ, Coombs JJ, Masa AN, Hansey CN, Hamilton JP, Veilleux RE, Buell CR, Douches DS (2012) Integration of two diploid potato linkage maps with the potato genome sequences. PLoS ONE 7:E36347

Friedman M (2006) Potato glycoalkaloids and metabolites: roles in the plant and the diet. J Agric Food Chem 54:8655–8681

Gebhardt C, Ritter E, Debener T, Schachtscchabel U, Walkemeier B, Uhrig H, Salamini F (1989) RFLP analysis and linkage mapping in *Solanum tuberosum*. Thero Appl Genet 78:65–75

Gebhardt C, Ritter E, Barone A et al (1991) RFLP maps of potato and their alignment with the homoeologous tomato genome. Theor Appl Genet 83:49–57

Hamilton JP, Hansey CN, Whitty BR, Stoffel K, Massa AN, Van Deynze A, De Jong WS, Douches DS, Buell CR (2011) Single nucleotide polymorphism discovery in elite North American potato germplasm. BMC Genom 12:3–2

Hardigan MA, Bamber J, Buell CR, Douches DS (2014) Taxonomy and genetic differentiation among wild and cultivated germplasm of *Solanum* sect. Petota. Plant Gen 8:1–16

Hardigan MA, Crisovan E, Hamilton JP, Kim J, Laimbeer P, Leisner CP, Manrique-Carpintero NC, Newton L, Pham GM, Vaillancourt B, Yang X, Zeng Z, Douches DS, Jiang J, Veilleux RE, Buell CR (2016) Genome reduction uncovers a large dispensable genome and adaptive role for copy number variation in asexually propagated *Solanum tuberosum*. Plant Cell 28:388–405

Hardigan MA, Laimbeer FPE, Newton L, Crisovan E, Hamilton JP, Vaillancourt B, Wiegert-Rininger K, Wood JC, Douches DS, Farre EM, Veilleux RE, Buell CR (2017) Genome diversity of tuber-bearing Solanum uncovers complex evolutionary history and targets of domestication in the cultivated potato. Proc Natl Acad Sci USA 114:E9999–E10008

Hawkes JG (1958) Significance of wild species and primitive forms for potato breeding. Euphytica 7:257–270

Hirsch CN, Hirsch CD, Felcher K et al (2013) Retrospective view of North American potato (*Solanum tuberosum* L.) breeding in the 20th and 21st centuries. Genes Genom Gen 3:1003–1013

Hosaka K, Hanneman RE (1998a) Genetics of self-compatibility in a self-incompatible wild diploid potato species *Solanum chacoense*. 1. Detection of an S locus inhibitor (Sli) gene. Euphytica 99:191–197

Hosaka K, Hanneman RE (1998b) Genetics of self-compatibility in a self-incompatible wild diploid potato species *Solanum chacoense*. 2. Localization of an S locus inhibitor (Sli) gene on the potato genome using DNA markers. Euphytica 103:265–271

Hosaka K, Sanetomo R (2020) Creation of highly homozygous diploid potato using the S locus inhibitor (*Sli*) gene. Euphytica 126:169

Howard HW (1970) Genetics of the potato *Solanum tuberosum*. Logos Press, London

Jacobs JME, Van Eck HJ, Arens P et al (1995) A genetic map of potato (*Solanum tuberosum*) integrating molecular markers, including transposons, and classical markers. Theor Appl Genet 91:289–300

Jansky SH, Chung YS, Kittipadukal P (2014) M6: A diploid potato inbred line for use in breeding and genetics research. J Plant Reg 8:195–199

Jansky SH, Charkowski AO, Douches DS et al (2016) Reinventing potato as a diploid inbred line-based crop. Crop Sci 56:1412

Kaiser N, Manrique-Carpintero NC, DiFonzo C, Coombs J, Douches D (2020) Mapping *Solanum chacoense* mediated Colorado potato beetle (*Leptinotarsa decemlineata*) resistance in a self-compatible F_2 diploid population. Theor Appl Genet 133:2538–2603

Kao T-H, McCubbin AG (1996) How flowering plants discriminate between self and non-self pollen to prevent inbreeding. Proc Natl Acad Sci USA 93:12059–12065

Kyriakidou M, Anglin NL, Ellis D, Tai HH, Stromvik MV (2020) Genome assembly of six polyploid potato genomes. Sci Data 7:1–6

Kyriakidou M, Achakkagari SR, Lopez JHG, Zhu X, Tang CY, Tai HH, Anglin NL, Ellis D, Stromvik MV (2019) Structural genome analysis in cultivated potato taxa. Theor Appl Genet 133:951–966

Leisner CP, Hamilton JP, Crisovan E, Manrique-Carpintero NC, Marand AP, Newton L, Pham GM, Jiang J, Douches DS, Jansky SH, Buell CR (2018) Genome sequence of M6, a diploid inbred clone of the high-glycoalkaloid-producing tuber-bearing potato species *Solanum chacoense*, reveals residual heterozygosity. Plant J 94:562–570

Li XQ, De Jong H, De Jong DM, De Jong WS (2005) Inheritance and genetic mapping of tuber eye depth in cultivated diploid potatoes. Theor Appl Genet 110:1068–1073

Li Y, Li G, Li C, Qu D, Huang S (2013) Prospects of diploid hybrid breeding in potato. Chin Potato J 27:96–99

Li Y, Colleoni C, Zhang J, Lian Q, Hu Y, Ruess H, Simon R et al (2018) Genomic analyses yield markers for identifying agronomically important genes in potato. Molec Plant 11:473–484

Lindhout P, Meijer D, Schotte T, Hutten RCB, Visser RGF, Eck HJ (2011) Towards F_1 hybrid seed potato breeding. Potato Res 54:301–312

Marand AP, Jansky SH, Gage JL, Hamernik AJ, de Leon N, Jiang J (2019) Residual heterozygosity and epistatic interactions underlie the complex genetic architecture of yield in diploid potato. Genetics 212:317–332

Mendoza HA, Haynes FL (1974) Genetic relationship among potato cultivars grown in the United States. HortSci 9:328–330

Miller JT, Spooner DM (1996) Introgression of *Solanum chacoense* (*Solanum* sect. *Petota*): Upland populations reexamined. Syst Bot 21:461–475

Navarre, DA, Goyer A, Shakya R (2009) Nutritional value of potatoes: vitamin, phytonutrient, and mineral content. In: Singh J, Kaur L (eds) Advances in potato chemistry and technology. Elsevier, Burlington MA

Pandey KKK (1962) Interspecific incompatibility in *Solanum* species. Am J Bot 49:874–882

Pham GM, Hamilton JP, Wood JC, Burke JT, Zhano H, Vaillancourt B, Ou S, Jiang J, Buell CR (2020) Construction of a chromosome-scale long-read reference genome assembly for potato. Gigascience 9:1–11

Phumichai C, Mori M, Kobayashi A, Kamijima O, Hosaka K (2005) Toward the development of highly homozygous diploid potato lines using the self-compatibility controlling *Sli* gene. Genome 48:977–984

Prashar A, Hornyik C, Young V, McLean K, Sharma SK, Dale MFB, Bryan GJ (2014) Construction of a dense SNP map of a highly heterozygous diploid potato population and a QTL analysis of tuber shape and eye depth. Theor Appl Genet 127:2159–2171

Pushkarnath P (1942) Studies on sterility in potatoes: 1. The genetics of self- and cross-incompatibilities. Indian J Genet Plant Breed 2:11–36

Ramulu KS, Dijkuis P, Roest S (1983) Phenotypic variation and ploidy level of plants regenerated from protoplasts of tetraploid potato (*Solanum tuberosum* L. cv. 'Bintje'). Theor Appl Genet 65: 329

Rivard SR, Cappadocia R, Landry BS (1996) A comparison of RFLP maps based on anther cultuver derived, selfed and hybrid progenies of *Solanum chacoense*. Genome 39:611–621

Sanford LL, Kobayashi RS, Deahl KL, Sinden SL (1996) Segregation of leptines and other glycoalkaloids in *Solanum tuberosum* (4x) x *S. chacoense* (4x) crosses. Am Potato J 73:21–33

Simão FA, Waterhouse RM, Ioannidis P, Kriventseva EV, Zdobnov (2015) BUSCO: assessing genome assembly

and annotation completeness with single-copy orthologs. Bioinformatics 31: 3210–3212
Sinden SL, Webb RE (1974) Effect of environment on glycalkaloid content of six potato varieties at 39 locations. USDA-ARS Tech Bull 1742, US Government. Washington DC
Spooner DM, Ghislain M, Simon R, Jansky SH, Gavrilenko T (2014) Systematics, diversity, genetics, and evolution of wild and cultivated potatoes. Bot Rev 80:283–383
The Potato Genome Sequencing Consortium (2011) Genome sequence and analysis of the tuber crop potato. Nature 475:189–197
Uribe P, Jansky SH, Halterman D (2014) Two CAPS markers predict Verticillium wilt resistance in wild *Solanum* species. Mol Breed 33:465–476
van Eck HJ, Jacobs JME, Stam P, Ton J, Stiekema WJ, Jacobsen E (1994) Multiptle alleles for tuber shape in diploid potato detected by qualitative and quantitative genetic analysis using RFLPS. Genetics 137: 303–309; G3 10: 3489–3495
van Lieshout N, van der Burgt A, de Vries ME, ter Maat M, Eickholt D, Esselink D, van Kaauwen MPW, Kodde LP, Visser RGF, Lindhout P, Finkers R (2020) Solyntus, the new highly contiguous reference genome for potato (*Solanum tuberosum*)
Visser RGF, Bachem CWB, Boer JM, Byran GJ, Chakrabati SK, Feingold S, Gromadka R, Ham RCHJ, Huan S, Jacobs JME, Kuznetsov B, Mel PE, Milbourne DM, Orjeda G, Sagredo B, Tang X (2009) Sequencing the potato genome: outline and first results to come from the elucidation of the sequence of the world's third most important food crop. Am J Potato Res 86:417–429
Ye M, Peng Z, Tang D, Yang Z, Li D, Xu Y, Zhang C, Huang S (2018) Generation of self-comptability diploid potato by knockout of *S-RNase*. Nature Plants. https://doi.org/10.1038/s41477-018-0218-6
Zhang C, Wang P, Tang D, Yang Z, Lu F, Qi J, Tawari NR, Shang Y, Li C, Huang S (2019) The genetic basis of inbreeding depression in potato. Nat Gen https://doi.org/10.1038/s41588-018-0319-1
Zhou Q, Tang D, Huang W, Yang Z, Zhang Y, Hamilton JP, Visser RGF, Bachem CWB, Buell CR, Zhang Z, Zhang C, Huang S (2020) Haplotype-resolved genome analyses of heterozygous diploid potato. Nat Genet 52:1018–1023

Genomic Resources in the Eggplant Wild Genepool

10

Pietro Gramazio, Jaime Prohens, and Santiago Vilanova

Abstract

Access to high-throughput next-generation sequencing is now becoming affordable also for non-model crops and small laboratories. In addition, for the most important economic crops, their wild relatives are being sequenced and exploited for the introgression of important traits and to unravel the crops domestication history. Despite the importance of the three cultivated species in the eggplant genepool, common (*Solanum melongena*), scarlet (*S. aethiopicum*), and gboma eggplant (*S. macrocarpon*), genomic studies are still sparse. In this chapter, we reviewed the few genomic studies performed in the eggplant relatives, which are the primary source for the introgression of traits present in these wild relatives to develop resilient cultivated varieties to address the challenges of the present and future agriculture. In addition, we present the preliminary results of the first resequencing study of the eggplant wild relative *S. incanum* and compare it to the other seven accessions from the cultivated *S. melongena*. Among the set of these eight accessions, over 10 million polymorphisms were identified, most of them in the wild *S. incanum*, supporting the hypothesis of multiple genetic bottlenecks occurred during the domestication process that has led to narrowing the genetic diversity of the cultivated common eggplant compared to its wild counterparts. The distribution of the identified polymorphisms along the chromosomes revealed the footprints of ancient interspecific hybridizations events that occurred at a different chronology. The main goal of this chapter is to provide relevant genomics information for the enhancement and utilization of eggplant wild relatives in order to encourage eggplant breeders to develop and use new genomic tools as occurred in other economically important Solanaceae.

10.1 Introduction

The sequencing efforts of the last decade have fostered an unprecedented understanding of genome architecture and complexity, gene and repetitive elements composition and function, phylogenetic relationships and domestications events, among others (Goodwin et al. 2016). In addition, the impressive advances in the

P. Gramazio
Faculty of Life and Environmental Sciences, University of Tsukuba, 1-1-1 tennodai, Tsukuba 305-8572, Japan

J. Prohens (✉) · S. Vilanova
Instituto de Conservación y Mejora de la Agrodiversidad Valenciana (COMAV), Universitat Politècnica de València, Camino de Vera 14, 46022 Valencia, Spain
e-mail: jprohens@btc.upv.es

© Springer Nature Switzerland AG 2021

D. Carputo et al. (eds.), *The Wild Solanums Genomes*, Compendium of Plant Genomes,
https://doi.org/10.1007/978-3-030-30343-3_10

technology of the new sequencing platforms and the plummeting cost per megabase have allowed the assembly of a reference genome for many plant species, including non-model crops (Li et al. 2018). However, despite an easier access to genome sequencing, even for small laboratories and neglected crops, a huge gap still exists for most crops compared to model species and the most economically important cultivated species. In fact, in these crops, more than one reference genome is available, including some of their wild relatives genomes, or even pangenomes (Khan et al. 2020). For example, in tomato (*S. lycopersicum*), apart from a few different de novo reference genome assemblies, some of its wild relatives have been already sequenced, like *S. pennellii* (Bolger et al. 2014), *S. pimpinellifolium* (Razali et al. 2018), *S. lycopersicoides* (Powell et al. 2020) or are in progress such as *S. habrochaites*. Along with these reference genomes, hundreds of cultivated and wild relatives accessions have been resequenced in tomato genepool as well the first pangenomes have been release (Aflitos et al. 2014; Lin et al. 2014; Gao et al. 2019; Alonge et al. 2020; Gramazio et al. 2020a).

On the other hand, the genomic resources available in the eggplant genepool are still sparse despite the efforts done recently to shorten the gap with other Solanaceae and staple crops (Gramazio et al. 2018). Although eggplant is an economically important crop, ranking fifth among vegetables in total worldwide production (Faostat, 2019), and is considered as one of the most important crops to guarantee food security, and so was included in Annex 1 of the International Treaty on Plant Genetic Resources for Food and Agriculture (www.fao.org), until recently the only eggplant reference genome released was a preliminary draft (Hirakawa et al. 2014). This first draft genome, from the Asian cultivar "Nakate-Shinkuro," covers 833.1 Mb (ca. 74% of the eggplant genome) and is assembled in 33,873 scaffolds, thus quite far to be assembled in pseudomolecules like in the case of the tomato genome. The "67/3" reference genome (Barchi et al. 2019a), which was assembled using a combination of Illumina sequencing and optical mapping, encompassed 1.06 Gb (N50 of 2.9 Mb) across 12 pseudomolecules anchored using a high-resolution genetic map. The number of proteins coding genes identified in "67/3" was 34.916, very similar to those described in tomato and much less and more precise compared to the over 42,000 genes predicted by Hirakawa et al. (2014) in their draft genome. Similarly, the high-quality reference genome "HQ-1315" (Wei et al. 2020), which was assembled implementing a combined strategy using Illumina, Nanopore, 10X, and Hi-C technologies, retrieved 36,582 protein-coding genes (1.17 Gb, N50 of 5.26 Mb) as well 35,018 protein-coding genes were annotated for "GUIQIE-1" high-quality reference genome (Li et al. 2021), assembled combining PacBio and Hi-C data (1.15 Gb, N50 of 93.9 Mb). Surprisingly, a draft genome sequence (1.02 Gb, N50 of 516.1 kb) was recently released also for the second most economically important cultivated eggplant, the hypervariable *S. aethiopicum*, along with the resequencing of 65 accessions from the "Gilo" and "Shum" groups and the wild ancestor *S. anguivi*, where 34,906 protein-coding genes were annotated (Song et al. 2019). As it happened for other crops, the availability of the first draft genomes is encouraging genomic studies also in eggplants.

In the eggplant genepool, the wild relatives display a plethora of valuable and beneficial breeding traits that are worth to introgress in the genetic background of the three eggplant cultivated species (Knapp et al. 2013; Syfert et al. 2016). On this basis, our group together with partners from Ivory Coast, Sri Lanka, and Taiwan have been doing a great effort in this direction in the framework of two projects on eggplant pre-breeding for adaptation to climate change within the initiative "Adapting Agriculture to Climate Change: Collecting, Protecting and Preparing Crop Wild Relatives" managed by the Global Crop Diversity Trust (Dempewolf et al. 2014). As a result of these projects (http://eggplantprebreeding.upv.es/), many interspecific hybrids have been obtained with CWRs from the primary, secondary, and tertiary genepools, as well as advanced backcross (AB) generations of these hybrids toward the cultivated eggplant *S.*

melongena (Kouassi et al. 2016; Plazas et al. 2016; García-Fortea et al. 2019). In addition, the development of three introgression lines (ILs) populations with CWRs, and the refining another IL population already developed (Gramazio et al. 2017), as well the first multi-parent advanced generation inter-cross (MAGIC) population (Arrones et al. 2020) are underway (García-Fortea et al. 2019). Most of the CWRs used in this project have been described as a source of variation for a large number of traits of interest in breeding, due to their distribution in desertic and dryland areas (Knapp et al. 2013; Rotino et al. 2014).

10.2 Genomic Studies in Eggplant CWRs

Eggplant is related to many wild species from subgenus *Leptostemonum* (Syfert et al. 2016; Vorontsova and Knapp 2016), many of which can be crossed with *S. melongena* (Daunay and Hazra 2012; Rotino et al. 2014; Kouassi et al. 2016; Plazas et al. 2016). Similarly to other Solanaceae and major crops, the genomic investigation on eggplant wild relatives has lagged behind that of cultivated eggplant (Gramazio et al. 2017). In fact, even though the importance of the CWRs has been clearly defined as essential to broaden the narrow genetic diversity of the crops and to develop resilient varieties (Dempewolf et al. 2017), in some staple crops like eggplant, the generation of informative data for introgression breeding is still very limited. Up to now, a few RNA-based studies have been performed in eggplant relatives (Table 10.1), due to the fact that RNA-Seq is faster and more cost-effective compared to genome assembly and WGRS for non-model crops. The first study was the whole transcriptome assembly of *S. torvum*, an American species from the tertiary genepool (Syfert et al. 2016), along with the first transcriptome of common eggplant (Yang et al. 2014). *Solanum torvum* is usually used as rootstock for being resistant to many soil-borne diseases such as root-knot nematodes, bacterial wilt, Verticillium wilt, and Fusarium wilt, among other stresses (Gousset et al. 2005; Yamaguchi et al. 2010), and is widely consumed in tropical and subtropical countries (Jaiswal 2012). Due to its large phylogenetic distance, up to now, no backcross materials with *S. torvum* have been obtained with the common eggplant although many efforts are being performed in that direction (Plazas et al. 2016), which would result in materials highly valuable for breeding (Daunay and Hazra 2012). In this study (Yang et al. 2014), the two de novo transcriptome assemblies were built on 38,185 and 34,174 unigenes for *S. torvum* and common eggplant, respectively, that were functionally annotated and identified 815 and 621 plant resistance genes. In addition, a comparison of these two transcriptomes along with 11 plant proteomes identified 276 high-confidence single-copy orthologous clusters (Yang et al. 2014).

In 2016, the de novo transcriptome of *S. aculeatissimum* (Zhou et al. 2016), another CWR from the tertiary genepool of common eggplant, was released. *Solanum aculeatissimum* is used for medicinal purposes and has been reported as resistant to root-knot nematode and *Verticillium dahliae* (Handique 1986; Borua 1990; Zhuang et al. 2012). Introgression of the verticillium wilt resistance, which causes important economic losses, in the genetic background of common eggplant is a breeding priority objective. Interspecific hybrids and an amphidiploid between *S. aculeatissimum* and *S. melongena* have been obtained after embryo culture (Rattan et al. 2015; Zhou et al. 2018). The assembly was built on 69,824 unigenes that were functionally annotated in protein, GO term, Clusters of Orthologous Groups (COG), and KEGG databases. Differentially expressed genes (DEGs) analysis found 11,696 up-regulated and 5949 down-regulated genes identifying putative resistance (R-genes) and pathogenesis-related proteins (PRs) genes.

Solanum aculeatissimum is also considered tolerant to low temperatures (Yang et al. 2017). In order to elucidate the molecular mechanism of its cold resistance, Yang et al. (2017) investigated the role of microRNAs (miRNAs), which facilitate chilling tolerance (Zhao et al. 2012), in the transcriptome of *S. aculeatissimum* using

Table 10.1 Statistics for RNA-seq studies genomic studies performed in eggplant CWRs

Species	Plant tissue	Library	Raw reads (M)	Final assembly (unigenes)	NCBI accession	Source
S. torvum	Leaves, root, stem	PE (72 bp)	54	34,174	SRR1104128	Yang et al. (2014)
S. melongena			30	38,185	SRR1104129	
S. aculeatissimum	Root	–	56	69,824	SRS1383901	Zhou et al. (2016)
					SRS1383902	
S. aculeatissimum	Leaf	–	425	95,642	–	Yang et al. (2017)
S. incanum	Leaf, floral bud, fruit	PE (300 bp)	105	83,905	SRR2289250	Gramazio et al. (2016)
S. aethiopicum			114	87,084	SRR2229192	
S. viarum	Stem	–	–	156,926	–	Pandey et al. (2018)
S. sisymbriifolium	Bud, root, stem, leaf	PE (250–450 bp)	31	41,189	GGFC00000000	Wixom et al. (2018)
S. sisymbriifolium	Root	PE (150–200 bp)	480	221,695	PRJNA495012	Wu et al. (2019)

small RNA sequencing under low-temperature stress. The raw reads were assembled in 95,642 unigenes that were functionally annotated for protein, GO terms, COG, and KEGG classification. The DEG analysis found 56 down-regulated and 28 up-regulated miRNAs involved to cold-stress tolerance.

The de novo transcriptome assemblies of *S. incanum* and *S. aethiopicum* were released in 2016 by Gramazio et al. (2016). As reported above, this specific accession of *S. incanum*, MM577, was used in many studies. *Solanum aethiopicum*, whose cultivation is widespread in Africa and in some areas of Caribbean, Brazil, and south Italy (Schippers 2002; Maundu et al. 2009; Sunseri et al. 2010), is usually used as a rootstock (Gisbert et al. 2011). In fact, since some scarlet eggplant accessions have been found resistant to several biotic stresses like fungi (*Fusarium oxysporum*, *F. solani*, *Pythium vexans*, *Phytophthora parasitica*), bacteria (*Ralstonia solanacearum*), insects (*Leucinodes orbonalis*), and to root-knot nematodes (*Meloidogyne incognita*) (Cappelli et al. 1995; Rizza et al. 2002; Collonnier et al. 2001). These resistances could be introgressed in common eggplant since interspecific hybrids and ABs have been already obtained, although with different degrees of fertility (Mennella et al. 2010; Prohens et al. 2012). The de novo transcriptomes were assembled in 83,905 and in 87,084 unigenes for *S. incanum* and *S. aethiopicum*, respectively, which were deeply annotated for many structural and functional categories. Using also the transcriptomes of *S. torvum* and *S. melongena* published by Yang et al. (2014), thousands of intraspecific and interspecific polymorphisms (SNPs, INDELs, and SSRs) were identified and several subsets of markers were established to characterize eggplant CWR germplasm. In this regard, a broader study was performed using the same *S. incanum* and *S. aethiopicum* accessions of the transcriptomes along with other 74 accessions from the three cultivated eggplants and 14 CWRs accessions from the three genepools, identifying 75,399 polymorphic sites and establishing comprehensive genetic relationships (Acquadro et al. 2017).

A phylogenetically distant eggplant CWRs, *S. viarum* Dunal, native to Argentina and Brasil (Coile 1993), was used to perform RNA-Seq analysis to gain insight into prickle development using a spontaneous prickleless mutant (Pandey et al. 2018). *Solanum viarum*, also known as "tropical soda apple," is a perennial noxious weed that causes ecologic and economic losses to agriculture and pasture systems infestating extensive areas worldwide like in the USA or Iran (Eskandari and Abdi Fouladkolae, 2020). However, *S. viarum* has a great commercial interest in the production of steroidal alkaloids and in phytoremediation (Pandey et al. 2018; Afonso et al. 2019). In the prickleless mutant, several defense regulators like ethylene, salicylic acid, PR-proteins as well development-related transcription factors were down-regulated suggesting a connection between prickle development and plant defense (Pandey et al. 2018). On the contrary, secondary metabolites related-genes like terpenoid, steroid, flavonoid, or glucosinolate were up-regulated in the mutant, suggesting that the loss of prickles was compensated with higher metabolites that are involved in biotic and abiotic stress defense.

The last CWRs genomic studies performed so far in eggplant genepool were de novo transcriptome (Wixom et al. 2018) and RNA-Seq (Wu et al. 2019) of *S. sisymbriifolium* another species from the tertiary genepool of common eggplant (Syfert et al. 2016). *Solanum sisymbriifolium* is a weedy plant that is considered a trap-crop for nematodes like *Globodera pallida*, a biotic stress for many Solanaceae (Timmermans 2005; Dandurand and Knudsen 2016). Despite many efforts in our projects on generating eggplant preebreding materials with introgressions with CWRs, no interspecific hybrids have been obtained so far between *S. sisymbriifolium* and common eggplant. Using a single molecule real-time (SMRT) sequencing approach, Wixom et al. (2018) built a high-quality de novo transcriptome of *S. sisymbriifolium* that consisted of 41,189 unigenes from different plant tissues. Analyzing the data, the authors suggested that this plant could have been experienced a recent genome duplication and re-diploidization and accumulate a large number of genes associated with photosynthesis and amino acid metabolism that may be involved in its nematode resistance. On the other hand, Wu et al. (2019) investigated the transcriptomic response of *S. sisymbriifolium* roots resistant to *Verticillium dahliae*, a devasting eggplant disease, under the fungal pathogen infection to dissect its complex mechanism defense. Even though some putative genes were identified, the authors suggested further studies to identify other characters that play important roles in the defense mechanism against this pathogen.

To our knowledge, no other genomic studies have been performed for eggplant CWRs.

10.3 Whole-Genome Resequencing Studies

Up to now, the genomic resources available in the eggplant genepool are sparse and just a few studies have been performed. To our knowledge, up to now no resequencing studies have been performed so far in eggplant, probably due to the lack, until recently, of high-quality reference genome assemblies available. This encouraged us to perform the whole-genome resequencing (WGRS) of seven accessions of the cultivated common eggplant plus the wild relative *S. incanum* (Gramazio et al. 2019). These eight accessions are the founder parents of a multi-parent advanced generation inter-cross (MAGIC) population that we are developing and they have been selected for their genetic, phenotypic, and geographical diversity (Arrones et al. 2020). The *S. incanum* accession used as a founder of the MAGIC population, MM577, has been selected for its considerable interest in eggplant breeding since exhibited a phenolic compounds contents several times higher than common eggplant (Stommel and Whitaker 2003; Prohens et al. 2013), and higher tolerance to drought (Knapp et al. 2013). The great interest in this species for breeding has been reflected in many studies and publications (Salas et al. 2011; Gisbert et al. 2011; Meyer et al. 2015; Acquadro et al. 2017; Gramazio et al. 2014).

The main objective of this WGRS study, among many others, was to generate a large set of reliable and dispersed molecular markers among the MAGIC founders to develop, with international collaborators, a high-throughput genotyping platform, the Single Primer Enrichment Technology (SPET) platform (Barchi et al. 2019b), to efficiently genotype the segregating population over the several cycles of selfing required to fix the lines. In addition, the study intended to unravel the genetics of complex traits of breeding importance and to shed light on the domestication and evolutionary history of eggplant through the detection of potentially shared introgressions and ultimately provide valuable tools for eggplant breeders.

The sequencing was performed with the Illumina HiSeq 4000 sequencer, after preparing the paired-end libraries of 300 bp, resulting in over 100 Gb of raw data with an average of over 180 million reads per sample (Fig. 10.1). The trimming and cleaning processes only removed less than 3% of the raw reads before mapping them against the "67/3" high-quality reference genome (Barchi et al. 2019a). Over 80% high-quality clean reads mapped, resulting in about 20× average coverage. However, while for the seven *S. melongena* accessions, the reads covered all the genome length, the *S. incanum* mapping coverage was nearly 95%, uniformly distributed across all the chromosomes.

After filtering with stringent criteria, the variant calling identified over 10 million high-quality polymorphisms among the MAGIC founders. Interestingly, while the common eggplant accessions showed a range of variants

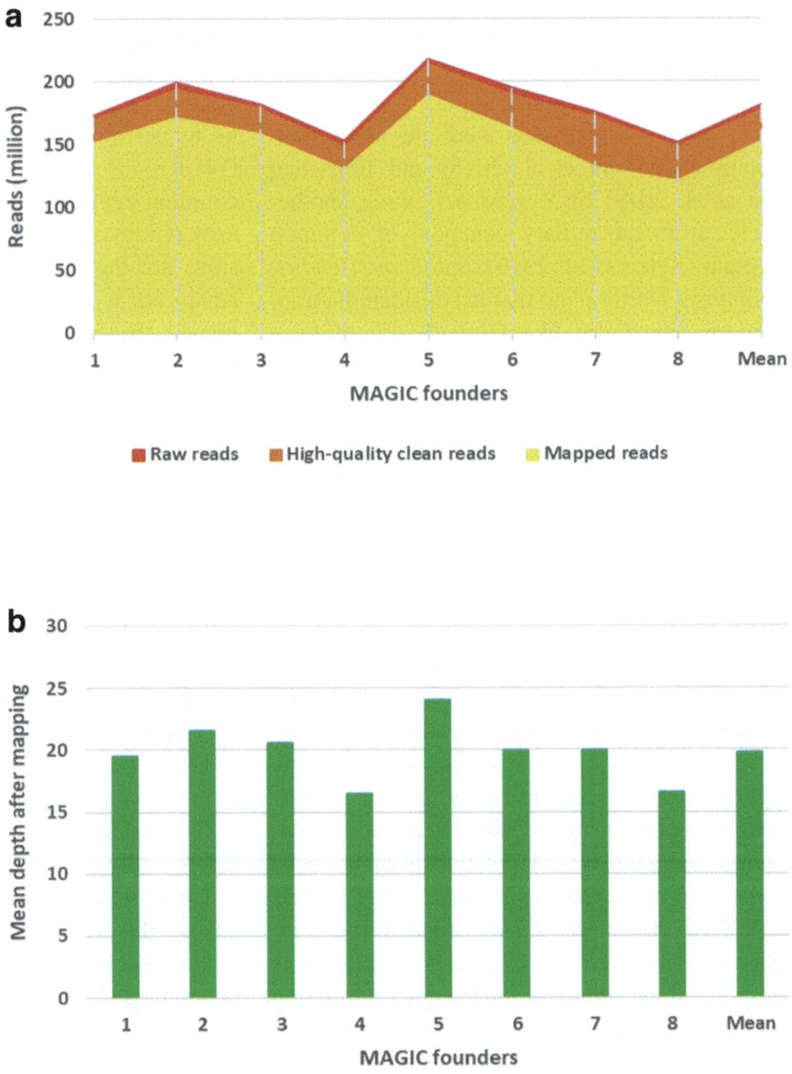

Fig. 10.1 **a** Sequencing and mapping statistics of the eight eggplant MAGIC founders. The codes 1–7 correspond to *S. melongena* accessions while the code 8 to *S. incanum*. **b** Average mapping depth of the MAGIC founders

identified between 800,000 and 1,200,000, in the *S. incanum* accessions were over nine million (Fig. 10.2). Large differences in polymorphisms identified between cultivated and wild relatives are not unusual and is more significant in species that have suffered an important genetic erosion due to artificial selection and breeding toward uniformity (Aflitos et al. 2014; Zhou et al. 2015). For this reason, there is growing interest in crop wild relatives (CWRs) to restore partially the genetic diversity lost during domestication to face the new biotic and biotic threats like those posed by climate change (Fita et al. 2015; Hulme et al. 2017). In fact, plant breeders are designing new strategies and approaches, like introgressiomics (Prohens et al. 2017), to introgress and exploit the novel beneficial variation of CWRs to develop resilient crop varieties.

In order to find footprints of potential common ancestral introgressions through the

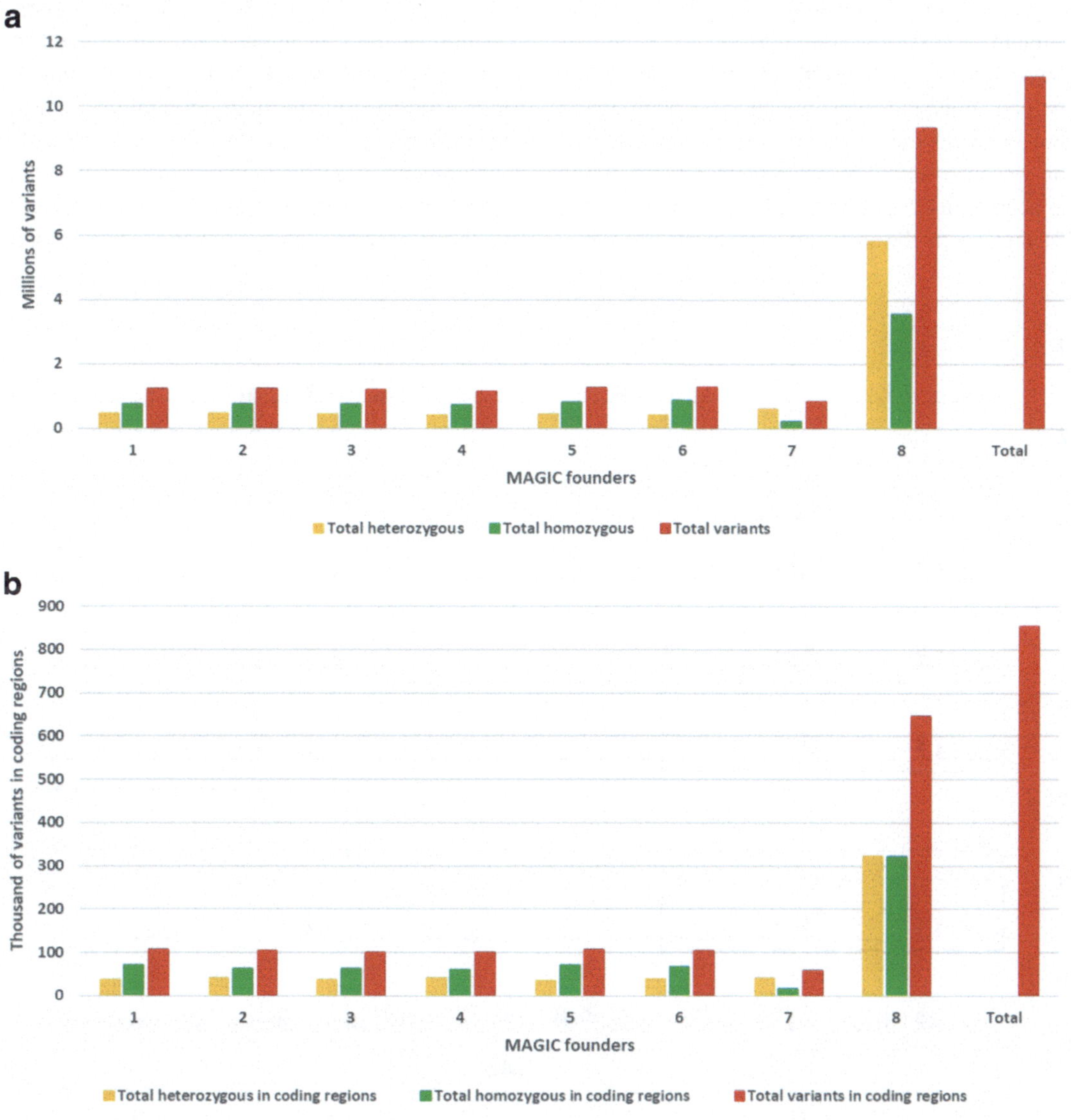

Fig. 10.2 **a** Polymorphisms identified among the eight eggplant MAGIC founders. The codes 1–7 correspond to *S. melongena* accessions while the code 8 to *S. incanum*. **b** Variants identified in coding regions among the eight MAGIC founders

identification of similar patterns of variations, the polymorphisms detected in the eight eggplant accessions subjected to WGRS were divided into 10 Mbp-sized bins. Using this approach, Aflitos et al. (2014) identified potential interspecific introgressions with CWRs in 54 tomato accessions compared to Heinz reference genome. Among our accessions, several potential introgressions were detected although not in all chromosomes. Figure 10.3 highlighted similar patterns of homozygous variations between the accessions for chromosome 6. The common eggplant accessions coded as 3, 4, 5, and 7 shared the same polymorphism distribution in the first part of the chromosome up to 25 Mbp, while until 60 Mbp the same polymorphism distribution was also shared by accessions 3 and 4, while the accessions 2 and 6 shared the same variation distribution in the last part of the chromosome. All the potential introgressions detected among the eight accessions may confirm the hypothesis about the domestication events of the eggplant from Africa to Asia and subsequently to Europe brought by Arab traders in the fourteenth century (Meyer et al. 2012). Furthermore, the set of polymorphisms identified was comprehensively structurally and functionally annotated.

The genomic information and variants identified in this first WGRS in eggplant, as well as the SPET platform developed, are being extremely valuable for breeders for the assisted selection of traits of interest and for researchers to unravel phylogenetic relationships and to foster conservation through integrative approaches (Gramazio et al. 2020b). This will foster the development of resilient eggplant varieties that could better face the present and future threats, like those posed by climate change. In addition,

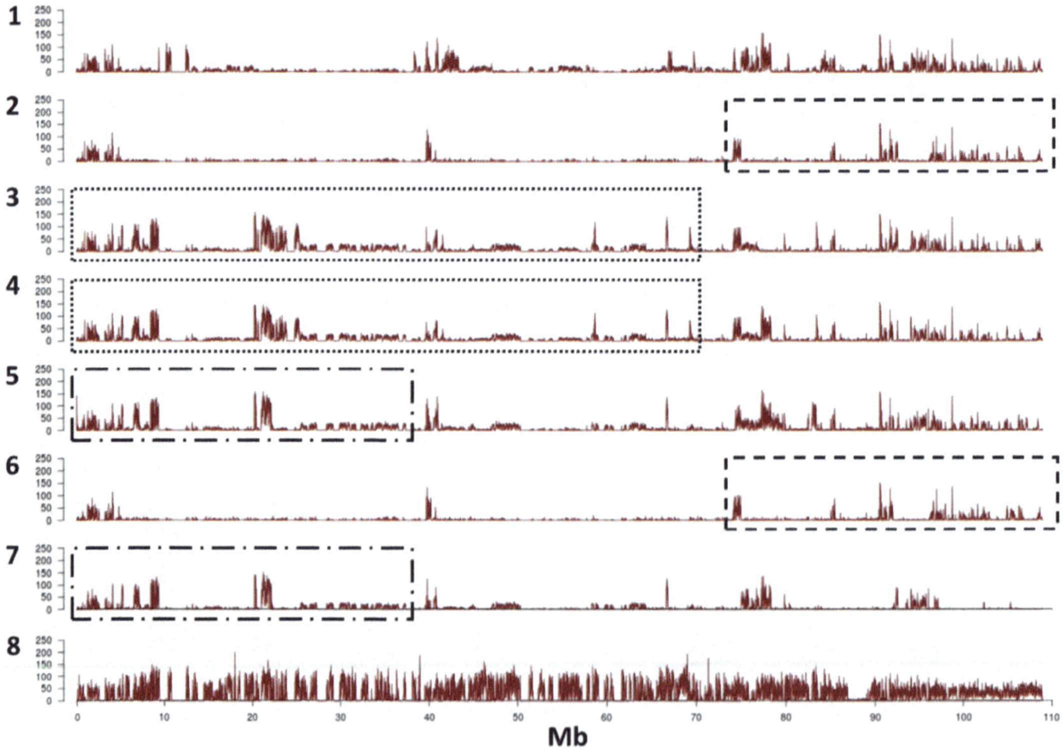

Fig. 10.3 a Polymorphisms distribution among the eight eggplant MAGIC population founders divided into 10 Mbp-sized bins (in red) across the chromosome 6. The x-axis represents the chromosome 6 length in Mb and y-axis the homozygous SNPs identified. The dashed line boxes indicate the similar SNP distribution pattern. The codes 1–7 correspond to *S. melongena* accessions while the code 8 to *S. incanum*

this study could stimulate further studies like fine-mapping of breeding traits or on domestication history in the eggplant genepool.

10.4 Conclusions

Next-generation sequencing platforms are allowing generating and providing an extraordinarily large amount of data unthinkable just a few years ago. While some years ago these technologies were economically available just for model plants and the most important economically cultivated crops, nowadays also non-model plants, wild relatives and small laboratories can benefit of them thanks to the technological improvements and the decreasing costs. In the eggplant genepool, although the importance of the three cultivated species in many geographical areas, genomic studies are still scarce and far compared to the other Solanaceae. In this book chapter, we reviewed the genomic data available in the eggplant wild genepool and introduced the first resequencing study of one wild relative (*S. incanum*) where the largest amount of polymorphisms in the eggplant genepool have been identified so far. This study could assist the development of new eggplant varieties more resilient to the current and future stresses and threats and with improved fruit quality along with shedding light on the genomic structure, gene functions, and breeding history. In addition, this study aimed at encouraging eggplant breeders to perform more genomic studies, especially in wild relatives, in light of the new high-quality eggplant reference genome recently available.

Acknowledgements This work was undertaken as part of the initiative "Adapting Agriculture to Climate Change: Collecting, Protecting and Preparing Crop Wild Relatives," which is supported by the Government of Norway. The project is managed by the Global Crop Diversity Trust with the Millennium Seed Bank of the Royal Botanic Gardens, Kew, and implemented in partnership with national and international gene banks and plant breeding institutes around the world. For further information see the project website: http://www.cwrdiversity.org/. Funding was also received from the European Union's Horizon 2020 Research and Innovation Programme under grant agreement No 677379 (G2P-SOL project: Linking genetic resources, genomes and phenotypes of Solanaceous crops) and from Spanish Ministerio de Economía, Industria y Competitividad and Fondo Europeo de Desarrollo Regional (grant AGL2015-64755-R from MINECO/FEDER). Pietro Gramazio is grateful to Japan Society for the Promotion of Science for a postdoctoral grant (P19015, FY2019 JSPS Postdoctoral Fellowship for Research in Japan [Standard]).

References

Acquadro A, Barchi L, Gramazio P, Portis E, Vilanova S, Comino C, Plazas M, Prohens J, Lanteri S (2017) Coding SNPs analysis highlights genetic relationships and evolution pattern in eggplant complexes. PLoS ONE 12(7):e0180774

Aflitos S, Schijlen E, Jong H, Ridder D, Smit S, Finkers R, Wang J, Zhang G, Li N, Mao L, Bakker F, Dirks R, Breit T, Gravendeel B, Huits H, Struss D, Swanson-Wagner R, Leeuwen H, van Ham RCHJ, Fito L, Guignier L, Sevilla M, Ellul P, Ganko E, Kapur A, Reclus E, Geus B, van de Geest H, Hekkert B, van Haarst J, Smits L, Koops A, Sanchez-Perez G, van Heusden AW, Visser R, Quan Z, Min J, Liao L, Wang X, Wang G, Yue Z, Yang X, Xu N, Schranz E, Smets E, Vos R, Rauwerda J, Ursem R, Schuit C, Kerns M, van den Berg J, Vriezen W, Janssen A, Datema E, Jahrman T, Moquet F, Bonnet J, Peters S (2014) Exploring genetic variation in the tomato (*Solanum* section *Lycopersicon*) clade by whole-genome sequencing. Plant J 80:136–148

Afonso TF, Demarco CF, Pieniz S, Camargo FA, Quadro MS, Andreazza R (2019) Potential of *Solanum viarum* Dunal in use for phytoremediation of heavy metals to mining areas, southern Brazil. Environ Sci Pollut Res Int 26:24132–24142

Alonge M, Wang X, Benoit M, Soyk S, Pereira L, Zhang L, Suresh H, Ramakrishnan S, Maumus F, Ciren D, Levy Y, Harel TH, Shalev-Schlosser G, Amsellem Z, Razifard H, Al C, Tieman DM, Klee H, Kirsche M, Aganezov S, Ranallo-Benavidez T, Lemmon ZH, Kim J, Robitaille G, Kramer M, Goodwin S, McCombie WR, Hutton S, Van Eck J, Gillis J, Eshed Y, Sedlazeck FJ, van der Knaap E, Schatz MC, Lippman ZB (2020) Major impacts of widespread structural variation on gene expression and crop improvement in tomato. Cell 182:145–161

Arrones A, Vilanova S, Plazas M, Mangino G, Pascual L, Díez MJ, Prohens J, Gramazio P (2020) The dawn of the age of multi-parent MAGIC populations in plant breeding: novel powerful next-generation resources for genetic analysis and selection of recombinant elite material. Biology 9:229

Barchi L, Delledonne M, Lanteri S, Dal Molin A, Minio A, Ferrarini A, Venturini L, Avanzato C, Toppino L, Sala T, Bassolino L, Comino C, Acquadro A, Portis E, Rinaldi R, Scaglione D,

Francese G, D'Alessandro A, Mennella G, Perrone D, Acciarri N, Pietrella M, Aprea G, Sulli M, Giuliano G, Rotino GL (2019) A chromosome-anchored eggplant genome sequence reveals key events in Solanaceae evolution. Sci Rep 9:11769

Barchi L, Acquadro A, Alonso D, Aprea G, Bassolino L, Demurtas O, Ferrante P, Gramazio P, Mini P, Portis E, Scaglione D, Toppino L, Vilanova S, Díez MJ, Rotino GL, Lanteri S, Prohens J, Giuliano G (2019) Single primer enrichment technology (SPET) for high-throughput genotyping in tomato and eggplant germplasm. Front Plant Sci 10:1005

Bolger A, Scossa F, Bolger ME, Lanz C, Maumus F, Tohge T, Quesneville H, Alseekh S, Sørensen I, Lichtenstein G, Fich EA, Conte M, Keller H, Schneeberger K, Schwacke R, Ofner I, Vrebalov J, Xu Y, Sonia O, Aflitos SA, Schijlen E, Jiménez-Goméz JM, Ryngajllo M, Kimura S, Kumar R, Koenig D, Headland LR, Maloof JN, Sinha N, van Ham RCHJ, Lankhorst RK, Mao L, Vogel A, Arsova B, Panstruga R, Fei Z, Rose JKC, Zamir D, Carrari F, Giovannoni JJ, Weigel D, Usadel B, Fernie AR (2014) The genome of the stress-tolerant wild tomato species *Solanum pennellii*. Nat Genet 46:1034–1038

Borua PK (1990) Failure in an interspecific cross between *Solanum khasianum* Clarke and *Solanum mammosum* L. Euphytica 46:1–6

Cappelli C, Stravato VM, Rotino GL, Buonaurio R (1995) Sources of resistance among *Solanum* spp. to an Italian isolate of *Fusarium oxysporum* f. sp. *melongenae*. In: Proceedings 9th Eucarpia meeting on genetics and breeding of capsicum & eggplant, pp 221–224

Coile NC (1993). Tropical soda apple, *Solanum viarum* Dunal: The plant from hell. Botany circular 27

Collonnier C, Mulya K, Fock I, Mariska I, Servaes A, Vedel F, Siljak-Yakovlev S, Souvannavong V, Ducreux G, Sihachakr D (2001) Source of resistance against *Ralstonia solanacearum* in fertile somatic hybrids of eggplant (*Solanum melongena* L.) with *Solanum aethiopicum* L. Plant Sci 160:301–313

Dandurand LM, Knudsen GR (2016) Effect of the trap crop *Solanum sisymbriifolium* and two biocontrol fungi on reproduction of the potato cyst nematode, *Globodera pallida*. Ann Appl Biol 169:180–189

Daunay M, Hazra P (2012) Eggplant. In: Peter K, Hazra P (eds) Handbook of vegetables. Studium Press LLC, Houston, TX, USA, pp 257–322

Dempewolf H, Eastwood RJ, Guarino L, Khoury CK, Müller JV, Toll J (2014) Adapting agriculture to climate change: a global initiative to collect, conserve, and use crop wild relatives. Agroecol Sustain Food Syst 38:369–377

Dempewolf H, Baute G, Anderson J, Kilian B, Smith C, Guarino L (2017) Past and future use of wild relatives in crop breeding. Crop Sci 57:1070–1082

Eskandari M, Abdi Fouladkolaei N (2020) *Solanum viarum* (Solanaceae), a new invasive plant for Iran. Rostaniha 21:299–302

Faostat (2019) http://www.fao.org/faostat. Accessed 28 Feb 2021

Fita A, Rodríguez-Burruezo A, Boscaiu M, Prohens J, Vicente O (2015) Breeding and domesticating crops adapted to drought and salinity: a new paradigm for increasing food production. Front Plant Sci 6:978

García-Fortea E, Gramazio P, Vilanova S, Fita A, Mangino G, Villanueva G, Arrones A, Knapp S, Prohens J, Plazas M (2019) First successful backcrossing towards eggplant (*Solanum melongena*) of a New World species, the silverleaf nightshade (*S. elaeagnifolium*), and characterization of interspecific hybrids and backcrosses. Sci Hortic 246:563–573

Gao L, Gonda I, Sun H, Ma Q, Bao K, Tieman DM, Burzynski-Chang EA, Fish TL, Stromberg KA, Sacks GL, Thannhauser TW, Foolad MR, Diez MJ, Blanca J, Canizares J, Xu Y, van der Knaap E, Huang S, Klee HJ, Giovannoni FZ (2019) The tomato pan-genome uncovers new genes and a rare allele regulating fruit flavor. Nat Genet 51:1044–1051

Gisbert C, Prohens J, Raigón MD, Stommel JR, Nuez F (2011) Eggplant relatives as sources of variation for developing new rootstocks: effects of grafting on eggplant yield and fruit apparent quality and composition. Sci Hortic 28:14–22

Goodwin S, McPherson JD, McCombie WR (2016) Coming of age: ten years of next-generation sequencing technologies. Nat Rev Genet 17:333–351

Gousset C, Collonnier C, Mulya K, Mariska I, Rotino GL, Besse P, Servaes A, Sihachakr D (2005) *Solanum torvum*, as a useful source of resistance against bacterial and fungal diseases for improvement of eggplant (*S. melongena* L.). Plant Sci 168:319–327

Gramazio P, Prohens J, Plazas M, Andújar I, Herraiz FJ, Castillo E, Knapp S, Meyer RS, Vilanova S (2014) Location of chlorogenic acid biosynthesis pathway and polyphenol oxidase genes in a new interspecific anchored linkage map of eggplant. BMC Plant Biol 14:350

Gramazio P, Blanca J, Ziarsolo P, Herraiz FJ, Plazas M, Prohens J, Vilanova S (2016) Transcriptome analysis and molecular marker discovery in *Solanum incanum* and *S. aethiopicum*, two close relatives of the common eggplant (*Solanum melongena*) with interest for breeding. BMC Genomics 17:300

Gramazio P, Prohens J, Plazas M, Mangino G, Herraiz FJ, Vilanova S (2017) Development and genetic characterization of advanced backcross materials and an introgression line population of *Solanum incanum* in a *S. melongena* background. Front Plant Sci 8:1477

Gramazio P, Prohens J, Plazas M, Mangino G, Herraiz FJ, García-Fortea E, Vilanova S (2018) Genomic tools for the enhancement of vegetable crops: a case in eggplant. Not Bot Horti Agrobot Cluj-Napoca 46 (1):1–13

Gramazio P, Yan H, Hasing T, Vilanova S, Prohens J, Bombarely A (2019). Whole-genome resequencing of seven eggplant (*Solanum melongena*) and one wild relative (*S. incanum*) accessions provides new insights and breeding tools for eggplant enhancement. Front Plant Sci 10:1220

Gramazio P, Pereira-Dias L, Vilanova S, Prohens J, Soler S, Esteras J, Garmendia A, Díez MJ (2020a)

Morphoagronomic characterization and whole-genome resequencing of eight highly diverse wild and weedy *S. pimpinellifolium* and *S. lycopersicum* var. cerasiforme accessions used for the first interspecific tomato MAGIC population. Hortic Res 7:1–6

Gramazio P, Jaén-Molina R, Vilanova S, Prohens J, Marrero Á, Caujapé-Castells J, Anderson GJ (2020b) Fostering conservation via an integrated use of conventional approaches and high-throughput SPET genotyping: A case study using the endangered Canarian endemics *Solanum lidii* and *S. vespertilio* (Solanaceae). Front Plant Sci 11:757

Handique AK (1986) Breeding behaviour of *Solanum khasianum* Clarke. Euphytica 35:631–632

Hirakawa H, Shirasawa K, Miyatake K, Nunome T, Negoro S, Ohyama A, Yamaguchi H, Sato S, Isobe S, Tabata S, Fukuoka H (2014) Draft genome sequence of eggplant (*Solanum melongena* L.): the representative *Solanum* species indigenous to the old world. DNA Res 21:649–660

Hulme PE (2017) Climate change and biological invasions: evidence, expectations, and response options. Biol Rev 92:1297–1313

Jaiswal BS (2012) *Solanum torvum*: A review of its traditional uses, phytochemistry and pharmacology. Int J Pharma Bio Sci 3:104–111

Khan AW, Garg V, Roorkiwal M, Golicz AA, Edwards D, Varshney RK (2020) Super-pangenome by integrating the wild side of a species for accelerated crop improvement. Trends Plant Sci 25:148–158

Knapp S, Vorontsova MS, Prohens J (2013) Wild relatives of the eggplant (*Solanum melongena* L.: Solanaceae): new understanding of species names in a complex group. PLoS One 8:e57039

Kouassi B, Prohens J, Gramazio P, Kouassi AB, Vilanova S, Galán-Ávila A, Herraiz FJ, Kouassi A, Seguí-Simarro JM, Plazas M (2016) Development of backcross generations and new interspecific hybrid combinations for introgression breeding in eggplant (*Solanum melongena*). Sci Hortic 213:199–207

Li D, Qian J, Li W, Yu N, Gan G, Jiang Y, Li W, Liang X, Chen R, Mo Y, Lian J (2021) A high-quality genome assembly of the eggplant provides insights into the molecular basis of disease resistance and chlorogenic acid synthesis. Mol Ecol Resour. https://doi.org/10.1111/1755-0998.13321

Li FW, Harkess A (2018) A guide to sequence your favorite plant genomes. Appl Plant Sci 6:e1030

Lin T, Zhu G, Zhang J, Xu X, Yu Q, Zheng Z, Zhang Z, Lun Y, Li S, Wang X, Huang Z, Li J, Zhang C, Wang T, Zhang Y, Wang A, Zhang Y, Lin K, Li C, Xiong G, Xue Y, Mazzucato A, Causse M, Fei Z, Giovannoni JJ, Chetelat RT, Zamir D, Städler T, Li J, Ye Z, Du Y, Huang S (2014) Genomic analyses provide insights into the history of tomato breeding. Nat Genet 46:1220–1226

Maundu P, Achigan-Dako E, Morimoto Y (2009) Biodiversity of African vegetables. In: Shackleton C, Pasquini M, Drescher A (eds) African Indig. Routledge, Veg. urban Agric, pp 65–104

Mennella G, Rotino GL, Fibiani M, D'Alessandro A, Franceses G, Toppino L, Cavallanti F, Acciarri N, Scalzo R (2010) Characterization of health-related compounds in eggplant (*Solanum Melongena* L.) lines derived from introgression of allied species. J Agric Food Chem 58:7597–7603

Meyer RS, Karol KG, Little DP, Nee MH, Litt A (2012) Phylogeographic relationships among Asian eggplants and new perspectives on eggplant domestication. Mol Phylogenet Evol 63:685–701

Meyer RS, Whitaker BD, Little DP, Wu SB, Kennelly EJ, Long CL, Litt A (2015) Parallel reductions in phenolic constituents resulting from the domestication of eggplant. Phytochemistry 115:194–206

Pandey S, Goel R, Bhardwaj A, Asif MH, Sawant SV, Misra P (2018). Transcriptome analysis provides insight into prickle development and its link to defense and secondary metabolism in *Solanum viarum* Dunal. Sci Rep 8(1):1–2

Plazas M, Vilanova S, Gramazio P, Rodríguez-Burruezo A, Fita A, Herraiz FJ, Ranil R, Fonseka R, Niran L, Fonseka H, Kouassi B, Kouassi A, Kouassi A, Prohens J (2016) Interspecific hybridization between eggplant and wild relatives from different genepools. J Am Soc Hortic Sci 141:34–44

Powell AF, Courtney LE, Schmidt MH, Feder A, Vogel A, Xu Y, Lyon DA, Dumschott KE, McHale M, Suplice R, Bao K, Duhan A, Hallab A, Denton AK, Mueller LA, Alseekh S, Lie J, Martin C, Fernie AR, Hind SR, Martin GB, Fei Z, Giovannoni JJ, Strickler SR, Usadel B (2020) A *Solanum lycopersicoides* reference genome facilitates biological discovery in tomato. BioRxiv. https://doi.org/10.1101/2020.04.16.039636

Prohens J, Plazas M, Raigón MD, Seguí-Simarro JM, Stommel JR, Vilanova S (2012) Characterization of interspecific hybrids and first backcross generations from crosses between two cultivated eggplants (*Solanum melongena* and *S. aethiopicum* Kumba group) and implications for eggplant breeding. Euphytica 186:517–538

Prohens J, Whitaker BD, Plazas M, Vilanova S, Hurtado M, Blasco M, Gramazio P, Stommel JR (2013) Genetic diversity in morphological characters and phenolic acids content resulting from an interspecific cross between eggplant, *Solanum melongena*, and its wild ancestor (*S. incanum*). Ann Appl Biol 162:242–257

Prohens J, Gramazio P, Plazas M, Dempewolf H, Kilian B, Díez MJ, Fita A, Herraiz FJ, Rodríguez-Burruezo A, Soler S, Knapp S, Vilanova S (2017) Introgressiomics: a new approach for using crop wild relatives in breeding for adaptation to climate change. Euphytica 213:158

Rattan P, Kumar S, Salgotra RK, Samnotra RK, Sharma F (2015) Development of interspecific F1hybrids (*Solanum melongena* × *Solanum khasianum*) in eggplant through embryo rescue technique. Plant Cell Tissue Organ Cult 120:379–386

Razali R, Bougouffa S, Morton MJL, Lightfoot DJ, Alam I, Essack M, Arold ST, Kamau AA, Schmöckel

SM, Pailles Y, Shahid M, Michell GT, Al-Babili S, Ho YS, Tester M, Bajic VB, Sónia Negrão S (2018) The genome sequence of the wild tomato *Solanum pimpinellifolium* provides insights into salinity tolerance. Front Plant Sci 2018:9

Rizza F, Mennella G, Collonnier C, Sihachakr D, Kashyap V, Rajam M, Prestera M, Rotino G (2002) Androgenic dihaploids from somatic hybrids between *Solanum melongena* and *S. aethiopicum* group gilo as a source of resistance to *Fusarium oxysporum* f. sp. *melongenae*. Plant Cell Rep 20:1022–1032

Rotino GL, Sala T, Toppino L (2014). Eggplant. In: Pratap A, Kumar J (eds) Alien gene transfer in crop plants, vol 2, Springer, NY, pp 381–409. https://doi.org/10.1007/978-1-4614-9572-7_16

Salas P, Prohens J, Seguí-Simarro JM (2011) Evaluation of androgenic competence through anther culture in common eggplant and related species. Euphytica 182:261–274

Schippers RR (2002) african indigenous vegetables, an overview of the cultivated species. University of Greenwich, Natural Resources Institute

Song B, Song Y, Fu Y, Kizito EB, Kamenya SN, Kabod PN, Liu H, Muthemba S, Kariba R, Njuguna J, Maina S, Stomeo F, Djikeng A, Hendre PS, Chen X, Chen W, Li X, Sun W, Wang S, Cheng S, Muchugi A, Jamnadass R, Shapiro HY, Van Deynze A, Yang H, Wang J, Xu X, Odeny DA, Liu X (2019) Draft genome sequence of *Solanum aethiopicum* provides insights into disease resistance, drought tolerance, and the evolution of the genome. GigaScience 8:giz115

Stommel JR, Whitaker BD (2003) Phenolic acid content and composition of eggplant fruit in a germplasm core subset. J Amer Soc Hort Sci 128:704–710

Sunseri F, Polignano GB, Alba V, Lotti C, Bisignano V, Mennella G, D'Alessandro A, Bacchi M, Riccardi P, Fiore MC, Ricciardi L (2010) Genetic diversity and characterization of African eggplant germplasm collection. African J Plant Sci 4:231–241

Syfert MM, Castañeda-Álvarez NP, Khoury CK, Särkinen T, Sosa CC, Achicanoy HA, Bernau V, Prohens J, Daunay M-C, Knapp S (2016) Crop wild relatives of the brinjal eggplant (*Solanum melongena*): poorly represented in genebanks and many species at risk of extinction. Am J Bot 103:635–651

Timmermans BGH (2005) *Solanum sisymbriifolium* (Lam.): a trap crop for potato cyst nematodes. PhD thesis, Wageningen University

Vorontsova M, Knapp S (2016) A revision of the "spiny *Solanums*" *Solanum* Subgenus *Leptostemonum* (Solanaceae), in Africa and Madagascar. Syst Bot Monogr 99:1–432

Wei Q, Wang J, Wang W, Hu T, Hu H, Bao C (2020) A high-quality chromosome-level genome assembly reveals genetics for important traits in eggplant. Horti Res 7:1–5

Wixom N, Casavant NC, Kuhl JC, Xiao F, Dandurand LM, Caplan AB (2018) Assessment of an organ-specific de novo transcriptome of the nematode trap-crop, *Solanum sisymbriifolium*. G3-Genes Genom Genet:g3–200327

Wu L, Du G, Bao R, Li Z, Gong Y, Liu F (2019) De novo assembly and discovery of genes involved in the response of *Solanum sisymbriifolium* to *Verticillium dahlia*. Physiol Mol Biol Plants 25:1009–1027

Yamaguchi H, Fukuoka H, Arao T, Ohyama A, Nunome T, Miyatake K, Negoro S (2010) Gene expression analysis in cadmium-stressed roots of a low cadmium-accumulating solanaceous plant, *Solanum torvum*. J Exp Bot 61:423–437

Yang X, Cheng YF, Deng C, Ma Y, Wang ZW, Chen XH, Xue LB (2014) Comparative transcriptome analysis of eggplant (*Solanum melongena* L.) and turkey berry (*Solanum torvum* Sw.): phylogenomics and disease resistance analysis. BMC Genom 15:412

Yang X, Liu F, Zhang Y, Wang L, Cheng Y (2017) Cold-responsive miRNAs and their target genes in the wild eggplant species *Solanum aculeatissimum*. BMC Genom 18:1000

Zhao Z, Tan L, Dang C, Zhang H, Wu Q, An L (2012) Deep-sequencing transcriptome analysis of chilling tolerance mechanisms of a subnival alpine plant. Chorispora Bungeana. BMC Plant Biol 12:222

Zhou Z, Jiang Y, Wang Z, Gou Z, Lyu J, Li W, Yu Y, Shu L, Zhao Y, Ma Y, Fang C, Shen Y, Liu T, Li C, Li Q, Wu M, Wang M, Wu Y, Dong Y, Wan W, Wang X, Ding Z, Gao Y, Xiang H, Zhu B, Lee S, Wang W, Tian Z (2015) Resequencing 302 wild and cultivated accessions identifies genes related to domestication and improvement in soybean. Nat Biotechnol 33:408–414

Zhou X, Bao S, Liu J, Zhuang Y (2016) De novo sequencing and analysis of the transcriptome of the wild eggplant species *Solanum aculeatissimum* in response to *Verticillium dahliae*. Plant Mol Biol Report 34:1193–1203

Zhou X, Bao S, Liu J, Yang Y, Zhuang Y (2018) Production and characterization of an amphidiploid derived from interspecific hybridization between *Solanum melongena* L. and *Solanum aculeatissimum* Jacq. Sci Hortic 230:102–106

Zhuang Y, Zhou X, Wang S (2012) Genetic diversity of NBS-LRR class disease-resistance gene analogs in cultivated and wild eggplants. Plant Syst Evol 298:1399–1406

11 Transposable Elements and Genome Expansion in Cultivated and Wild Potato and Tomato Species

M. Gantuz, C. F. Marfil, and R. W. Masuelli

Abstract

The sequence of potato and tomato genomes showed that repetitive sequences constitute high proportions of it. Transposable elements (TEs) and among them long terminal repeats (LTR) retrotransposons are the most abundant TEs in *Solanum* genomes. Biotic and abiotic stress induces the mobilization of TEs, possibly by the relaxation of epigenetic mechanisms, which alters gene expression due to the insertion of a TE inside or close to genes and generates novel phenotypes. This chapter highlights the importance of TEs as a source of phenotypic variation important to evolution and potato and tomato breeding.

11.1 Introduction

Solanum is a large and diverse genus of flowering plants, which include crops of high economic importance, like potato and tomato. Research in this genus has produced comparative genomic resources in crop and wild relative species that bring useful data for agronomic improvement. Crop species like potato (*Solanum tuberosum* L.) and tomato (*Solanum lycopersicum* L.) were first sequenced (The Potato Genome Sequencing Consortium 2011; The Tomato Genome Consortium 2012), followed by wild relative species. Among them, the genome of the wild tomatoes *S. pimpinellifolium* L. and *Solanum pennellii* Correll, endemic to Andean regions in South America (Bolger et al. 2014) and the wild potatoes *Solanum commersonii* Dunal and *Solanum chacoense* Bitter, native to South America (Argentina, Brazil and Paraguay) (Aversano et al. 2015; Leisner et al. 2018) were assembled (see details in Chaps. 7, 8, and 9). Sequence analysis in the above-mentioned taxa showed that, as in other plant species, nearly half or more of these genomes are constitute of repetitive elements like (i) interspersed repeats, which include TEs and (ii) tandem repeats, grouping satellite, minisatellite, and microsatellite repeats (Grover and Sharma 2017) (Table 11.1). TEs represent the major proportion of the repetitive elements in *Solanum* genomes. Particularly, we found a positive correlation between genome size and repeat content when all tomato and potato species reviewed in this chapter were analyzed ($r = 0.79$; Fig. 11.1). This result seems to be consistent with observations made in several plant genomes, and it is despite the fact that most genome sequencing projects fail to assemble a considerable proportion of the genome due to high copy number repeats that are hard (or impossible) to assemble with current sequencing technologies (Michael 2014).

M. Gantuz · C. F. Marfil · R. W. Masuelli (✉)
Facultad de Ciencias Agrarias, Instituto de Biología Agrícola de Mendoza, CONICET-Universidad Nacional de Cuyo, Alte. Brown 500, M5507 Chacras de Coria, Mendoza, Argentina
e-mail: rmasuelli@fca.uncu.edu.ar

© Springer Nature Switzerland AG 2021

D. Carputo et al. (eds.), *The Wild Solanums Genomes*, Compendium of Plant Genomes,
https://doi.org/10.1007/978-3-030-30343-3_11

TEs are mobile genetic elements present in practically all plant species that can move through the genome and generate new copies. TEs were discovered by McClintock (1984), through her genetic and cytogenetic work on maize (*Zea mays*), for which she was awarded the Nobel Prize in Physiology or Medicine in 1983. This discovery introduced the notion of genome fluidity, constituting a major shift in our concepts of heredity (Grandbastien 2015). Originally, the TEs were assumed to be junk DNA, but now they are considered drivers in evolution (Lanciano and Mirouze 2018). TE amplification and polyploidization are considered the main mechanisms to increase plant genome size and, more generally, for plant genome evolution (Wendel et al. 2016). In fact, TE amplification and polyploidization may not be completely independent mechanisms, and it is believed that these two phenomena greatly influence one another, reinforcing their potential to drive plant genome evolution (Negi et al. 2016).

TEs are extremely diverse in terms of their types and life cycles (Fig. 11.2). It can be subdivided in two main classes: class 1 elements (or retrotransposons) that move by copy and paste mechanisms and has an RNA intermediate; class 2 (or DNA transposons) that move by cut and paste with DNA as intermediate. These two classes are themselves divided into subclasses, each species containing a specific number of families representing the subclasses. At the same time, they can also be classified into autonomous and not autonomous (for a complete review of the TE classification system see Wicker et al. 2007).

TEs activation and their implication in species adaptation is an active field of study. It was reported that the activation can be induced by various stresses, extensible reviewed by Negi et al. (2016). As a consequence of the "genome shock", proposed by McClintock, the massive activation of TEs can bring about large-scale chromosomal rearrangements which might collectively shape the organization and expression of the genome and facilitate adaptive evolution (McClintock 1984; Sanmiguel and Bennetzen 1998; Du et al. 2009), as has been demonstrated for many plant genomes including *Solanum* species like tomato and pepper (Bolger et al. 2014; Qin et al. 2014; Kim et al. 2017). Also, biotic and abiotic stresses could trigger TEs activation, for example: salt (Naito et al. 2009; Woodrow et al. 2010), wounding (Mhiri et al.

Table 11.1 Representation of repetitive sequences in the potato and tomato genomes

Solanum species	Genotype	Genome size (Mb)	Total length repetitive sequences (Mb)	Proportion of the genome (%)	References
Potatoes					
S. tuberosum	DM1-3	727	452.4	62.2	The Potato Genome Sequencing Consortium (2011)
S. commersonii	PI243503	838	383	44.5	Aversano et al. (2015)
S. chacoense	M6	822	501	56.8	Leisner et al. (2018)
Tomatoes					
S. lycopersicum	Heinz 1706	760	466	61.3	The Tomato Genome Consortium (2012)
S. pimpinellifolium	LA1589	739	282	38.1	The Tomato Genome Consortium (2012)
S. pimpinellifolium	LA0480	811	482	59.5	Razali et al. (2018)
S. pennellii	LA0716	942	743	78.8	Bolger et al. (2014)

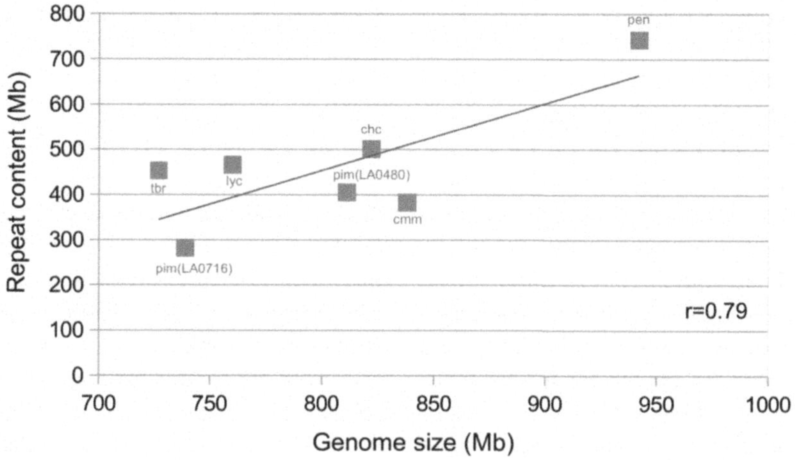

Fig. 11.1 Correlation between genome size and repeat content in potato and tomato species. Data from *Solanum tuberosum* (tbr), *S. commersonii* (cmm), *S. chacoense* (chc), *S. lycopersicum* (lyc), *S. penellii* (pen), and *S. pimpinellifolium* (pim) were taken from reference in Table 11.1. Pearson correlation coefficient (r) was performed

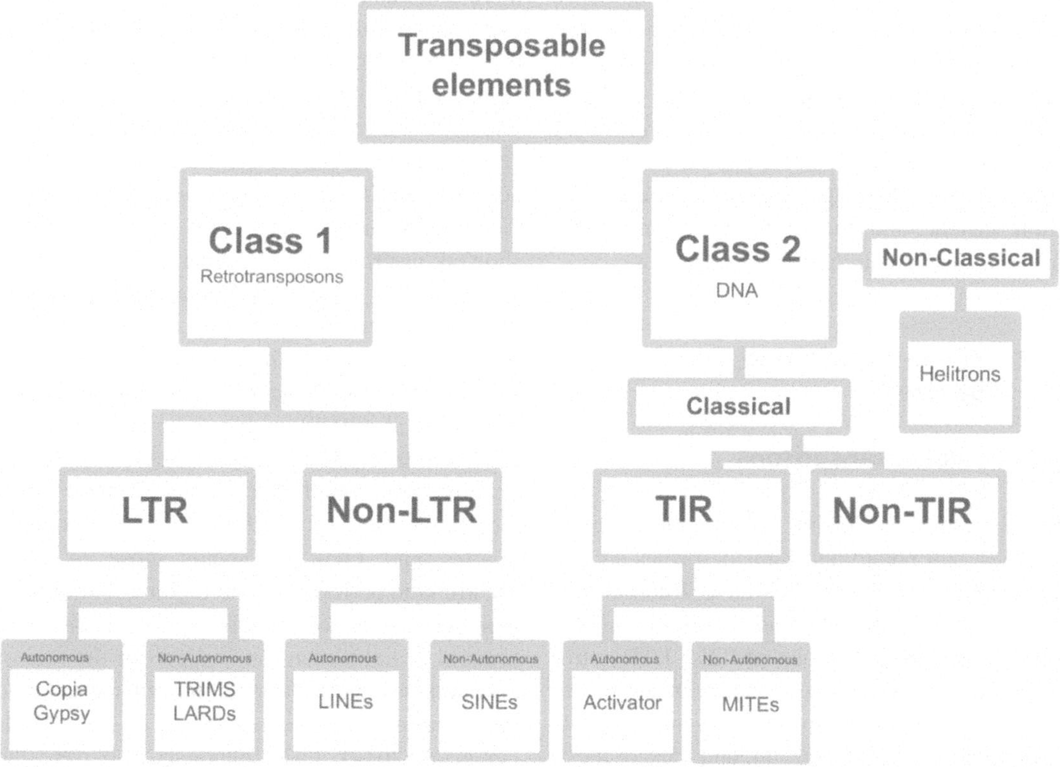

Fig. 11.2 Classification of plant transposable elements (TEs). TEs are divided into two major classes based in their life cycle: retrotransposons (class 1) and DNA transposons (class 2). Further classification is based on the classical /non-classical mode of transposition, and the presence/absence of long terminal repeat (LTR) and terminal inverted repeats (TIR) sequences. Subsequently are classified into autonomous and not autonomous. Modified from Negi et al. (2016)

1997), temperature (Ivashuta et al. 2002; Naito et al. 2009; Ito et al. 2011; Ishiguro et al. 2014), and pathogens (Grandbastien et al. 2005; Buchmann et al. 2009; Anca et al. 2014).

Besides their direct effect on genome size, TEs could affect gene structure and function. They can disrupt functional genes, affect chromatin status and expression of neighboring genes, and facilitate homology-driven non-allelic recombination leading to genomic rearrangements (reviewed by Lisch 2013). They are commonly silenced by epigenetic mechanisms, which are specific to subclasses and families of TEs. As TEs propagate in plant genomes and attract epigenetic marks, their neo-insertions can lead to the formation of new heritable epigenetic variants (epialleles) of genes and impact on gene regulatory networks. The epigenetic interplay between TEs and genes thus exerts a crucial role in the evolution of the genome (Lisch 2009; Bucher et al. 2012). So TEs contribute to genetic variability and play an important role in gene and genome evolution, although it is not yet clear whether TE activity is the cause or effect of speciation events. In this chapter, we analyze the importance of TEs in genome architecture, evolution, genetic, and phenotypic variation in wild and cultivated potato and tomato species.

11.2 TEs in Potato and Tomato Species

11.2.1 Retrotransposons

TEs represent the major component of plant genomes, and potato and tomato species are not an exception. Plant genomes have different proportion in TE contents and family members. Among the different types of TEs, LTR retrotransposons are the most abundant in the potato and tomato genomes so far sequenced: LTRs accounts for 29, 34, 43, 62, 44, and 45% of the *S. tuberosum*, *S. commersonii*, *S. chacoense*, *S. lycopersicum, S. penellii, and S. pimpinellifolium* genomes, respectively (The Potato Genome Sequencing Consortium 2011; The Tomato Genome Consortium 2012; Bolger et al. 2014; Aversano et al. 2015; Leisner et al. 2018) (Fig. 11.3). For example, the assembled potato genome, generated from the doubled monoploid *S. tuberosum* Group *Phureja* DM1-3, represents 727 Mb containing 39,031 annotated genes and 62.2% of the genome correspond to repetitive elements and LTRs comprising the majority of the TE classes, representing 29.4% of the genome (The Potato Genome Sequencing Consortium 2011) (Fig. 11.3). These proportions were similar in the wild potato *S. commersonii*, but they were more abundant in the *S. chacoense* genome (Aversano et al. 2015; Leisner et al. 2018). LTR retrotransposons are flanked by LTR sequences and are divided in Copia and Gypsy according to the coding sequences of capsid protein; PR protease; integrase; reverse transcriptase, and ribonuclease H genes (Parisod et al 2010). In particular, Gypsy superfamily is the most abundant. Comparing tomato and potato, they have a similar differential age distribution of the two LTR retrotransposon superfamilies, Gypsy and Copia. Copia elements are younger with a downward slope toward older insertions. Gypsy elements are older, with past maximum activities and few recent insertions (The Tomato Genome Consortium 2012). Recent articles showed insertion time estimations of approximately 2 and 4 Mya for Copia and Gypsy elements, respectively, for tomato (Paz et al. 2017), potato, and *S. commesonii* (Esposito et al. 2019; Zavallo et al. 2020) genomes. In a recent work, Zavallo et al. (2020) analyzed the distribution of LTR retrotransposon superfamilies across potato genome and found that Copia elements were located close to euchromatic subtelomeric regions and Gypsy elements close to heterochromatic pericentromeric regions. The whole-genome analysis of the wild tomato *S. pennellii* brought to light interesting aspects of the role of retrotransposons in the evolution of the genome architecture and genes associated to stress (Bolger et al. 2014). LTRs elements also were the most abundant repeats, comprising ~45% of the *S. pennellii* genome assembly. As in *S. lycopersicum* and other *Solanum* species, the genome of *S. pennellii* presented many more Gypsy than Copia LTRs. Since LTR retrotransposons could have an important role in genome size variation, it was proposed that the difference observed on genome size between *S. lycopersicum and S. pennellii,* representing 355 Mb of the 781 Mb and 428 Mb of 942 Mb, respectively, could be to a specific burst of LTR retrotransposons (Bolger et al. 2014). In fact, the analysis of insertion age of LTRs showed that *S. pennellii* has a higher abundance of young LTRs than *S. lycopersicum,* which is especially pronounced for Copia elements. These results point toward different genome dynamics in these two species since their separation from a common ancestor, probably driven by LTR amplification. At the same time, the authors observed a differential distribution of Gypsy and Copia-like elements, and a significant association between stress-related genes and Copia-like retrotransposons in *S. pennellii* and *S. lycopersicum* (Bolger et al. 2014). Gaiero et al. (2018) performed a global comparative analysis on repetitive genome fractions of wild and cultivated potato and tomato species and found a high correlation between repeat proportion and genome size differences. Also, the authors reported difference in the frequency of specific groups of LTRs (Gaiero et al. 2018). Consistently with previous reports, Gypsy elements were the most abundant, in particular the Chromovirus lineage. Although these elements are highly prolific in all species, they show significant variation in abundance, with some species having as much as twice the relative content as the others. In the case of tomato and its relatives, Jinling was the most frequently observed element, which was almost undetectable in the potato clade (Gaiero et al. 2018). Most Jinling elements were located in heterochromatin regions where the gene density is low, were highly methylated and rarely transcribed. Nonetheless, they have spread throughout the pericentromeric heterochromatin in cultivated and wild tomato species fairly recently-well after tomato diverged from potato and other related *Solanaceae* species (Wang et al. 2006).

The particular amplification of LTRs and its association with a plant-section or species point out that retroelements expansion within a genome is unique and depends on the evolutionary process within each plant species (Paz et al. 2017). Rider element in tomato is another example of specific amplification. Database searches and DNA blots using the LTR as probe indicate that the Rider element is present in all tomato species tested and absent from related species such as potato and tobacco (Cheng et al. 2009; Jiang et al. 2009). Rider is an autonomous Copia-like element. Unlike other LTR retrotransposons Rider element are located in both heterochromatic and euchromatic regions, so they are more likely to be surrounded by genes. These retrotransposons were also associated to phenotype changes related to fruit shape locus *SUN* (Xiao et al. 2008).

TEs can be a source of long noncoding RNAs (lncRNAs), molecules that regulate gene expression and biological processes. lncRNAs are present in all sequenced genomes of tomato and potato, but are diverse, only a small proportion (6.7%, 24 of 353) of them are considered to be conserved between tomato and potato (Wang et al. 2016). Sequence analysis of specific lncRNA loci in *S. lycopersicum* and *S. pennellii* showed that the origin of these molecule is associated with TEs (Wang et al. 2016). For example, lncRNA-314, a fruit-specific lncRNA expressed in *S. lycopersicum* and S. *pimpinellifolium,* but not in *S. pennellii,* was originated through two evolutionary events: speciation of *S. pennellii* that resulted in the insertion of an LTR retrotransposon into chromosome 10 and contributed to most of the transcribed region of lncRNA-314; a large deletion in *Lycopersicon* that generated the promoter region and part of the transcribed region of lncRNA-314. In this work, Wang et al. (2016) proposed that the genomic structural variations resulting from TE, are primary factors for the origin of lncRNAs in tomato.

SINEs are non-autonomous class 1 retrotrasposons (Fig. 11.2). In *Solanaceae,* an evolutionary model of the emergence of SINEs based on the structural relationship to LINEs has been described. *Solanaceae* SINEs are dispersed along chromosomes and distributed without clustering but with preferred integration into short A-rich motifs. They emerged more than 23 million years ago and were species-specific amplified during

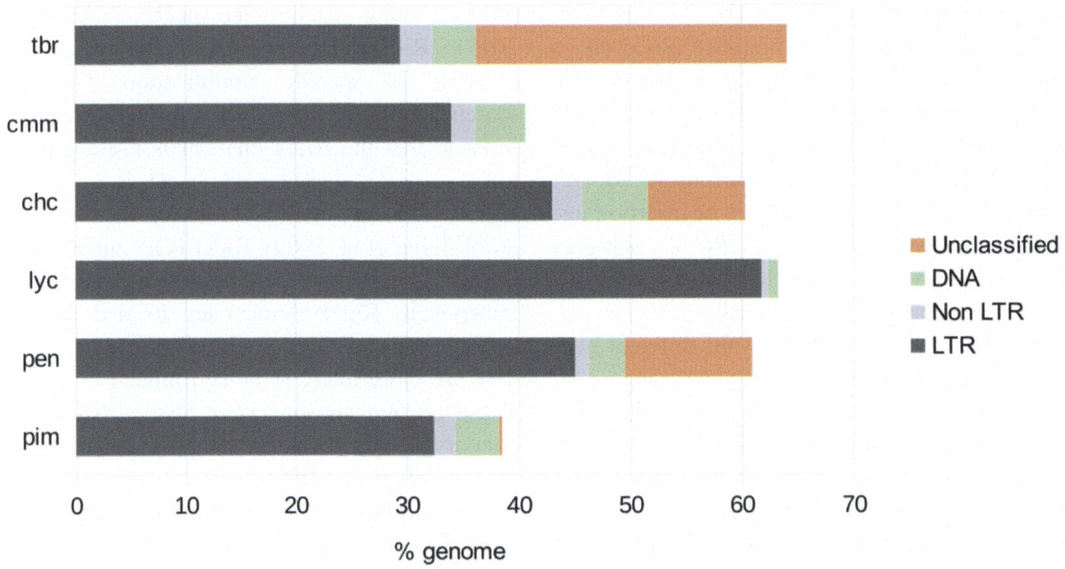

Fig. 11.3 Percentage of transposable elements in the potato and tomato genomes. Data for *S. tuberosum* (tbr), *S. commersonii* (cmm), *S. chacoense* (chc), S. *lycopersicum* (lyc), *S. pennellii* (pen), and *S. pimpinellifolium* (pim) were taken from references in Table 11.1. DNA corresponds to all Class 2 TEs. Unclassified proportions were not available for lyc and cmm

the radiation of potato, tomato, and tobacco (*Nicotiana tabacum* L.) (Wenke et al. 2011). The study of *Solanaceae* SINEs pointed out the importance of these elements on the structural variation of genes and genomes. Seibt et al. (2016) identified SINEs involved in post-insertional duplication leading to increasing UTR and intron lengths; providing splice sites, exons, start, and stop codons to genes; duplicated and forming tandem-like structures; and transducing flanking genomic regions upon transposition. In potato, SINEs contribute 0.32% to the genome, whereas in tomato and *S. pennellii* SINEs account for approximately 0.15%. Exclusively in potato, a strong amplification was reported for the families SolS-IIIa, IV, V, VI. This indicates most likely a recent amplification burst of these families limited to potato after its divergence from tomato, leading to high numbers of homogeneous copies (Seibt et al. 2016). Since genome shock is one factor triggering transpositional activity, the high SINE genome proportion and recent activity in potato might be caused by species hybridization, polyploidization, and introgression (Liu and Wendel 2000; Rodríguez et al. 2011). In comparison with *S. tuberosum*, *S. chacoense* presented a similar proportion of SINEs elements. In *S. commersonii,* characterization of SINE families allowed annotating 1925 SINEs with significant similarity to families previously described in *S. tuberosum* and in other *Solanaceae* (Aversano et al. 2015). The proportion of the genome also was similar to those observed in *S. tuberosum.*

11.2.2 DNA Transposons

Miniature inverted-repeat TEs (MITEs) are non-autonomous DNA transposon (Fig. 11.2). Generally, they are less than 600 bp in length and lack open reading frames, and their mobility depends on the activity in trans of transposases encoded by cognate full-length autonomous transposons (Feschotte et al. 2002). Based on their relations in sequences of the terminal inverted repeats (TIRs) and insertion site, MITEs can be classified in two major groups: Tourist-like MITEs (Zhang et al. 2004) and Stowaway-like MITEs (Bureau and Wessler 1994), subsequently they can be classified in superfamilies and families. These little DNA transposons predominate in plant genomes, the first report of MITE identification in *Solanum* species was performed by Kuang et al. (2009), who examined homeologous R-gene regions from different haplotypes of the same or closely related *Solanaceae* species and identified by small RNA sequencing 22 families of MITEs (MiS1–MiS22), including MITE-derived siRNAs. The analysis of whole-genome analysis in tomato and potato describes 104 and 171 families, respectively, (P-MITE: a Pant MITE database, http://pmite.hzau.edu.cn/mite_family_info/), belonging to five superfamilies (Tc1/Mariner, Mutator, hAT, PIF/Harbinger and CACTA). These two species share number of families but differ in proportion: in potato, Mutator and hAT superfamily are predominant, but in tomato, Mutator is more abundant. A specific study of potato Stowaway MITEs suggested that the chromatin status is associated with H3K4me3-modified nucleosome and the presence of Stowaway transposons (Marand et al. 2017). In this work, Marand and co-workers (2017) detected a significantly enriched MITE within their crossover data set of two independent populations. This result may reflect the preference of crossovers and Stowaway elements to occur within open chromatin given the preference of Stowaway elements to insert in TA rich sequences depleted of nucleosomes. The analysis of TEs distribution in potato genome showed that MITEs presented a concentration pattern around gene-rich subtelomeric regions, and these results suggest that MITEs are involved in gene regulation (Zavallo et al. 2020).

11.3 Dynamic of TEs in the *Solanum* genome and Stress-Induced TE Mobilization

The first active retrotransposon Tnt1 was described in *N. tabacum* by Grandbastien and co-workers (1989). Since then, Tnt1-like sequences have been described in several *Solanaceae* species: Retrolyc I, Retrolyc I-I, TLC-1, and Retrosol in *S. lycopersicum*, *S. peruvianum*, *S. chilense*, and in several *Solanum* species from section *Petota*, respectively (Costa et al. 1999; Yañez et al. 1998; Manetti et al. 2007). The Tnt1 superfamily was present early in the evolution of *Solanaceae* and evolved mainly through the diversification of the U3 region (Manetti et al 2007). This region has similar motifs to genes involved in pathogen resistance and abiotic stress (Casacuberta and González 2013). It was proposed that retrotransposons may contribute to the origin of the inducible promoters of plant defense genes (Grandbastien et al. 1997; Takeda et al. 1999).

McClintock (1984) proposed that many stressful sources such as hybridization, among others, induce unprogrammed genome restructuring. A consequence of these can be the activation of TEs, and it is not possible to know when a particular TE will be activated. The insertion of TEs may induce changes in gene expression, insertional inactivation, or neofunctionalization (Vicient and Casacuberta 2017).

Hybridization and polyploidization are two important forces of flowering plant genome evolution and speciation. Both mechanisms have a profound impact on genome expression and architecture. Interspecific hybridization is a source of adaptive variation and functional novelties (Lewontin and Birch 1966) and the advantage of polyploids over diploids, such as higher yields and heterosis were analyzed by several authors (Leitch and Bennett 1997; Wendel 2000; Comai 2005). Wild and cultivated potato species constitute an euploid series with chromosome numbers ranging from $2n = 2x = 24$ to $2n = 6x = 72$; and homoploid hybridization was recognized as the main mechanism involved in the origin and evolution of the diploid potato species (Masuelli et al. 2009). Also, interploidal hybridization was acknowledged as an additional source of genetic and epigenetic variation in wild potato populations (Cara et al. 2013, 2019). Investigations in various polyploid systems showed that polyploidization, principally allopolyploidization, was associated with changes in genome structure, DNA methylation, TEs activation, and gene expression of low and high copy coding (Yaakov and Kashkush 2010).

The mobilization of TEs could be analyzed by sequence-specific amplified polymorphism (SSAP). This technique is a modification of the AFLP adapted to detect polymorphism caused by TE activity. Using SSAP technique with primers specific to LTR, activation of Tnt1 and Tto1 Copia retrotransposons were detected in interspecific hybrids between the wild potato species *S. kurtzianum* and *S. microdontum*. Both elements showed moderate mobilization (0.9–7.8%) depending on the parental genotypes, and there was a negative correlation between Tnt1 activation and pollen viability (Paz et al. 2015). There were differences in activation among different TE families and genotypes. Another study uses SSAP to detect 10 LTR in *S. tuberosum* ($2n = 2x = 24$) x *S. kurtzianum* ($2n = 2x = 24$) interspecific hybrid, activation of Copia and Gypsy retrotransposons was observed as a result of interspecific hybridization. In the same work, retrotransposons were not activated in allopolyploids of *S. tuberosum* x *S. kurtzianum* hybrid and autopolyploids of *S. kurtzianum* obtained by chemical treatment (Gantuz and Masuelli 2017) (Fig. 11.4). In plants of a triploid hybrid between *S. tubersoum* and *S. phureja* regenerated by leaf protoplast fusion, activation of Stowaway MITESs was detected causing the variation of skin color in potato tubers (Momose et al. 2010).

11.4 Epigenetic Control of TE Activation

TEs were considered as parasitic elements, due to their increase in copy number and potential for gene disruption (Slotkin and Martienssen 2007). However, most of the TEs are heavily methylated and silenced. In this sense, it was proposed that the principal role of DNA methylation is TEs silencing (Yoder et al. 1997). In potato and tomato DNA methylation levels were positively correlated with TE density and CG methylation is higher in tomato than in potato in both genes and TEs. Tomato has a higher methylation level at CG and CHG sites than potato, while CHH sites are more methylated in the potato genome (Wang et al. 2016). The activation of the LTRs is influenced by several stimuli such as pathogen infection, environmental stress, tissue culture, polyploidization, or hybridization (Grandbastien 2015). Activation of TEs was detected after polyploidization events in several species (Comai et al. 2000). The activation of the TEs induced by hybridization or polyploidization should be accompanied by changes in the methylation status that can release the transcriptional repression of the TEs (Parisod et al. 2010). Genome-wide DNA methylation status of 5'-CCGG-3' sites flanking transposons could be analyzed by transposon methylation display (TMD) technique (Kashkush and Khasdan 2007). This technique was used by Paz et al. (2015) to analyze changes in methylation status of interspecific hybrids between *S. kurtzianum* and *S. microdontum*, respect to parental genomes. They found that for Tnt1 and Tto1, there was a positive correlation between retrotransposon activity and hypomethylation in the hybrids.

Recently, the prevailing view considering the evolution of the epigenetic mechanisms as a defense against the TEs has been challenged in his favor as a driver of evolutionary innovation and phenotypic plasticity (Federoff 2012). New evidence showed that TE mobilization induces changes in gene expression and generate novel phenotypes, such as stress tolerance, fruit shape and color, among others (Lisch 2013; Negi et al. 2016).

11.5 TEs Induced Novel Phenotypes

TEs induce genotypic and phenotypic changes in plants. Barbara McClintock (1984), who conducted her pioneering experiments in maize, described the activation of TEs by the entrance of the newly ruptured end of a chromosome into a telophase nucleus. The newly activated elements were identified because, after its insertion, affected the regulations of a gene required for the anthocyanin production inducing variegated phenotypes: dots of red or purple pigment in the aleurone layer. Then, McClintock acknowledged that the genomic modification induced by TE

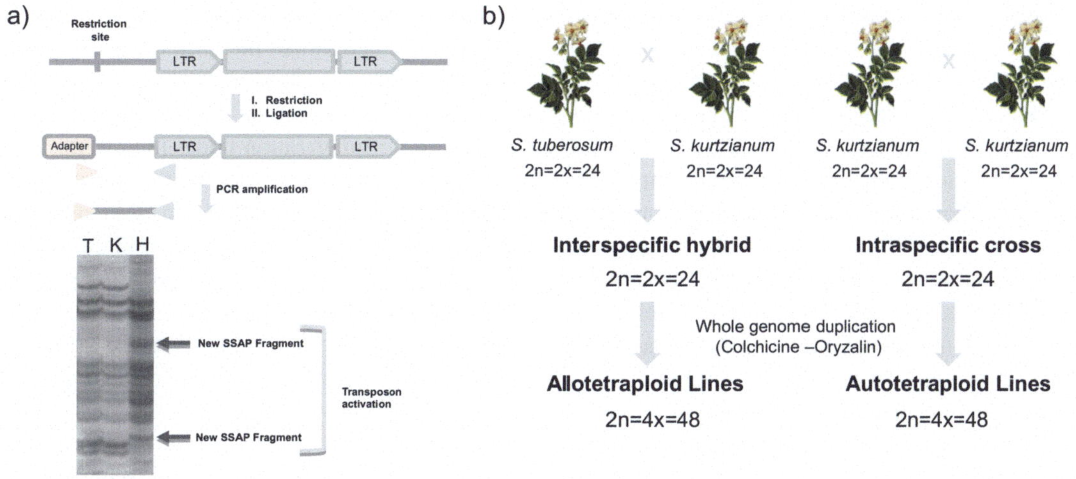

Fig. 11.4 Molecular tools and experimental models to study TE mobilization. **a** Determination of TEs activation (sequence-specific amplification polymorphisms, SSAP), briefly whole genome was digested with restriction enzymes, adapters were ligated to the generated fragments, and PCR amplification with specific LTR sequence was performed. Fragments were resolved in polyacrylamide denaturing gels. New SSAP fragments are new transposed events. **b** Experimental model, parental genotypes are dihaploid of *S. tuberosum* (T) and *S. kurtzianum* (K), the interspecific hybrid (H) was obtained by cross, allotetraploid, and autotetraploid lines were obtained by treatment with colchicine and oryzaline, respectively

could have a relevant role in plant diversification and be potent sources for selection by the plant breeders.

TEs have contributed to tomato diversity. *Solanum pennellii* is a wild relative that diverged from the cultivated tomato *S. lycopersicum* around 2.7 million years ago (Kamenetzky et al. 2010). By comparing the recent activity of LTR retrotransposon activity in these two tomato genomes, Bolger and co-workers (2014) demonstrated differential genome dynamics since their separation from a common ancestor: the wild tomato presented higher abundance of young LTR retrotransposon than the cultivated one. In addition, results obtained from the genome sequencing suggested that TE had a role in the evolution of stress tolerance in *S. pennellii* (Bolger et al. 2014). This wild tomato is an Andean species adapted to arid environments and characterized by its abiotic stress tolerance. Nonrandom association of drought and salt tolerance-related genes with TEs could explain the high stress tolerance of *S. pennellii*. The colocalization of stress-related genes with Copia and Gypsy elements along with the correlation between the distribution of Copia elements and gene expression in *S. pennellii* led authors to propose that LTR retrotransposon could regulate gene expression and that Copia elements might represent conditional regulatory sequences co-opted by the *S. pennellii* upon exposure to different environmental stresses (Bolger et al. 2014).

The outstanding role of retrotransposons on genome dynamics and evolution also includes the emergence of new genes. Sequence analysis of lncRNA loci in *S. lycopersicum*, *S. pimpinelifollium*, and *S. pennellii* reveals that TEs played important roles in the origin of these sequences in the tomato genomes (Wang et al. 2016). The authors found that several lncRNAs specific of the *Lycopersycon* group (*S. lycopersicum* and *S. pimpinellifolium*) obtained the promoter and transcribed region from TEs during tomato species divergence. The study also demonstrated that one TEs-derived lncRNAs (lncRNA-314) is tissue-specific and highly expressed in the ripening fruit of tomatoes only (Wang et al. 2016), indicating a potential role during tomato domestication.

During ancestral domestication and subsequent tomato improvement emerged a vast diversity in fruit shape and color (Paran and van der Knaap 2007). *SUN* is one of the major genes controlling the elongated fruit shape in tomato. *SUN* mutations arose after a gene duplication event mediated by the retrotransposons Rider (Xiao et al. 2008), and most of the long and oxheart cultivated tomatoes carry the mutant allele of this gene (Rodríguez et al. 2011).

Also, a nutritional trait has been associated with retrotransposons activity: Vitamin E in tomato is controlled by *VTE3(1)* gene coding a 2-methyl-6-phytylquinol methyl transferase, a key enzyme for α- and γ-tocopherols synthesis (Quadrana et al. 2014). By comparing the sequence and expression differences between the *S. lycopersicum* and *S. pennellii VTE3(1)* alleles in introgressed lines, a SINE retrotransposon located in the promoter region of the cultivated allele was characterized. In addition, it was demonstrated that this TE derived genomic restructuration induced the hypermethylation of the proximal promoter region in the *S. lycopersicum* allele and its transcriptional silencing; while the *S. pennellii* allele showed low levels of DNA methylation and higher expression levels (Quadrana et al. 2014).

In the cultivated potato was described the first active Stowaway MITE in dicotyledons (Momose et al. 2010). The mobilization of this TE in potato induced a change of tuber skin color by affecting the gene encoding for a flavonoid 3′,5′-hydroxylase, which is involved in purple anthocyanin synthesis (Momose et al. 2010).

It has been proposed that primitive indigenous cultivated potatoes were domesticated from wild relatives in the current territory of Peru (Spooner et al. 2005). Then, this Andean landraces (*S. tuberosum* Andigenum group), which tuberize under short-day conditions, migrated southward up to south-central Chile, where the Chilean landraces (*S. tuberosum* Chilotanum group) were adapted to form tubers in the long days of southern latitudes. Similar day length of Europe, likely conditioned the selection of this trait in the modern potato cultivars (*S. tuberosum* Tuberosum group). Nowadays, tubers ancestrally domesticated in Andean valleys around 10,000 years ago turned the third more important crop for human consumption because this germplasm went through a crucial process of adaptation for tuberization under long days. This adaptation was possible due to the presence of TE-induced genetic variability in wild relatives species and its further introgression in cultivated ones during the domestications process (Kloosterman et al. 2013; Hardigan et al. 2017). Tuberization is a reproductive strategy photoperiod-regulated, and the locus *StCDF1* is a central regulator of this process in potato (Kloosterman et al. 2013). These authors found that *StCDF1* has different variants with premature stop codons induced by TEs. These structural variations resulting in truncated StCDF1 proteins with dominant-allele effects determined the long day tuberization phenotype. Exploring the StCDF1 haplotype diversity in a panel that included wild diploid species, South American landraces (groups Andigena, Phureja, Stenotomum and Chilotanum) and North American modern cultivars (group Tuberosum), Hardigan and co-workers (2017) revealed introgression of truncated alleles from wild species, particularly from *S. microdontum*, in long day adapted cultivars. This is an extraordinary example of the essential role of potato wild relatives as source of untapped adaptive potential (Hardigan et al. 2017) and a confirmation of the McClintock thought about that TEs induced phenotypic change provides selection opportunities for plant breeding.

11.6 Conclusion

TEs constitute a high proportion of the *Solanum* genomes, where they account for approximately 40–70% of the genome based on the species considered. More than "parasites" as was considered before, they play an essential role in the evolution of *Solanum* genomes and stress responses. Several studies showed the importance of TEs in genome diversity, inducing for example high stress tolerance in *S. pennelli* or adaptation for tuberization in potato genomes.

TEs are active elements that mobilize under different stresses, such as hybridization or polyploidization, changing and reshaping the genomes. Epigenetic mechanisms control the activity of the TEs; changes in the genome methylation status under stress induce the mobilization of TEs. On the other hand, mobilized TEs spread their epigenetic marks to new loci inducing changes in the phenotype, such as modification in the tuber skin color in potato (Momose et al. 2010). Therefore, we must considered TEs as part of a complex mechanism of *Solanum* genomes where TEs activity is controlled by the plant genome and, at the same time, its activation is a source of genetic and epigenetic diversity. New phenotypes generated by TEs activity are an important source of variation not only for the evolution of *Solanum* genomes but for utilization in crop breeding programs.

References

Anca I-A, Fromentin J, Bui QT, Mhiri C, Grandbastien M-A, Simon-Plas F (2014) Different tobacco retrotransposons are specifically modulated by the elicitor cryptogein and reactive oxygen species. J Plant Physiol 171:1533–1540. https://doi.org/10.1016/j.jplph.2014.07.003

Aversano R, Contaldi F, Ercolano MR, Grosso V, Iorizzo M, Tatino F, Xumerle L, Molin AD, Avanzato C, Ferrarini A, Delledonne M, Sanseverino W, Cigliano RA, Capella-Gutierrez S, Gabaldón T, Frusciante L, Bradeen JM, Carputo D (2015) The solanum commersonii genome sequence provides insights into adaptation to stress conditions and genome evolution of wild potato relatives. Plant Cell 27:954–968. https://doi.org/10.1105/tpc.114.135954

Bolger A, Scossa F, Bolger ME, Lanz C, Maumus F, Tohge T, Quesneville H, Alseekh S, Sørensen I, Lichtenstein G, Fich EA, Conte M, Keller H, Schneeberger K, Schwacke R, Ofner I, Vrebalov J, Xu Y, Osorio S, Aflitos SA, Schijlen E, Jiménez-Goméz JM, Ryngajllo M, Kimura S, Kumar R, Koenig D, Headland LR, Maloof JN, Sinha N, van Ham RC, Lankhorst RK, Mao L, Vogel A, Arsova B, Panstruga R, Fei Z, Rose JK, Zamir D, Carrari F, Giovannoni JJ, Weigel D, Usadel B, Fernie AR (2014) The genome of the stress-tolerant wild tomato species *Solanum pennellii*. Nat Genet 46:1034–1038. https://doi.org/10.1038/ng.3046

Bucher E, Reinders J, Mirouze M (2012) Epigenetic control of transposon transcription and mobility in Arabidopsis. Curr Opin Plant Biol 15:503–510. https://doi.org/10.1016/j.pbi.2012.08.006

Buchmann RC, Asad S, Wolf JN, Mohannath G, Bisaro DM (2009) Geminivirus AL2 and L2 proteins suppress transcriptional gene silencing and cause genome-wide reductions in cytosine methylation. J Virol 83:5005–5013. https://doi.org/10.1128/JVI.01771-08

Bureau TE, Wessler SR (1994) Stowaway: a new family of inverted repeat elements associated with the genes of both monocotyledonous and dicotyledonous plants. Plant Cell 6:907–916. https://doi.org/10.1105/tpc.6.6.907

Casacuberta E, González J (2013) The impact of transposable elements in environmental adaptation. Mol Ecol 22:1503–1517. https://doi.org/10.1111/mec.12170

Cheng X, Zhang D, Cheng Z, Keller B, Hong-qing L (2009) A new family of Ty1-copia-like retrotransposons originated in the tomato genome by a recent horizontal transfer event. Genetics 181:1183–1193. https://doi.org/10.1534/genetics.108.099150

Cara N, Marfil CF, Masuelli RW (2013) Epigenetic patterns newly established after interspecific hybridization in natural populations of *Solanum*. Ecol Evol 3:3764–3779

Cara N, Ferrer MS, Masuelli RW, Camadro EL, Marfil CF (2019) Epigenetic consequences of interploidal hybridisation in synthetic and natural interspecific potato hybrids. New Phyto 222:1981–1993

Comai L, Tyagi AP, Winter K, Holmes-Davis R, Reynolds SH, Stevens Y, Byers B (2000) Phenotypic instability and rapid gene silencing in newly formed arabidopsis allotetraploids. Plant Cell 12:1551–1568

Comai L (2005) The advantages and disadvantages of being polyploids. Nat Rev Genet 6:836–846; Costa APP, Scortecci KC, Hashimoto RY, Araujo PG, Grandbastien M-A, Van Sluys M-A (1999) Retrolyc-1, a member of the Tnt1 retrotransposon super-family in the *Lycopersicum peruvianum* genome. Genetica. 197:65–72

Costa AP, Scortecci KC, Hashimoto RY, Araujo PG, Grandbastien MA, Van Sluys MA (1999) Retrolyc1-1, a member of the Tntl retrotransposon super-family in the *Lycopersicon peruvianum* genome. Genetica 107:65–72. https://doi.org/10.1023/A:1004028002883

Du C, Fefelova N, Caronna J, He L, Dooner HK (2009) The polychromatic Helitron landscape of the maize genome. Proc Natl Acad Sci 106:19916–19921. https://doi.org/10.1073/pnas.0904742106

Esposito S, Barteri F, Casacuberta J, Mirouze M, Carputo D, Aversano R (2019) LTR-TE abundance, timing and mobility in Solanum commersonii and S. tuberosum genomes following cold-stress conditions. Planta 250:1781–1787

Federoff NV (2012) Transposable elements, epigenetics and genome evolution. Science 338:758–767

Feschotte C, Jiang N, Wessler SR (2002) Plant transposable elements: where genetics meets genomics. Nat Rev Genet 3:329–341. https://doi.org/10.1038/nrg793

Gaiero P, Vaio M, Peters SA, Schranz EM, de Jong H, Speranza PR (2018) Comparative analysis of repetitive sequences among species from the potato and the tomato clades. Ann Bot. https://doi.org/10.1093/aob/mcy186

Gantuz M, Masuelli RW (2017) Characterization and phylogenetic analysis of retrotransposons families present in the genome of *Solanum tuberosum* L. (*Solanaceae*). In: Abstract of the XXXVI Jornadas Argentinas de Botánica, Mendoza 18–22 Sept 2017. Bol Soc Argent Bot 52 (Supl.), p 258

Grandbastien MA (2015) LTR retrotransposons, handy hitchhikers of plant regulation and stress response. Biochim Biophys Acta 1849:403–416. https://doi.org/10.1016/j.bbagrm.2014.07.017

Grandbastien MA, Spielmann A, Caboche M (1989) Tnt1, a mobile retroviral-like transposable element of tobacco isolated via plant cell genetics. Nature 337:376–380

Grandbastien MA, Lucas H, Morel JB, Mhiri C, Vernhettes S, Casacuberta JM (1997) The expression of the tobacco Tnt1 retrotransposon is linked to plant defense responses. Genetica 100:241–252

Grandbastien MA, Audeon C, Bonnivard E et al (2005) Stress activation and genomic impact of Tnt1 retrotransposons in *Solanaceae*. Cytogenet Genome Res 110:229–241. https://doi.org/10.1159/000084957

Grover A, Sharma PC (2017) Repetitive sequences in the potato and related genomes. In: Kumar Chakrabarti S, Xie C, Kumar Tiwari J (eds) The potato genome. Compendium of plant genomes. Springer, Cham

Hardigan MA, Laimbeer FPE, Newton L, Crisovan E, Hamilton JP, Vaillancourt B, Wiegert-Rininger K, Wood JC, Douches DS, Farré EM, Veilleux RE, Buell CR (2017) Genome diversity of tuber-bearing *Solanum* uncovers complex evolutionary history and targets of domestication in the cultivated potato. PNAS 114:E9999–E10008. https://doi.org/10.1073/pnas.1714380114

Ishiguro S, Ogasawara K, Fujino K, Sato Y, Kishima Y (2014) Low temperature-responsive changes in the anther transcriptome's repeat sequences are indicative of stress sensitivity and pollen sterility in rice strains 1,2[W]. Plant Physiol 164:671–682. https://doi.org/10.1104/pp.113.230656

Ito H, Gaubert H, Bucher E et al (2011) An siRNA pathway prevents transgenerational retrotransposition in plants subjected to stress. Nature 472:115–119. https://doi.org/10.1038/nature09861

Ivashuta S, Naumkina M, Gau M, Uchiyama K, Isobe S, Mizukami Y, Shimamoto Y (2002) Genotype-dependent transcriptional activation of novel repetitive elements during cold acclimation of alfalfa (*Medicago sativa*). Plant J 31:615–627. https://doi.org/10.1046/j.1365-313X.2002.01383.x

Jiang N, Gao D, Xiao H, van der Knaap E (2009) Genome organization of the tomato sun locus and characterization of the unusual retrotransposon rider. Plant J Cell Mol Biol 60:181–193. https://doi.org/10.1111/j.1365-313X.2009.03946.x

Kamenetzky L, Asís R, Bassi S, de Godoy F, Bermúdez L, Fernie AR, Van Sluys M-A, Vrebalov J, Giovannoni JJ, Rossi M, Carrari F (2010) Genomic analysis of wild tomato introgressions determining metabolism- and yield-associated traits. Plant Physiol 152:1772–1786. https://doi.org/10.1104/pp.109.150532

Kashkush K, Khasdan V (2007) Large-scale survey of cytosine methylation of retrotransposons and the impact of readout transcription from long terminal repeats on expression of adjacent rice genes. Genetics 177:1975–1985. https://doi.org/10.1534/genetics.107.080234

Kim S, Park J, Yeom S-I, Kim Y-M, Seo E, Kim K-T, Kim M-S, Lee JM, Cheong K, Shin H-S, Kim S-B, Han K, Lee J, Park M, Lee H-A, Lee H-Y, Lee Y, Oh S, Lee JH, Choi E, Choi E, Lee SE, Jeon J, Kim H, Choi G, Song H, Lee J, Lee S-C, Kwon J-K, Lee H-Y, Koo N, Hong Y, Kim RW, Kang W-H, Huh JH, Kang B-C, Yang T-J, Lee Y-H, Bennetzen JL, Choi D (2017) New reference genome sequences of hot pepper reveal the massive evolution of plant disease-resistance genes by retroduplication. Genome Biol 18:210. https://doi.org/10.1186/s13059-017-1341-9

Kloosterman B, Abelenda JA, Gomez M del MC, Oortwijn M, Boer JM de, Kowitwanich K, Horvath BM, Eck HJ van, Smaczniak C, Prat S, Visser RGF, Bachem CWB (2013) Naturally occurring allele diversity allows potato cultivation in northern latitudes. Nature 495:246–250.https://doi.org/10.1038/nature11912

Kuang H, Padmanabhan C, Li F et al (2009) Identification of miniature inverted-repeat transposable elements (MITEs) and biogenesis of their siRNAs in the *Solanaceae*: new functional implications for MITEs. Genome Res 19:42–56. https://doi.org/10.1101/gr.078196.108

Lanciano S, Mirouze M (2018) Transposable elements: all mobile, all different, some stress responsive, some adaptive? Curr Opin Genet Dev 49:106–114. https://doi.org/10.1016/j.gde.2018.04.002

Leisner CP, Hamilton JP, Emily C, Manrique-Carpintero NC, Marand AP, Linsey N, Pham GM, Jiming J, Douches DS, Jansky SH, Robin BC (2018) Genome sequence of M6, a diploid inbred clone of the high glycoalkaloid producing tuber bearing potato species Solanum chacoense, reveals residual heterozygosity. Plant J 94:562–570. https://doi.org/10.1111/tpj.13857

Leitch IJ, Bennett MD (1997) Polyploidy in angiosperms. Trends Plant Sci 2:470–476. https://doi.org/10.1016/S1360-1385(97)01154-0

Lewontin RC, Birch LC (1966) Hybridization as a source of variation for adaptation to new environments. Evolution 20:315–336. https://doi.org/10.1111/j.1558-5646.1966.tb03369.x

Lisch D (2009) Epigenetic regulation of transposable elements in plants. Annu Rev Plant Biol 60:43–66. https://doi.org/10.1146/annurev.arplant.59.032607.092744

Lisch D (2013) How important are transposons for plant evolution? Nat Rev Genet 14:49–61. https://doi.org/10.1038/nrg3374

Liu B, Wendel JF (2000) Retrotransposon activation followed by rapid repression in introgressed rice plants. Genome 43:874–880

Manetti ME, Rossi M, Costa A, Clausen AM, Van Sluys MA (2007) Radiation of the Tnt1 retrotransposon superfamily in three *Solanaceae* genera. BMC Evol Biol 7:34. https://doi.org/10.1186/1471-2148-7-34

Marand AP, Jansky SH, Zhao H et al (2017) Meiotic crossovers are associated with open chromatin and enriched with Stowaway transposons in potato. Genome Biol 18:203. https://doi.org/10.1186/s13059-017-1326-8

Masuelli RW, Camadro EL, Erazzu LE, Bedogni MC, Marfil CF (2009) Homoploid hybridization in the origin and evolution of wild diploid potato species. Plant Syst Evol 277:143–151. https://doi.org/10.1007/s00606-008-0116-x

McClintock B (1984) The significance of responses of the genome to challenge. Science 226:792–801. https://doi.org/10.1126/science.15739260

Mhiri C, Morel J-B, Vernhettes S, Casacuberta JM, Lucas H, Grandbastien M-A (1997) The promoter of the tobacco Tnt1 retrotransposon is induced by wounding and by abiotic stress. Plant Mol Biol 33:257–266. https://doi.org/10.1023/A:1005727132202

Michael TP (2014) Plant genome size variation: bloating and purging DNA. Brief Funct Genomics 13:308–317. https://doi.org/10.1093/bfgp/elu005

Momose M, Abe Y, Ozeki Y (2010) Miniature inverted-repeat transposable elements of Stowaway are active in potato. Genetics 186:59–66. https://doi.org/10.1534/genetics.110.117606

Naito K, Zhang F, Tsukiyama T, Saito H, Hancock CN, Richardson AO, Okumoto Y, Tanisaka T, Wessler SR (2009) Unexpected consequences of a sudden and massive transposon amplification on rice gene expression. Nature 461:1130–1134. https://doi.org/10.1038/nature08479

Negi P, Rai AN, Suprasanna P (2016) Moving through the stressed genome: emerging regulatory roles for transposons in plant stress response. Front Plant Sci 7:1448. https://doi.org/10.3389/fpls.2016.01448

Paran I, van der Knaap E (2007) Genetic and molecular regulation of fruit and plant domestication traits in tomato and pepper. J Exp Bot 58:3841–3852

Parisod C, Alix K, Just J, Petit M, Sarilar V, Mahiri C, Ainouche M, Chalhoub B, Grandbastien MA (2010) Impact of transposable elements on the organization and function of allopolyploid genomes. New Phytol 186:37–45. https://doi.org/10.1111/j.1469-8137.2009.03096.x

Paz RC, Kozaczek ME, Rosli HG, Andino NP, Sanchez-Puerta MV (2017) Diversity, distribution and dynamics of full-length Copia and Gypsy LTR retroelements in *Solanum lycopersicum*. Genetica 145:417–430. https://doi.org/10.1007/s10709-017-9977-7

Paz RC, Rendina González A, Ferrer S, Masuelli RW (2015) Short-term hybridisation activates Tnt1 and Tto1 Copia retrotransposons in wild tuber-bearing *Solanum* species. Plant Biol 17:860–869. https://doi.org/10.1111/plb.12301

Qin C, Yu C, Shen Y, Fang X, Chen L, Min J, Cheng J, Zhao S, Xu M, Luo Y, Yang Y, Wu Z, Mao L, Wu H, Ling-Hu C, Zhou H, Lin H, González-Morales S, Trejo-Saavedra DL, Tian H, Tang X, Zhao M, Huang Z, Zhou A, Yao X, Cui J, Li W, Chen Z, Feng Y, Niu Y, Bi S, Yang X, Li W, Cai H, Luo X, Montes-Hernández S, Leyva-González MA, Xiong Z, He X, Bai L, Tan S, Tang X, Liu D, Liu J, Zhang S, Chen M, Zhang L, Zhang L, Zhang Y, Liao W, Zhang Y, Wang M, Lv X, Wen B, Liu H, Luan H, Zhang Y, Yang S, Wang X, Xu J, Li X, Li S, Wang J, Palloix A, Bosland PW, Li Y, Krogh A, Rivera-Bustamante RF, Herrera-Estrella L, Yin Y, Yu J, Hu K, Zhang Z (2014) Whole-genome sequencing of cultivated and wild peppers provides insights into **Capsicum** domestication and specialization. PNAS 111:5135–5140. https://doi.org/10.1073/pnas.1400975111

Quadrana L, Almeida J, Asís R, Dominguez PG, Bermúdez L, Conti G, Corrêa da Silva JV, Peralta IE, Colot V, Asurmendi S, Fernie AR, Rossi M, Carrari F (2014) Natural occurring epialleles determine vitamin E accumulation in tomato fruits. Nat Commun 5:4027

Razali R, Bougouffa S, Morton MJL, Lightfoot DJ, Alam I, Essack M, Arold ST, Kamau AA, Schmöckel SM, Pailles Y, Shahid M, Michell CT, Al-Babili S, Ho YS, Tester M, Bajic VB, Negrão S (2018) The genome sequence of the wild tomato *Solanum pimpinellifolium* provides insights into salinity tolerance. Front Plant Sci 9:1402. https://doi.org/10.3389/fpls.2018.01402

Rodríguez GR, Muños S, Anderson C, Sim SC, Michel A, Causse M, Gardener BB, Francis D, van der Knaap E (2011) Distribution of *SUN*, *OVATE*, *LC* and *FAS* in the tomato germplasm and the relationship to fruit shape diversity. Plant Physiol 156:275–285

Sanmiguel P, Bennetzen JL (1998) Evidence that a recent increase in maize genome size was caused by the massive amplification of intergene retrotransposons. Ann Bot 82:37–44. https://doi.org/10.1006/anbo.1998.0746

Seibt KM, Wenke T, Muders K, Truberg B (2016) Short interspersed nuclear elements (SINEs) are abundant in *Solanaceae* and have a family-specific impact on gene structure and genome organization. Plant J 86:268–285. https://doi.org/10.1111/tpj.13170

Slotkin RK, Martienssen R (2007) Transposable elements and the epigenetic regulation of the genome. Nat Rev Genet 8:272–285. https://doi.org/10.1038/nrg2072

Spooner DM, McLean K, Ramsay G, Waugh R, Bryan GJ (2005) A single domestication for potato based on multilocus amplified fragment length polymorphism genotyping. Proc Natl Acad Sci USA 102:14694–14699. https://doi.org/10.1073/pnas.0507400102

Takeda S, Sugimoto K, Otsuki H, Hirochika H (1999) A 13-bp cis-regulatory element in the LTR promoter of the tobacco retrotransposon Tto1 is involved in responsiveness to tissue culture, wounding, methyl jasmonate and fungal elicitors. Plant J 18:383–393

The Potato Genome Sequencing Consortium (2011) Genome sequence and analysis of the tuber crop potato. Nature 475:189–195. https://doi.org/10.1038/nature10158

The Tomato Genome Consortium (2012) The tomato genome sequence provides insights into fleshy fruit evolution. Nature 485:635–641. https://doi.org/10.1038/nature11119

Vicient CM, Casacuberta JM (2017) Impact of transposable elements on polyploid plant genomes. Ann Bot 120:195–207. https://doi.org/10.1093/aob/mcx078

Wang Y, Tang X, Cheng Z, Mueller L, Giovannoni J, Tanksley SD (2006) Euchromatin and pericentromeric heterochromatin: comparative composition in the tomato genome. Genetics 172:2529–2540. https://doi.org/10.1534/genetics.106.055772

Wang X, Ai G, Zhang C, Cui L, Wang J, Li H, Zhang J, Ye Z (2016) Expression and diversification analysis reveals transposable elements play important roles in the origin of *Lycopersicon*-specific lncRNAs in tomato. New Phytol 209:1442–1455. https://doi.org/10.1111/nph.13718

Wendel JF (2000) Genome evolution in polyploids. Plant Mol Biol 42:225–249

Wendel JF, Jackson SA, Meyers BC, Wing RA (2016) Evolution of plant genome architecture. Genome Biol 17:37. https://doi.org/10.1186/s13059-016-0908-1

Wenke T, Döbel T, Sörensen TR, Junghans H, Weisshaar B, Schmidt T (2011) Targeted identification of short interspersed nuclear element families shows their widespread existence and extreme heterogeneity in plant genomes[W]. Plant Cell 23:3117–3128. https://doi.org/10.1105/tpc.111.088682

Wicker T, Sabot F, Hua-Van A, Bennetzen JL, Capy P, Chalhoub B, Flavell A, Leroy P, Morgante M, Panaud O, Paux E, SanMiguel P, Schulman AH (2007) A unified classification system for eukaryotic transposable elements. Nat Rev Genet 8:973–982

Woodrow P, Pontecorvo G, Fantaccione S, Fuggi A, Kafantaris I, Parisi D, Carillo P (2010) Polymorphism of a new Ty1-copia retrotransposon in durum wheat under salt and light stresses. Theor Appl Genet 121:311–322. https://doi.org/10.1007/s00122-010-1311-z

Xiao H, Jiang N, Schaffner E, Stockinger EJ, van der Knaap E (2008) A retrotransposon-mediated gene duplication underlies morphological variation of tomato fruit. Science 319:1527–1530. https://doi.org/10.1126/science.1153040

Yaakov B, Kashkush K (2010) Massive alterations of the methylation patterns around DNA transposons in the first four generations of a newly formed wheat allohexaploid. Genome 54:42–49. https://doi.org/10.1139/G10-091

Yañez M, Verdugo I, Rodríguez M, Prat S, Ruiz-Lara S (1998) Highly heterogeneous families of Ty1/copia retrotransposons in the *Lycopersicon chilense* genome. Gene 222:223–228. https://doi.org/10.1016/s0378-1119(98)00486-7

Yoder JA, Walsh CP, Bestor TH (1997) Cytosine methylation and the ecology of intragenomic parasites. Trends Genet 13:335–340

Zavallo D, Crescente JM, Gantuz M, Leone M, Vanzetti LS, Masuelli RW, Asurmendi S (2020) Genomic re-assessment of the transposable elements landscape of the potato genome. Plant Cell Rep 39:1161–1174

Zhang X, Jiang N, Feschotte C, Wessler SR (2004) PIF- and Pong-like transposable elements: distribution, evolution and relationship with Tourist-like miniature inverted-repeat transposable elements. Genetics 166:971–986

Cultivar Improvement with Exotic Germplasm: An Example from Potato

12

S. H. Jansky, W. S. De Jong, D. S. Douches, K. G. Haynes, and D. G. Holm

Abstract

Potato breeders have abundant germplasm resources at their disposal in wild relatives and landrace varieties. Exotic germplasm has been used mainly for disease resistance and processing quality genes. Linkage drag is common when introgressing wild germplasm, with poor adaptation and high glycoalkaloid levels commonly observed in hybrid offspring. Poor tuber type, late maturity, and short tuber dormancy accompany desirable traits introduced from landraces. More targeted introgression of exotic genes for potato improvement is needed in future breeding efforts. Potato programs worldwide are exploring the conversion of potato into a diploid inbred-hybrid crop. Backcrossing in this system would allow valuable genes from exotic sources to be fixed in high-performing cultivated potato backgrounds. This offers exciting new possibilities for contributions of wild and cultivated potato relatives to potato cultivar improvement.

Have wild potatoes made a significant contribution to modern cultivars? A wide range of opinions can be found in the literature. Some suggest that wild relative germplasm is abundant in cultivated potato due to the efforts of breeders and it has been critical for the development of varieties. Others claim that the contributions from wild relatives have been minimal. This chapter attempts to unpack the complex answer to this question and suggest ways in which modern genomics technologies may contribute to the introgression of exotic germplasm into new potato cultivars.

There is no question that the potato germplasm resource is immense and highly accessible to breeders. This resource is described in detail in Chap. 4. Reviews of introgression strategies are provided by (Jansky 2006; Ortiz et al. 2009; Spooner et al. 2014; Bethke et al. 2017). The question to be answered in this chapter is "To what extent has exotic germplasm had an impact on the development of successful modern cultivars?".

S. H. Jansky (✉)
USDA-ARS and University of Wisconsin-Madison, Madison, USA
e-mail: shelley.jansky@ars.usda.gov

W. S. De Jong
Cornell University, Ithaca, USA

D. S. Douches
Michigan State University, East Lansing, USA

K. G. Haynes
USDA-ARS, Beltsville, MD, USA

D. G. Holm
Colorado State University, Fort Collins, USA

© This is a U.S. government work and not under copyright protection in the U.S.; foreign copyright protection may apply 2021
D. Carputo et al. (eds.), *The Wild Solanums Genomes*, Compendium of Plant Genomes,
https://doi.org/10.1007/978-3-030-30343-3_12

12.1 What Exotic Germplasm Has Been Introgressed into Potato Cultivars?

The potato germplasm resource consists of 107 wild relatives, four landrace cultivated species (*S. tuberosum* Andigenum Group, including the tropical zone Andean tetraploids and diploids commonly referred to as Andigena and Phureja, respectively), and temperate zone modern cultivars (*S. tuberosum* Chilotanum Group) (Spooner et al. 2007, 2014; Ovchinnikova et al. 2011). It is striking to note that, while most of these species are sexually compatible with cultivated potato, only a few are found in the pedigrees of cultivars. Four review articles list only 12 wild species in the pedigrees of European and North American cultivars—*S. acaule*, *S. chacoense*, *S. demissum*, *S. fendleri*, *S. kurtizianum*, *S. microdontum*, *S. multidissectum*, *S. raphanifolium*, *S. spegazzinii*, *S. stenotomum*, *S. toralapanum*, and *S. vernei* (Ross 1986; Plaisted and Hoopes 1989; Love 1999; Bradshaw 2009). Nearly 30 years ago, *S. sparsipilum* and *S. chacoense* were mentioned as species that should be used in breeding programs due to their "broad range of desirable attributes" (Plaisted and Hoopes 1989). However, new introgressions of these species are not apparent in pedigrees. In fact, new introgressions of any additional species are rare. This observation is in line with Bradshaw (2009), who concluded that "with a few exceptions, it has proved to be difficult to successfully utilize wild species in potato breeding" and Pavek and Corsini (2001), who suggests that "there is a great disparity between the large number of wild species that showed promise when evaluated and those actually used in breeding."

An examination of cultivar pedigrees reveals that a few clones have been repeatedly used as parents by breeders. This interrelatedness among clones in breeding programs has led to the broadly-held belief that cultivated potato has a narrow genetic base (Glendinning 1983; Plaisted and Hoopes 1989; Mohan et al. 1990; Love 1999). However, a recent genomic analysis based on resequencing of North American cultivars, South American landraces, and wild relatives has revealed higher levels of genetic diversity than in any other crop evaluated to date (Hardigan et al. 2017). Amazingly, 73% of the alleles from a set of 20 wild potato species were detected in potato cultivars and 31.2% of 5 kb genome sequences in North American cultivars carried wild species alleles. Wild alleles presumably accumulated in cultivated potato even after domestication. As potato migrated from its origin in the Peruvian Andes to lowland Chile, it hybridized with wild potatoes and accumulated new genes, including those for photoperiod adaptation and stress tolerance. Introgressions of *S. microdontum* were found in most of the cultivars, even though this species is not present in their pedigrees. *Solanum microdontum* likely contributed to the genetic variation needed to tuberize under increasingly longer days as potato migrated from tropical to temperate zones. Furthermore, Hardigan et al. (2017) found that, when wild species alleles were preferentially retained in cultivars, those genes were often associated with abiotic and biotic stress tolerance.

In addition to the natural introductions of wild species genes during the evolution of South American cultivated potatoes, breeders have introgressed exotic genes. With a few exceptions, such as *S. demissum* for late blight resistance (discussed below), exotic germplasm was not extensively used by breeders until the mid-twentieth century (Love 1999; Bradshaw 2009). This infusion of exotic germplasm is apparent when comparing genetic markers in cultivars released before 1945 with those found in cultivars released after 1945 (Vos et al. 2015). They found new genetic variants clustered in the genome in a subset of cultivars. These regions are likely introgressions of exotic germplasm.

Early attempts to systematically introgress wild germplasm focused on bringing late blight resistance (R) genes into cultivars from the wild hexaploid species *S. demissum* (Grünwald et al. 2002; Bradshaw and Ramsay 2005). The first cultivars carrying these resistance genes were released in the 1950s, but resistance was quickly overcome by the pathogen (Bradshaw et al. 2006). The use of this exotic germplasm was not without challenges. Stevenson and Clark (1937) noted that "Owing to sterility, incompatibility, etc., the results thus far have been simply what could be got, not what it was planned to get." Nevertheless, *S. demissum* germplasm is in the pedigrees of a large proportion of North American and European cultivars. The Mexican National Potato Program, initiated in 1949 and supported by the Rockefeller Foundation, produced several cultivars with late blight resistance, mostly derived from *S. demissum* (Grünwald et al. 2002). Some products of this breeding effort, such as Tollocan, are still used as sources of durable resistance to late blight. For example, Tollocan is found in the pedigrees of Jacqueline Lee (Douches et al. 2001) and Missaukee (Douches et al. 2009).

Unlike wild relatives, South American landrace germplasm has undergone selection, so it is enhanced for traits of interest to breeders. Consequently, it could be used for both base broadening and as a source of specific genes. Hybrid vigor has been reported at the tetraploid level in crosses between Chilotanum and Andigena (Plaisted 1973; Tarn and Tai 1973, 1983; Hoopes et al. 1980; Maris 1989), in the tetraploid progeny of crosses between tetraploid Chilotanum and dihaploid x Phureja hybrids (Hanneman and Peloquin 1967; Mendiburu 1971; Quinn and Peloquin 1973; Kidane-Mariam and Peloquin 1975; Mok and Peloquin 1975; De Jong and Tai 1977), and at the diploid level (McHale and Lauer 1981; Hilali et al. 1987; Buso et al. 2000). Andigena (Andean highland tropical zone cultivated potato) and Chilotanum (Chilean lowland temperate zone cultivated potato) may represent two heterotic groups. In fact, 80% of genes that have likely been under selection are population-specific (Hardigan et al. 2017). That is, regional differences rather than conserved processes led to superior performance of Andigena in highland tropical and Chilotanum in lowland temperate regions. However, there are some challenges associated with the use of this germplasm for breeding (discussed below). Recurrent selection efforts have improved tetraploid Andigena, creating so-called Neotuberosum (Glendinning 1975a, b, c, 1976; Rasco et al. 1980; Pavek and Corsini 2001; Bradshaw 2009) and diploid Phureja (Haynes and Christ 1999; Christ and Haynes 2001; Haynes 2008; Santa Cruz et al. 2009; Haynes et al. 2014). The cultivar Eva, produced by crossing Steuben with bulk pollen of a collection of Neotuberosum x Tuberosum hybrids, has shallow eyes and exceptional tuber appearance (Plaisted et al. 2001).

12.2 How Has Exotic Germplasm Been Introgressed into Potato Cultivars?

Most wild potato species are diploid, so ploidy manipulations must be used to introgress them into tetraploid potato cultivars. One option is to create dihaploids from potato cultivars, cross them to diploid wild relatives (or cultivated diploids), and then use sexual polyploidization to return to the tetraploid level. This strategy has been used in recent decades in some breeding programs in North America and Europe (Buso et al. 2000; Hayes and Thill 2002; Buso et al. 2003; Carputo and Barone 2005; Sterrett et al. 2006; Ortiz et al. 2009; Haynes et al. 2011). Introgression and selection at the diploid level allows for recombination and selection against undesirable contributions from the wild parent (Bradshaw and Ramsay 2005).

The alternative is to cross wild potatoes directly to tetraploid cultivars. Diploid wild species must be somatically doubled or produce 2n gametes in order to create tetraploid offspring. For example, the nematode-resistant diploid wild species *S. vernei* was somatically doubled and then crossed to tetraploids to introduce nematode resistance (Bradshaw and Ramsay 2005). Alternatively, 2n gamete production in wild potato germplasm allows for direct polyploidization (DenNijs and Peloquin 1977a, b; Watanabe and Peloquin 1991; Werner and Peloquin 1991). Examples of direct hybridization with cultivars can be found when examining the pedigrees of cultivars, where crosses to wild relatives can be seen, followed by several generations of backcrossing to cultivated potato. This was the standard method for wild species germplasm utilization until the late twentieth century. A classic example is found in the pedigree of Lenape (Akeley et al. 1968). The diploid wild species *S. chacoense* was crossed to the tetraploid cultivar Menominee to produce a tetraploid hybrid. That hybrid is a grandparent of Lenape. As will be discussed later, Lenape has been an important cultivar in the development of North American chip processing cultivars. A significant challenge with this strategy is that it is much more difficult to remove undesirable alleles at the tetraploid level than in diploids. Introgression increases genetic load with respect to alleles of interest to the potato industry. Consequently, the use of this germplasm in breeding decreases the probability of finding good phenotypic combinations in the progeny.

The use of exotic germplasm for potato improvement in the northern hemisphere began in earnest after most cultivars were destroyed by late blight epidemics in Europe and North America in the mid-nineteenth century (Hawkes 1990). The Chilean cultivar Rough Purple Chile was introduced into the U.S. in 1851 and features prominently in the pedigrees of many North American and European cultivars. As discussed above, the best-known early example of the systematic use of a wild species in breeding is the hexaploid species *S. demissum*, used as a source of major genes for late blight resistance (Black 1970). Although *S. demissum* is hexaploid, it is sexually compatible with tetraploid cultivars and the pentaploid offspring are fertile. Resistance genes were introgressed but were ultimately overcome by the late blight pathogen, *Phytophthora infestans*. New efforts are underway to introduce resistance genes that are not likely to be easily overcome (Song et al. 2003; Colton et al. 2006; Kuhl et al. 2007; Bradeen et al. 2009). The *R8* gene from *S. demissum* confers broad spectrum resistance and is found in European (Sarpo Mira), North American (Jacqueline Lee, Missaukee), and Chinese (PB-06, S-60) cultivars (Vossen et al. 2016). R8 is also present in Tollocan, the source of late blight resistance in several breeding programs, and C88, the most widely-grown cultivar in China and the world. North American breeders find that late maturity is often introduced when using resistant cultivars from other regions. However, breeding efforts have been successful in selecting for earlier maturity while retaining disease resistance. New sources of late blight resistance have been identified and are being mapped (Liu and Halterman 2006; Islam et al. 2016; Vossen et al. 2017). These recent R-gene mapping studies will likely be paying dividends soon. Durable resistance will depend on the stacking of R genes. However, resistance has been shown to break down even when three R genes are stacked (Malcolmson 1969). Breeders should be searching for germplasm with host resistance genes that recognize different pathogen effectors (Halterman et al. 2010). Stacking these R genes will likely lead to more durable resistance. Marker-assisted selection is being used to create germplasm with stacked R genes, but cultivar deployment is further downstream. If potato is converted into a diploid inbred-hybrid crop (discussed below), then pyramiding R genes in inbred lines will be very efficient.

Dominant genes for resistance to potato virus Y and nematodes provide the other main success stories for the introgression of wild germplasm with resistance genes. The most widely used PVY resistance gene is from Andigenum (Ry_{adg}). It has been introgressed into the North American cultivars Eva (Plaisted et al. 2001), Saginaw Chipper, and Mackinaw. Several cultivars with PVY resistance genes introgressed from *S. stoloniferum* (Ry_{sto}) are listed by Ross (1986). However, the Ry_{sto} germplasm introduces male sterility, so backcross hybrids can only be used only as female parents (Song and Schwarzfischer 2008; Vales et al. 2010). Nematode resistance has been introgressed from Andigenum and *S. vernei* and is found in the pedigrees of many European and North American cultivars (Ross 1986; Plaisted and Hoopes 1989; Bradshaw and Ramsay 2005). The breeding clone CPC (Commonwealth Potato Collection) 1673 is the Andigena donor of the *H1* nematode resistance gene and is found in nearly all cultivars released by Cornell in the past half-century.

Wild germplasm has also contributed to tuber processing quality. The release of the cultivar

Lenape (Akeley et al. 1968), which has *S. chacoense* as a great-grandparent, marked the beginning of an era of breeding progress toward improved processing quality, including high tuber solids and low reducing sugar levels (Love et al. 1998). Lenape is considered a landmark cultivar for the development of processing germplasm, leading to steady gains from selection. In fact, Love et al. (1998) noted that, if Lenape and its progeny were removed from the data set, gains for improved dry matter content would disappear. An analysis of SNPs in 25 genes associated with carbohydrate metabolism revealed that breeders have significantly enriched alleles that contribute positively to high tuber solids and low reducing sugars in processing cultivars (Hirsch et al. 2013). Lenape is no longer grown commercially due to the production of high levels of tuber glycoalkaloids, which can cause bitter taste and gastric distress (Love et al. 1998). High glycoalkaloids continue to pose a challenge to breeders who use germplasm derived from Lenape. Three additional exotic sources of processing quality have been used in North American breeding programs. Phureja contributed to cold storage ability in NorValley (Novy et al. 1998). *Solanum berthaultii* is a grandparent of the clones S438 and S440. These two breeding clones are in the pedigrees of the chip processing cultivars Accumulator, Kalkaska, Lelah, Nicolet, Pinnacle, Tundra, and White Pearl. A clone in the Cornell breeding program (E48-2) produces offspring with exceptional chip color. E48-2 has *S. berthaultii* in its background and carries the Ry_{adg} PVY resistance gene. Finally, the clone AH66-4, with *S. raphanifolium* in its pedigree, is a parent of the recently released fry processing cultivar Dakota Russet.

After a wild or landrace potato is crossed to a cultivar, three to seven generations of backcrossing to other cultivars are typically required before commercially acceptable clones are generated (Bradshaw et al. 2006). Bradshaw and Ramsay (2005) provide numerous examples of this practice and suggest that additional generations of backcrossing lead to more successful cultivars. For example, Pentland Ace is the product of three generations of backcrossing after the introgression of *S. demissum*, but four generations of backcrossing produced Pentland Dell, which was a more successful cultivar. Similarly, resistance to potato cyst nematode was introduced from the wild diploid species *S. vernei*. Four generations of backcrossing produced Morag and Glenna, which were commercially acceptable, but five generations of crosses to cultivated potato produced the successful cultivar Lady Balfour. The number of generations of backcrossing to conventional cultivars appears to be similar whether the exotic germplasm is wild or landrace germplasm. Four backcrosses were necessary to generate Vales Everest, a cultivar with resistance to potato cyst nematode from Andigena (Bradshaw and Ramsay 2005). In North America, the cultivar Peconic is three generations removed from CPC 1673. It was not popular with growers, but was planted on farms with golden nematode infestations. One more generation of breeding produced Kanona (Plaisted et al. 1989), which was at the time a more desirable cultivar.

Backcrosses are carried out to eliminate as much exotic germplasm as possible while retaining the major gene(s) contributed by the exotic germplasm. Linkage drag is a problem and multiple cycles of meiosis are necessary to break linkages between desirable and undesirable genes in introgressions of exotic germplasm. Most new genetic variants introduced into potato cultivars since 1945 are found in a low frequency and are likely moving toward extinction (Vos et al. 2015). However, exotic germplasm sometimes contributes positive genes other than those for which they were selected. Vos et al. (2015) found genomic regions that underwent positive selection by breeders even though they do not appear to carry major resistance genes. For example, the *S. demissum R3a/R3b* late blight resistance locus was quickly overcome by *P. infestans*, so it is not important for the development of new varieties. However, its allele frequency in the cultivar gene pool has increased over time, perhaps because beneficial alleles are linked to this R gene. Ross (1986) noted that a heterotic benefit was observed when *S. demissum* was used in breeding programs.

Table 12.1 Major U.S. cultivars, year of release, percent certified seed acreage in 2017, wild species in pedigrees, and number of generations back to wild species parent

Cultivar	Year	Acreage	Species	Generations
Russet Burbank	1902	18.75	None	Not applicable
Russet Norkotah	1987	11.21	None	Not applicable
Norland	1958	6.84	None	Not applicable
Ranger Russet	1992	5.48	None	Not applicable
Umatilla Russet	2000	4.98	None	Not applicable
Atlantic	1978	2.55	*S. chacoense*, Andigena	4
Lamoka	2011	2.50	Andigena; *S. chacoense*	6-Andigena; 7-*S. chacoense* and Andigena
Clearwater Russet	2010	1.78	None	Not applicable
Red La Soda	1948	1.75	None	Not applicable
Snowden	1990	1.68	*S. chacoense*	4
Alturas	2002	1.52	*S. chacoense*, Andigena	6
Dakota Pearl	1999	1.49	Phureja	4
Shepody	1983	1.47	None	Not applicable
Canela Russet	2012	1.18	None	Not applicable
Bannock Russet	2002	1.11	None	Not applicable
Chieftain	1968	1.01	*S. acaule*	4

List is sorted based on acreage and includes all cultivars that occupied at least 1% of acreage

A positive side effect of the introduction of exotic germplasm into breeding programs is often an increase in fertility. Male sterility is a serious problem in potato breeding programs, limiting the availability of clones for use as male parents (Salaman 1910; Krantz 1924; Rieman et al. 1956; Birhman and Kaul 1989; Gopal 1993). Although most dihaploids are male sterile, fertility is often restored when they are crossed to wild germplasm (Hermundstad and Peloquin 1985).

In the U.S., cultivars are tracked by certified seed acres. These data can be used as a proxy for commercial production trends. Figure 1 shows production across time for all cultivars that occupied at least 1% of certified seed acres in 2017. There is a greater array of varieties now (16) than 20 years ago (10). This is partly due to increased specialization, with new varieties being developed for specific regions and growing seasons. Among russet varieties, the major cultivars Russet Burbank and Russet Norkotah are gradually being displaced by new cultivars such as Umatilla Russet and Clearwater Russet. Acreage of the chip processing cultivars Atlantic and Snowden is decreasing while that of Lamoka is increasing. However, these trends cannot be tied to the introgression of exotic germplasm (Table 12.1). In fact, only four sources of exotic germplasm are apparent in the pedigrees of these top 16 U.S. cultivars and only two (*S. chacoense* and Andigena) appear in more than one cultivar.

12.3 What Are the Challenges with Introgression of Exotic Germplasm into Cultivated Potato?

The two most significant challenges with the introduction of wild germplasm into a breeding program are poor adaptation and high glycoalkaloid levels. Most wild potatoes are adapted to tropical latitudes, so they tuberize under a short (12 h) photoperiod. However, most major production areas are in temperate regions, where that photoperiod occurs too late in the growing season to allow for adequate tuber growth. Crosses between Tuberosum dihaploids and wild potatoes result in segregation for photoperiod response, so breeders can select for adapted individuals even in first-generation hybrids (Hermundstad and Peloquin 1986; Yerk and Peloquin 1989; Kittipadukal et al. 2012). The *CDF1* gene on chromosome 5 codes for a transcription factor that plays a major role in the transition from vegetative growth to tuber initiation (Kloosterman et al. 2013). Many agronomic traits are associated with this region, presumably due to the strong effects of this maturity locus on plant growth and tuber production (Manrique-Carpintero et al. 2015). As discussed above, most long-day cultivars carry a *CDF1* allele derived from *S. microdontum*, even though that species has not been used by breeders (Hardigan et al. 2017). This allele was likely introgressed into cultivated potato as it migrated southward from Peru, allowing it to adapt to long photoperiods. Cultivated potato from Chile and Peru was then transported to Europe, but only clones from Chile, adapted to high latitude photoperiods, tuberized (Ríos et al. 2007).

The second negative trait commonly introduced from wild germplasm is high glycoalkaloid levels. As mentioned above, the cultivar Lenape was removed from commercial production due to high glycoalkaloid levels. Glycoalkaloids are secondary metabolites that can have toxic effects on animals if produced at high levels (200 mg/kg) in tubers (Hall 1992). Wild species vary widely in levels of glycoalkaloids and may produce unacceptably high concentrations in tubers (Van Gelder et al. 1988; Kozukue et al. 2008). Resistance to some major potato pests and pathogens is associated with high glycoalkaloid levels (Sinden et al. 1980; Paudel et al. 2017). Consequently, breeders who introduce wild relatives into cultivated germplasm must monitor glycoalkaloid levels in their programs. As discussed later, new breeding efforts in North America (Jansky et al. 2016) and Europe (Lindhout et al. 2011) are focusing on developing diploid inbred-hybrid cultivars. This strategy will allow for the systematic elimination of negative traits through backcrossing.

There are challenges when introgressing exotic cultivated (landrace) germplasm as well. Andigena and Phureja germplasm is adapted to the tropics, so photoperiod adaptation is a challenge, as was mentioned above with regard to wild germplasm. In addition, rough tubers and high set are common in landrace germplasm (Simmonds 1964; Glendinning, 1975a; McHale and Lauer 1981; Carroll and DeMaine 1989; DeMaine 1996; Jansky and Peloquin 2005; Bradshaw et al. 2006). Finally, Phureja exhibits short dormancy, a trait that is difficult to improve through breeding (DeMaine 1996; Jansky and Peloquin 2005; Ritter et al. 2008).

In a typical breeding program, clones carrying wild species introgressions are crossed to an advanced tetraploid clone or cultivar, and then the segregating population is evaluated for adaptation and agronomic performance in the field. During this first evaluation in the field, the breeder will discard 90% or more of the genotypes, based on appearance (Ross 1986; Tarn et al. 1992; Bradshaw and MacKay 1994; MacKay 2005). This strong selection pressure likely eliminates genotypes carrying wild germplasm. Even when breeders make a concerted effort to broaden the genetic base of parental clones and relax selection criteria in families carrying exotic germplasm, few interspecific hybrids advance beyond the first generation. Consequently, the "requirement that external appearance, uniformity, internal tuber quality, and yield conform to specific, high standards leads to the occurrence of limited parentage and genetic background in successful varieties" (Pavek and Corsini 2001). It is difficult to identify plants with desirable genes from wild relatives because they are masked by wild species genes for poor adaptation (Bonierbale et al. 1993). Furthermore, recombination between wild and cultivated genomes may be reduced in interspecific hybrids, hindering the separation of desirable from undesirable wild species genes (Naess et al. 2001). This may be especially relevant in tetraploid germplasm, where multiple pairing partners are available at meiosis. Population improvement through backcrossing and/or recurrent selection has not been widely practiced in potato germplasm enhancement programs. This breeding strategy offers the opportunity to create genetic combinations carrying the desired introgressions in commercially acceptable phenotypes. Two examples of successful breeding programs that have focused on recurrent selection of diploid germplasm include that of K. Haynes (USDA-Beltsville) and E. Zimnoch-Guzowska (Mlochów, Poland). The Haynes' Phureja-Stenotomum population has been improved for resistance to late blight, early blight, common scab, and internal heat necrosis, as well as yield and dry matter content (Haynes et al. 1995, 2009, 2014; Haynes and Christ 1999; Christ and Haynes 2001; Haynes 2001; Santa Cruz et al. 2009). In Poland, Dr. Zimnoch-Guzowska's program has carried out multiple cycles of recurrent selection on diploid populations containing cultivated potato and numerous wild relatives. This germplasm has been developed with the goals of improving disease resistance, starch content, tuber yield, cooking quality, and late blight resistance (Zimnoch-Guzowska and Dziewonska 1989; Zimnoch-Guzowska 1993; Jakuczun et al. 1995; Swiezynski and Zimnoch-Guzowska 1996; Zimnoch-Guzowska and Sieczka 1999; Domanski et al. 2004; Lebecka 2004; Śliwka et al. 2016). More recently, in the U.S., a recurrent selection program with a cultivated and wild germplasm base, has been carried out for five cycles (Alsahlany et al. 2016, 2017).

In addition to selecting for tuber quality, some breeders have been selecting for self-compatibility in diploids in anticipation of the production of hybrid cultivars from inbred lines. The main source of self-compatibility has been the *S. chacoense* clone M6, which is homozygous for a dominant self-incompatibility inhibitor (Jansky et al. 2014; Leisner et al. 2018). Initially, less than 20% of the individuals in the population were self-compatible. After five generations of selection, over 70% of the clones can now be self-pollinated (Douches, unpublished). Another promising source of self-compatibility is cultivated diploid germplasm. Haynes and Guedes (2018) discovered that nine of 42 long-day adapted Phureja-Stenotomum hybrids are self-compatible. These clones are being self-pollinated and used for diploid inbred-hybrid breeding.

Marker-assisted selection (MAS) provides an opportunity to track introgressions of major genes from exotic germplasm. Recent reviews of MAS in potato include De Koeyer et al. (2011) and Veilleux and Boluarte-Medina (2014). MAS is too expensive to carry out on first-generation plants, because a breeder typically evaluates 50,000 to 100,000 individuals. MAS is typically employed only on plants that pass this initial selection step and are carried to the next cycle in the breeding program (Ortega and Lopez-Vizcon 2012). An example of this scenario is provided by Vales et al. (2010) for a small group of breeding populations under development for potato virus Y resistance. Visual selection of 3698 clones in 28 families resulted in 98% discards. Of the 73 clones that were retained, 42 carried potato virus Y resistance markers.

With a dense set of DNA markers, it is possible to accelerate breeding progress by selecting against the exotic genome during the backcrossing steps. For example, marker-assisted introgression of the potato cyst nematode resistance gene from *S. vernei* can be achieved in three backcrosses across a span of six years (Bradshaw et al. 2006). Without MAS, this germplasm would require five generations across 30 years. The Douches breeding program (Michigan State University) has backcrossed Alca Tarma to a potato virus Y resistant parent in order to combine potato virus Y and potato leaf roll virus resistance. MAS has been used to identify double resistant clones and the program is now using genome-wide SNPs to accelerate the introgression of these traits.

Despite its potential power for selecting desirable segregants, MAS is not widely used in potato breeding programs (De Koeyer et al. 2011; Ortega and Lopez-Vizcon 2012). Linkage analysis of agronomic traits in potato is often carried out in experimental mapping populations, rather than breeding populations (Ortega and Lopez-Vizcon 2012). The incorporation of experimental mapping clones into breeding programs is typically not successful. They fail to be retained in the pedigrees of advanced selections due to poor adaptation or because they have other agronomic problems. Consequently, trait-specific markers developed in this experimental germplasm are not useful to breeders. In addition, many traits for which markers have been developed are not a high priority for breeders or are easy to score visually. MAS contributes to introgression efforts only when genotypic selection is more effective than phenotypic selection because "the number of generations and population sizes required for a successful introgression depend upon the frequencies of the desirable products of meiosis during sexual hybridization, which in turn depends upon the number of chromosomes and the number and distribution of chiasmata" (Bradshaw 2009). In the future, efforts to enhance meiotic recombination may be beneficial. When cultivars are used as sources of exotic genes, MAS may be more successful than when wild genes are tracked directly. For example, late blight resistance from *S. demissum* was incorporated into Tollocan by the Mexican National Potato Program. A marker for resistance derived from Tollocan has been developed and is being successfully implemented as a tool to select for late blight resistance in modern cultivars (Massa et al. 2015).

De Koeyer et al. (2011), suggest that the major limiting factor for the use of MAS in breeding programs is the lack of validated, user-friendly markers. Nie et al. (2018) has developed an improved marker for selection of potato virus X resistance. In many cases, the exotic sources of resistance genes are not known (Ortega and Lopez-Vizcon 2012). For example, Ortega and Lopez-Vizcon (2012) revealed that in their breeding program, there are 32 parents with PVY resistance, but only six and nine of those clones have the Ry_{adg} or Ry_{sto} gene, respectively, allowing them to be selected using species-specific markers. Similarly, of the 50 parents with nematode resistance, only 16 amplify known resistance markers.

Another challenge with introgression efforts is that the goal has generally been to improve disease resistance. However, while disease resistance is a desirable trait, it is not required for a cultivar's success. For example, Pentland Dell was released as a cultivar with resistance to late blight. Its resistance genes were overcome by *P. infestans*, but it is still a significant cultivar in Britain (Bradshaw 2009). Its agronomic and quality factors are sufficiently high to justify the use of fungicides when necessary to control late blight. According to Bradshaw (2009), "High levels of resistance in cultivars occurred largely by chance and susceptibility was accepted if the cultivar had other desirable traits." In some cases, breeding for disease resistance may be difficult to sell. Managers of some seed certification programs have felt that they can adequately control virus pathogens without the need for resistance genes. In fact, there may be a fear that the introduction of virus resistance into cultivars will lead to an increase in commercial production from farm-saved seed (Bradshaw 2009).

On the other hand, cultivars with high levels of disease resistance are not successful unless they exhibit superior quality traits. Several late blight resistant cultivars (Teena, Shelagh, Torridon, Brodick, Stirling, Cramond) were released by the Scottish Crop Breeding Institute, but none was commercially successful because they lacked required quality traits (Bradshaw 2009). In addition, these cultivars lacked resistance to an emerging pest, white potato cyst nematode, highlighting the risk of developing a cultivar in which the main selling point is resistance to a disease. In the U.S., several cultivars have been released with disease resistance as a main feature, but they did not become major cultivars because they did not meet other required standards. Examples include Kalkaska, Missaukee, Jacqueline Lee, Liberator, and Marcy.

One final challenge is the long-term commitment required by breeders/programs to introgress valuable genes from exotic relatives. As discussed above, many generations and decades of development efforts are typically required. The Cornell University cultivar Eva is 25% Andigena. However, 40 years passed between the first introduction of Andigena to Cornell and Eva's ultimate release. Five generations of crossing and selection for adaptation were needed first.

12.4 How Can Genomics Advances Support the Introgression of Exotic Germplasm?

Decades ago, four generations of backcrossing were generally enough to produce breeding clones worthy of consideration as cultivar. As quality expectations have increased, more generations, perhaps five or six, are likely necessary. Once wild germplasm is successfully introgressed into a successful cultivar, that germplasm is in a well-adapted background and can be transmitted to additional offspring through crosses between the successful cultivar and other clones in a breeding program. This pattern is obvious when evaluating introgressions in North American cultivars. Almost all exotic introgressions have arisen through only a few clones, including Allegany, Atlantic, Lenape, Snowden, Wauseon, Rose Gold, S440, and Krantz.

The main contribution of exotic germplasm, disease resistance, has not risen to the top of the priority list in the potato industry. In the future, as tighter regulations on pesticide use are implemented, this may change. According to Bradshaw (2009) disease resistance in cultivars is "unlikely to improve dramatically unless end-users demand it because they can see economic advantage, or governments legislate on it because they want environmental benefits. This is because breeders do not want to devote limited resources to traits that will not lead to commercially successful cultivars and hence financial rewards." In Chap. 5, genetic resistance to nematodes and viruses studied in potato species is reviewed. In addition, exotic germplasm may eventually contribute resistance to abiotic stresses, including water deficits, temperature extremes, and limited mineral availability. As these traits become more important due to climate change and the need to more tightly control the application of fertilizers, the contributions of exotic germplasm may become more significant. Some breeding programs are now beginning to focus on improving abiotic stress resistance and input use efficiencies. For example, the Haynes breeding program is using *S. chacoense* to improve nitrogen use efficiency and *S. candolleanum* as a source of heat tolerance.

We are still trying to unravel the genetic structure of potato relatives. Currently, the genome sequences of two of them, *S. commersonii* and *S. chacoense*, are available (see Chaps. 8 and 9). They should provide new insights to solve problems related to the efficient exploitation of these species. As expected, breeding system influences genotype frequencies, with outcrossing species having a higher proportion of heterozygous loci than those that can self-pollinate (Bamberg and Del Rio 2004). Recent studies based on the Illumina SNP array of wild and cultivated potatoes have reported that heterozygosity estimates in diploid wild potatoes are half of those of tetraploid cultivated potatoes (Hirsch et al. 2013). However, ascertainment bias may have led to an underestimate of heterozygosity in wild germplasm, since the SNP chip was developed using cultivated germplasm. In addition, for any given minor allele frequency, it is more likely that a tetraploid will be heterozygous than a diploid. Hardigan et al. (2017), using whole genome sequencing, showed that the current cultivated North American potato germplasm has high levels of allele sharing in coding sequences of wild species and landraces. As mentioned above, 73% of wild alleles were also found in cultivated germplasm. On the other hand, SSR markers revealed heterozygosity estimates ranging from 33 to 87% among *S. chacoense* accessions (Haynes et al. 2017). Similarly, heterozygosity estimates of 51% (SNPs) to 70% (RAPDs) were reported in *S. jamesii* (Bamberg and Del Rio 2004).

Extensive intra-genome variation exists in cultivated potato, with multiple alleles that exhibit differential expression and an abundance of copy number variants (Pham et al. 2017). Breeders have long believed that the opportunity to capitalize on this complex genetic variability in tetraploids is critical for cultivar development. However, despite the expectation that maximum heterozygosity leads to enhanced yields, levels of heterozygosity in potato cultivars have not changed across 150 years of breeding efforts, assuming no ascertainment bias with the SNP array (Hirsch et al. 2013). However, these are genome-wide estimates. Breeders may have increased heterozygosity in specific regions, so haplotype-based estimates of heterozygosity may tell a different story. Bonierbale et al. (1993) suggest that the presence of certain alleles in a heterozygous state is more important than overall heterozygosity for improving yield, especially when exotic germplasm is the source of new genetic diversity. Indeed, genome-wide SNP evaluations of a large array of cultivars revealed that "potato breeding during the last century resulted in changes to allele composition at select loci to address the needs of specific market classes rather than genome-scale alterations of heterozygosity" (Hirsch et al. 2013).

As mentioned above, breeding programs in Europe and North America are exploring the possibility of developing diploid inbred-hybrid potato cultivars (Lindhout et al. 2011; Jansky et al. 2016). *Solanum chacoense* is a major donor for self-compatibility (Jansky et al. 2014; Leisner et al. 2018). Other important sources of self-compatibility are found in diploid cultivated potato (De Jong 1977; Peterson et al. 2016). Diploid breeding with self-compatible germplasm can overcome many limitations of using wild germplasm and can also employ more efficient breeding methods. One advantage of inbred-hybrid breeding is that new genes can be incorporated into inbred lines through repeated backcrossing. While conventional potato literature often mentions backcrossing, the term is used loosely in that context. Backcrossing in conventional potato breeding refers to crossing back to cultivars. However, a different parent is used in every generation of backcrossing (Lauer 1959). It is impossible to breed a new cultivar that retains the characteristics of an existing cultivar, but that adds new traits. "It is difficult and time-consuming to introgress a resistance gene into an old cultivar by means of traditional breeding while retaining the desirable characteristics of the old cultivar. The outcome, in a highly heterozygous crop, does not recover the recurrent parent phenotype, but rather an approximation of it" (Zhang et al. 2007). By moving to the diploid level and generating inbred lines, we can take advantage of dense marker arrays to select for positive and against negative contributions from exotic germplasm. In tomato, after an inbred line is crossed to a wild species, the hybrid is repeatedly crossed back to the inbred line (using backcross in the strict sense). During the backcrossing steps, molecular markers are used to select against undesirable wild introgressions. This allows breeders to retain 99% of the cultivated genome after only three backcross generations (Tanksley et al. 1989).

It is relatively easy to create fertile, adapted hybrids between wild and cultivated potato (Hermundstad and Peloquin 1985, 1986; Yerk and Peloquin 1990; Watanabe et al. 1995). The next steps are not as clear. When hybrids are intercrossed, the so-called F2 families often show a range of phenotypes, many of which are unthrifty. This may be an example of hybrid breakdown resulting from the breakup of adapted gene complexes that have evolved independently in the cultivated and wild parents (Hawkes 1990). Hybrid breakdown has been widely reported in other organisms (Li et al. 1997; Ellison and Burton 2008; Burton et al. 2013; Barreto et al. 2015; Matsubara et al. 2015; Dai et al. 2016; Hwang et al. 2016). Continued intercrossing would not be productive, as it would continue to break up these complexes. However, backcrossing could eventually restore vigorous plants. In germplasm enhancement efforts, it will be important to determine which species/groups exhibit hybrid breakdown. These may be useful as sources of major genes, but not for base broadening. A recent example is long tuber dormancy contributed by *S. chacoense* and *S. berthaultii* (Bisognin et al. 2018). QTL on three chromosomes have been identified and may be used in marker-assisted breeding to select for or against long tuber dormancy.

While it may be surprising that a broader array of wild germplasm is not found in successful cultivars, efforts by breeders to introduce exotic germplasm have been successful. In fact, all major North American cultivars released between 1965 and 1999 contain exotic germplasm (Love 1999). Bradshaw (2009) suggests that "to ensure continued progress, our commercial breeding programmes will need new parental material developed by pre-breeding from our germplasm collections…" Likewise, according to Pavek and Corsini (2001), "transfer of useful characters from wild species to cultivate type parents, also referred to as 'pre-breeding' must continue if the long-term breeding of new varieties is to be efficient and to achieve its full potential." Perhaps germplasm enhancement efforts should focus more on the development of well-adapted, high-quality parents through backcrossing and selection following introgressions of exotic germplasm.

References

Akeley RV, Mills WR, Cunningham CE, Watts J (1968) Lenape: a new potato variety high in solids and chipping quality. Am Potato J 45:142–145

Alsahlany M, Zarka D, Coombs J, Douches D (2017) Recurrent selection in diploid potato for self-compatibility. Am J Potato Res 94:211–250

Alsahlany M, Zarka D, Coombs J, Jansky S, Douches D (2016) Redesigning diploid potato breeding with self-compatability. Am J Potato Res 93:120–149

Bamberg JB, Del Rio AH (2004) Genetic heterogeneity estimated by RAPD Polymorphism of four tuber-bearing potato species differing by breeding system. Am J Potato Res 81:377–383

Barreto FS, Pereira RJ, Burton RS (2015) Hybrid dysfunction and physiological compensation in gene expression. Mol Biol Evol 32:613–622

Bethke P, Halterman D, Jansky S (2017) Are we getting better at using wild potato species in light of new tools? Crop Sci 57:1–18

Birhman RK, Kaul MLH (1989) Flower production, male sterility and berry setting in andigena potato. Theor Appl Genet 78:884–888

Bisognin D, Manrique-Carpintero N, Douches D (2018) QTL analysis of tuber dormancy and sprouting in potato. Am J Potato Res. https://doi.org/10.1007/s12230-018-9638-0QTL

Black W (1970) Researches on potatoes at the Scottish Plant Breeding Station. Scott Plant Breed Stat Ann Rep 1970–71:52–60

Bonierbale M, Plaisted R, Tanksley S (1993) A test of the maximum heterozygosity hypothesis using molecular markers in tetraploid potatoes. Theor Appl Genet 86:481–491

Bradeen JM, Iorizzo M, Mollov DS, Raasch J, Kramer LC, Millett BP, Austin-Phillips S, Jiang J, Carputo D (2009) Higher copy numbers of the potato RB transgene correspond to enhanced transcript and late blight resistance levels. Mol Plant Microbe Interact 22:437–446

Bradshaw JE (2009) Potato breeding at the Scottish plant breeding station and the Scottish crop research institute: 1920–2008. Potato Res 52:141–172

Bradshaw JE, Bryan GJ, Ramsay G (2006) Genetic resources (including wild and cultivated *Solanum* species) and progress in their utilisation in potato breeding. Potato Res 49:49–65

Bradshaw JE, MacKay GR (1994) Breeding strategies for clonally propagated crops. In: Bradshaw JE, MacKay GR (eds) Potato genetics. CAB International, Wallington, pp 467–497

Bradshaw JE, Ramsay G (2005) Utilisation of the commonwealth potato collection in potato breeding. Euphytica 146:9–19

Burton RS, Pereira RJ, Barreto FS (2013) Cytonuclear genomic interactions and hybrid breakdown. Annu Rev Ecol Evol Syst 44:281–302

Buso JA, Boiteux LS, Peloquin SJ (2003) Tuber yield and quality of 4x-2x (FDR) potato progenies derived from the wild diploid species *Solanum berthaultii* and *Solanum tarijense*. Plant Breed 122:229–232

Buso JA, Boiteux LS, Peloquin SJ (2000) Heterotic effects for yield and tuber solids and type of gene action for five traits in 4x potato families derived from interploid (4x–2x) crosses. Plant Breed 119:111–117

Carputo D, Barone A (2005) Ploidy level manipulations in potato through sexual hybridisation. Ann Appl Biol 146:71–79

Carroll C, DeMaine M (1989) The agronomic value of tetraploid F1 hybrids between potatoes of Group Tuberosum and Group Phureja/Stenotomum. Potato Res 32:447–456

Christ BJ, Haynes KG (2001) Inheritance of resistance to early blight disease in a diploid potato population. Plant Breed 120:169–172

Colton L, Groza H, Wielgus S, Jiang J (2006) Marker-assisted selection for the broad-spectrum potato late blight resistance conferred by gene *RB* derived from a wild potato species. Crop Sci 46:589–594

Dai B, Guo H, Huang C, Zhang X, Lin Z (2016) Genomic heterozygosity and hybrid breakdown in cotton (*Gossypium*): Different traits, different effects. BMC Genet 17:58

De Jong H (1977) Self-compatibility in inbred cultivated diploid potatoes. Incompatibility Newslett 8:16–17

De Jong H, Tai GCC (1977) Analysis of tetraploid-diploid hybrids in cultivated potatoes. Potato Res 20:111–121

De Koeyer D, Chen H, Gustafson V (2011) Molecular breeding for potato improvement. In: Bradeen J, Kole C (eds) Genetics, genomics, and breeding of potato. CRC Press, Enfield, NH, pp 41–67

DeMaine MJ (1996) An assessment of true potato seed families of *Solanum phureja*. Potato Res 39:323–332

DenNijs TPM, Peloquin SJ (1977a) 2n gametes in potato species and their function in sexual polyploidization. Euphytica 26:585–600

DenNijs TPM, Peloquin SJ (1977b) Polyploid evolution via 2n gametes. Am Potato J 54:377–386

Domanski L, Zimnoch-Guzowska E, Domanska M, Zgorska K, Paczowska M (2004) The germplasm release of M-62774 and M-62805, two potato clones with cold-sweetening resistance. Folia Hortic 16:33–40

Douches DS, Coombs J, Felcher K, Kirk WW, Long C, Bird G (2009) Missaukee: a round white potato variety combining chip-processing with resistance to late blight, Verticillium wilt and golden cyst nematode. Am J Potato Res 87:10–18

Douches D, Jastrzebski K, Coombs J, Kirk W, Felcher K, Hammerschmidt R, Chase R (2001) Jacqueline Lee: a late-blight-resistant tablestock variety. Am J Potato Res 78:413–419

Ellison CK, Burton RS (2008) Interpopulation hybrid breakdown maps to the mitochondrial genome. Evolution 62:631–638

Glendinning DR (1975a) Neo-Tuberosum: new potato breeding material. 2. A comparison of Neo-Tuberosum with unselected Andigena and Tuberosum. Potato Res 18:343–350

Glendinning D (1975b) Neo-Tuberosum: new potato breeding material. 3. Characteristics Variability Neo-Tuberosum Potential Value Breed 18:351–362

Glendinning DR (1975c) Neo-Tuberosum: new potato breeding material. 1. The origin, composition, and development of the Tuberosum and Neo-Tuberosum gene pools. Potato Res 18:343–350

Glendinning DR (1976) Neo-Tuberosum: new potato breeding material. 4. The breeding system of Neo-Tuberosum, and the structure and composition of the Neo-Tuberosum. Potato Res 19:27–36

Glendinning DR (1983) Potato introductions and breeding up to the early 20th century. New Phytol 94:479–505

Gopal J (1993) Flowering behaviour, male sterility and berry setting in tetraploid *Solanum tuberosum* germplasm. Euphytica 72:133–142

Grünwald N, Cadena Hinojosa M, Covarrubias OR, Peña AR, Niederhauser JS, Fry WE (2002) Potato cultivars from the Mexican National program: Sources and durability of resistance against late blight. Phytopathology 92:688–693

Hall RL (1992) Toxicological burdens and the shifting burden of toxicology. Food Technol 46:109–112

Halterman DA, Chen Y, Sopee J, Berduo-Sandoval J, Sanchez-Perez A (2010) Competition between *Phytophthora infestans* effectors leads to increased aggressiveness on plants containing broad-spectrum late blight resistance. PLoS ONE 5(5):e10536

Hanneman RE, Peloquin SJ (1967) Crossability of 24-chromosome potato hybrids with 48-chromosome cultivars. Eur Potato J 10:62–73

Hardigan MA, Laimbeer FPE, Newton L, Crisovan E, Hamilton JP, Vaillancourt B, Wiegert-Rininger K, Wood JC, Douches DS, Farré EM, Veilleux RE, Buell CR (2017) Genome diversity of tuber-bearing *Solanum* uncovers complex evolutionary history and targets of domestication in the cultivated potato. Proc Natl Acad Sci 114:E9999–E10008

Hawkes JG (1990) The potato: evolution, biodiversity, and genetic resources. Smithsonian Institution Press, Washington, D.C.

Hayes RJ, Thill CA (2002) Introgression of cold (4 C) chipping from 2x (2 Endosperm Balance Number) potato species into 4x (4EBN) cultivated potato using sexual polyploidization. Am J Potato Res 79:421–431

Haynes KG (2001) Variance components for yield and specific gravity in a diploid potato population after two cycles of recurrent selection. Am J Potato Res 78:69–75

Haynes KG (2008) Heritability of chip color and specific gravity in a long-day adapted *Solanum phureja-S. stenotomum* population. Am J Potato Res 85:361–366

Haynes K, Guedes ML (2018) Self-compatibility in a diploid hybrid population of *S. tuberosum* Groups Phureja and Stenotomum. Am J Potato Res: In Presss

Haynes K, Christ B (1999) Heritability of resistance to foliar late blight in a diploid hybrid potato population of *Solanum phureja* x *Solanum stenotomum*. Plant Breed 118:431–434

Haynes K, Christ B, Burkhart C, Vinyard B (2009) Heritability of resistance to common scab in diploid potatoes. Am J Potato Res 81:165–170

Haynes KG, Clevidence BA, Rao D, Vinyard BT (2011) Inheritance of carotenoid content in tetraploid x diploid potato crosses. J Am Soc Hortic Sci 136:265–272

Haynes KG, Qu X, Christ BJ (2014) Two cycles of recurrent maternal half-sib selection reduce foliar late blight in a diploid hybrid *Solanum phureja-S. stenotomum* population by two-thirds. Am J Potato Res 91:254–259

Haynes KG, Wilson DR, Kang MS (1995) Genotype x environment interactions for specific gravity in diploid potatoes. Crop Sci 35:977–981

Haynes KG, Zaki HEM, Christensen CT, Ogden E, Rowland LJ, Kramer M, Zotarelli L (2017) High levels of heterozygosity hound for 15 SSR loci in *Solanum chacoense*. Am J Potato Res 94:638–646

Hermundstad SA, Peloquin S (1985) Male fertility and 2n pollen production in haploid-wild species hybrids. Am J Potato Res 62:479–487

Hermundstad SA, Peloquin SJ (1986) Tuber yield and tuber traits of haploid-wild species F1 hybrids. Potato Res 29:289–297

Hilali A, Lauer FI, Veilleux RE (1987) Reciprocal differences between hybrids of *Solanum tuberosum* Groups Tuberosum (haploid) and Phureja. Euphytica 36:631–639

Hirsch CN, Felcher K, Coombs J, Zarka D, VanDeynze A, DeJong W, Veilleux RE, Jansky S, Bethke P, Douches DS, Buell CR (2013) Retrospective view of North American potato (*Solanum tuberosum* L.) breeding in the 20th and 21st centuries. G3 Genes Genomes Genet 3:1003–1013

Hoopes R, Plaisted R, Cubillos A (1980) Yield and fertility of reciprocal-cross Tuberosum-Andigena hybrids. Am Potato J 57:275–284

Hwang, A.S., V.L. Pritchard, and S. Edmands. 2016. Recovery from hybrid breakdown in a marine invertebrate is faster, stronger and more repeatable under environmental stress. Journal of Evolutionary Biology: 1–11.

Islam S, Coombs J, Manrique-Carpintero N, Kirk W, Douches D (2016) Mapping Solanum berthaultii based late blight resistance in a diploid potato population. Am J Potato Res 93:133–134

Jakuczun H, Zgórska K, Zimnochguzowska E (1995) An investigation of the level of reducing sugars in diploid potatoes before and after cold storage. Potato Res 38:331–338

Jansky S (2006) Overcoming hybridization barriers in potato. Plant Breed 125:1–12

Jansky SH, Charkowski AO, Douches DS, Gusmini G, Richael C, Paul C, Spooner DM, Novy RG, De Jong H, De Jong WS, Bamberg JB, Thompson L, Bizimungu B, Holm DG, Brown CR, Haynes KG, Vidyasagar R, Veilleux RE, Miller JC, Bradeen JM, Jiang JM (2016) Reinventing potato as a diploid inbred line–based crop. Crop Sci 11:1–11

Jansky SHS, Chung YSY, Kittipadukal P (2014) M6: A diploid potato inbred line for use in breeding and genetics research. J Plant Registrations 8:195–199

Jansky SH, Peloquin SJ (2005) Advantages of wild diploid Solanum species over cultivated diploid relatives in potato breeding programs. Genet Resour Crop Evol 53:669–674

Kidane-Mariam H-M, Peloquin SJ (1975) Method of diplandroid formation and yield of progeny from reciprocal (4x–2x) crosses. J Am Soc Hortic Sci 100:602–603

Kittipadukal P, Bethke PCC, Jansky SHH (2012) The effect of photoperiod on tuberisation in cultivated x wild potato species hybrids. Potato Res 55:27–40

Kloosterman B, Abelenda JA, Gomez MDMC, Oortwijn M, de Boer JM, Kowitwanich K, Horvath BM, van Eck HJ, Smaczniak C, Prat S, Visser RGF, Bachem CWB (2013) Naturally occurring allele diversity allows potato cultivation in northern latitudes. Nature 495:246–250

Kozukue N, Yoon K-S, Byun G-I, Misoo S, Levin CE, Friedman M (2008) Distribution of glycoalkaloids in potato tubers of 59 accessions of two wild and five cultivated *Solanum* species. J Agric Food Chem 56:11920–11928

Krantz F (1924) Potato breeding methods III. A suggested procedure for potato breeding. Minn Agric Extension Stat Technical Bull 25:3–32

Kuhl J, Zarka K, Coombs J (2007) Late blight resistance of RB transgenic potato lines. J Am Soc Hortic Sci 132:783–789

Lauer FI (1959) Recovery of recurrent parent characters from crosses of *Solanum demissum* x *S. tuberosum* in successive backcrosses. Am Potato J 36:345–357

Lebecka R, Zimnoch-Guzowska E (2004) The inheritance of resistance to soft rot (*Erwinia carotovora* subsp. *atroseptica*) in diploid potato families. Am J Potato Res 81:395–401

Leisner CP, Hamilton JP, Crisovan E, Manrique-Carpintero NC, Marand AP, Newton L, Pham GM, Jiang J, Douches DS, Jansky SH, Buell CR (2018) Genome sequence of M6, a diploid inbred clone of the high glycoalkaloid-producing tuber-bearing potato species *Solanum chacoense,* reveals residual heterozygosity. Plant J

Li Z, Pinson SRM, Paterson AH, Park WD, Stansel JW (1997) Genetics of hybrid sterility and hybrid breakdown in an intersubspecific rice (*Oryza sativa* L.) population. Genetics 145:1139–1148

Lindhout P, Meijer D, Schotte T, Hutten RCB, Visser RGF, Eck HJ (2011) Towards F1 hybrid seed potato breeding. Potato Res 54:301–312

Liu Z, Halterman D (2006) Identification and characterization of RB-orthologous genes from the late blight resistant wild potato species *Solanum verrucosum*. Physiol Mol Plant Pathol 69:230–239

Love SL (1999) Founding clones, major contributing ancestors, and exotic progenitors of prominent North American potato cultivars. Am J Potato Res 76:263–272

Love SL, Pavek JJ, Thompson-Johns A, Bohl W (1998) Breeding progress for potato chip quality in North American cultivars. Am J Potato Res 75:27–36

MacKay G (2005) Propagation by traditional breeding methods. In: Razdan M, Mattoo A (eds) Genetic improvement of solanaceae crops volume 1 potato. Science Publishers Inc., Enfield, NH, pp 65–81

Malcolmson JF (1969) Races of *Phytophthora infestans* occurring in Great Britain. Trans Br Mycol Soc 53:417–423

Manrique-Carpintero N, Coombs JJ, Cui Y, Veilleux RE, Buell CR, Douches D (2015) Genetic map and QTL analysis of agronomic traits in a diploid potato population using single nucleotide polymorphism markers. Crop Sci 55:2566–2579

Maris M (1989) Analysis of an incomplete diallel cross among three ssp. tuberosum varieties and seven long-day adapted ssp. andigena clones of the potato (*Solanum tuberosum* L.). Euphytica 41:163–182

Massa AN, Manrique-Carpintero NC, Coombs JJ, Zarka DG, Boone AE, Kirk WW, Hackett CA, Bryan GJ, Douches DS (2015) Genetic linkage mapping of economically important traits in cultivated tetraploid potato (*Solanum tuberosum* L.). G3 Genes Genomes Genet 5:2357–2364

Matsubara K, Yamamoto E, Mizobuchi R, Yonemaru JI, Yamamoto T, Kato H, Yano M, Olsen K (2015) Hybrid breakdown caused by epistasis-based recessive incompatibility in a cross of rice (*Oryza sativa* L.). J Hered 106:113–122

McHale NA, Lauer FI (1981) Inheritance of tuber traits from phureja in diploid phureja - tuberosum hybrids. Am Potato J 58:93–102

Mendiburu AO (1971) Significance of 2n gametes in potato breeding and genetics. Ph.D. Thesis. University of Wisconsin-Madison

Mohan SK, Davis JR, Corsini DL, Sorensen LH, Pavek JJ (1990) Reaction of potato clones and accessions of *Solanum* spp. to *Verticllium dahliae* Kleb. and its toxin. Potato Res 33:449–458

Mok DW, Peloquin SJ (1975) Breeding value of 2n pollen (diplandroids) in tetraploid x diploid crosses in potatoes. Theor Appl Genet 46:307–314

Naess S, Bardeen J, Wielgus S, Haberlach G, McGrath J, Helgeson J (2001) Analysis of the introgression of *Solanum bulbocastanum* DNA into potato breeding lines. Mol Genet Genomics 265:694–704

Nie X, Dickison VL, Brooks S, Nie B, Singh M, De Koeyer DL, Murphy AM (2018) High resolution DNA melting assays for detection of *Rx1* and *Rx2* for high-throughput marker-assisted selection for extreme resistance to Potato virus X in tetraploid potato. Plant Dis 102:382–390

Novy RG, Secor GA, Farnsworth BL, Lorenzen JH, Holm ET, Preston DA, Gudmestad NC, Sowokinos JR (1998) NorValley: a white-skinned chipping cultivar with cold-sweetening resistance. Am J Potato Res 75:101–105

Ortega F, Lopez-Vizcon C (2012) Application of molecular marker-assisted selection (MAS) for disease resistance in a practical potato breeding programme. Potato Res 55:1–13

Ortiz R, Simon P, Jansky S, Stelly D (2009) Ploidy manipulation of the gametophyte, endosperm and sporophyte in nature and for crop improvement: a tribute to Professor Stanley J. Peloquin (1921–2008). Ann Bot 104:795–807

Ovchinnikova A, Krylova E, Gavrilenko T, Smekalova T, Zhuk M, Knapp S, Spooner DM (2011) Taxonomy of cultivated potatoes (*Solanum* section Petota: Solanaceae). Bot J Linn Soc 165:107–155

Paudel JR, Davidson C, Song J, Maxim I, Aharoni A, Tai HH (2017) Pathogen and pest responses are altered due to RNAi-mediated knockdown of *GLYCOALKALOID METABOLISM 4* in *Solanum tuberosum*. Mol Plant Microbe Interact 30:876–885

Pavek JJ, Corsini DL (2001) Utilization of potato genetic resources in variety development. Am J Potato Res 78:433–441

Peterson BA, Holt SH, Laimbeer FPE, Doulis AG, Coombs J, Douches DS, Hardigan MA, Buell CR, Veilleux RE (2016). Self-fertility in a cultivated diploid potato population examined with the Infinium 8303 potato single-nucleotide polymorphism array. Plant Genome 9(3)

Pham GM, Newton L, Wiegert-Rininger K, Vaillancourt B, Douches DS, Buell CR (2017) Extensive genome heterogeneity leads to preferential allele expression and copy number-dependent expression in cultivated potato. Plant J 92:624–637

Plaisted R (1973) Components of yield in potato crosses involving andigena and tuberosum germplasm. Am Potato J 50:336

Plaisted RL, Halseth DE, Brodie BB, Slack SA, Sieczka JB, Christ BJ, Paddock KM, Peck MW (2001) Eva: A midseason golden nematode- and virus-resistant variety for use as tablestock or chipstock. Am J Potato Res 78:65–68

Plaisted R, Halseth D, Thurston H, Brodie B, Loria R, Jones E (1989) Kanona: a golden nematode resistant potato variety for the chipping industry. Am Potato J 66:145–150

Plaisted RL, Hoopes RW (1989) The past record and future prospects for the use of exotic potato germplasm. Am J Potato Res 66:603–627

Quinn AA, Peloquin S (1973) Use of experimental tetraploids in potato breeding. Am Potato J 50:415–420

Rasco ET, Plaisted RL, Ewing EE (1980) Photoperiod response and earliness of S. tuberosum ssp. Andigena after six cycles of recurrent selection for adaptation to long days. Am Potato J 57:435–447

Rieman G, Hooker W, Krantz F, Werner H (1956) Potato improvement through parental line breeding. Am Potato J 33:319–323

Ríos D, Ghislain M, Rodríguez F, Spooner DM, Rios D, Rodriguez F (2007) What is the origin of the European potato? Evidence from Canary Island landraces. Crop Sci 47:1271–1280

Ritter E, Barandalla L, López R, De Galarreta JIR (2008) Exploitation of exotic, cultivated Solanum germplasm for breeding and commercial purposes. Potato Res 51:301–311

Ross H (1986) Potato breeding—Problems and perspectives. Verlag Paul Parey, Berlin and Hamberg

Salaman R (1910) Male sterility in potatoes, a dominant Mendelian character; with remarks on the shape of the pollen in wild and domestic varieties. J Linn Soc London, Bot 39:301–312

Santa Cruz JH, Haynes KG, Christ BJ (2009) Effects of one cycle of recurrent selection for early blight resistance in a diploid hybrid *Solanum phureja-S. stenotomum* population. Am J Potato Res 86:490–498

Simmonds N (1964) Studies of the tetraploid potatoes I. The variability of the Andigena Group. J Linn Soc (bot) 58:461–474

Sinden S, Sanford L, Osman S (1980) GLycoalkaloids and resistance to the Colorado potato beetle in Solanum chacoense Bitter. Am Potato J 57:331–343

Śliwka J, Sołtys-Kalina D, Szajko K, Wasilewicz-Flis I, Strzelczyk-Żyta D, Zimnoch-Guzowska E, Jakuczun H, Marczewski W (2016) Mapping of quantitative trait loci for tuber starch and leaf sucrose contents in diploid potato. Theor Appl Genet 129:131–140

Song JQ, Bradeen JM, Naess SK, Raasch JA, Wielgus SM, Haberlach GT, Liu J, Kuang HH, Austin-Phillips S, Buell CR, Helgeson JP, Jiang JM (2003) Gene RB cloned from *Solanum bulbocastanum* confers broad spectrum resistance to potato late blight. In: Proceedings of the national academy of sciences of the United States of America vol 100, pp 9128–9133

Song Y-S, Schwarzfischer A (2008) Development of STS markers for selection of extreme resistance (*Ry sto*) to PVY and maternal pedigree analysis of extremely resistant cultivars. Am J Potato Res 85:159–170

Spooner DMDM, Ghislain M, Simon R, Jansky SHSH, Gavrilenko T (2014) Systematics, diversity, genetics, and evolution of wild and cultivated potatoes. Bot Rev 80:283–383

Spooner DM, Núñez J, Trujillo G, Herrera MDR, Guzmán F, Ghislain M, Nunez J, del Rosario Herrera M, Guzma F (2007) Extensive simple sequence repeat genotyping of potato landraces supports a major reevaluation of their gene pool structure and classification. Proc Natl Acad Sci USA 104:19398–19403

Sterrett SB, Haynes KG, Yencho GC, Henninger MR, Vinyard BT (2006) 4x–2x potato clones with resistance or susceptibility to internal heat necrosis differ in tuber mineral status. Crop Sci 46:1471–1478

Stevenson FJ, Clark CF (1937) Breeding and genetics in potato improvement. Yearbook of Agriculture, USDA, pp 405–444

Swiezynski K, Zimnoch-Guzowska E (1996) Development of parental lines for Polish potato breeding. Genet Pol 37A:15–23

Tanksley SD, Young ND, Paterson AH, Bonierbale MW (1989) RFLP mapping in piant breeding: New tools for an old science. Biotechnology 7:257–264

Tarn T, Tai G (1973) Heterosis in F1 hybrids between group Andigena and group Tuberosum potatoes. Am Potato J 50:337

Tarn TR, Tai GCC (1983) Tuberosum x Tuberosum and Tuberosum x Andigena potato hybrids: Comparisons of families and parents, and breeding strategies for Andigena potatoes in long-day temperate environments. Theor Appl Genet 66:87–91

Tarn TR, Tai G, De Jong H, Tarn R, Murphy A, Seabrook J (1992) Breeding potatoes for long-day, temperate climates. Plant Breed Rev 9:217–232

Vales MI, Ottoman RJ, Ortega JA, Yilma S, Karaagac E (2010) Marker-assisted selection for PVY resistance in tetraploid potatoes. Acta Hort 859:409–416

Van Gelder WMJ, Vinke JH, Scheffer JJC (1988) Steroidal glycoalkaloids in tubers and leaves of *Solanum* species used in potato breeding. Euphytica 39:147–158

Veilleux R, Boluarte-Medina T (2014) Molecular breeding in the postgenomic era. In: Navarre R, Pavek M (eds) The potato: Botany, production, and uses. CABI International, Boston, MA, pp 290–309

Vos PG, Uitdewilligen JGAML, Voorrips RE, Visser RGF, van Eck HJ (2015) Development and analysis of a 20K SNP array for potato (*Solanum tuberosum*): An insight into the breeding history. Theor Appl Genet 128:2387–2401

Vossen JH, van Arkel G, Bergervoet M, Jo KR, Jacobsen E, Visser RGF (2016) The *Solanum demissum* R8 late blight resistance gene is an *Sw-5* homologue that has been deployed worldwide in late blight resistant varieties. Theor Appl Genet 129:1785–1796

Vossen J, Nijenhuis M, Aren-De Reuver M, van der Vossen E, Jacobsen E, Visser R (2017) Cloning and exploitation of a functional R-gene from *Solanum chacoense*. U.S. Patent 9,551,007 B2

Watanabe KN, Orrillo M, Vega S, Iwagana M, Ortiz R, Freyre R, Yerk G, Peloquin SJ, Ishiki K (1995) Selection of diploid potato clones from diploid (haploid x wild species) F1 hybrid families for short day conditions. Breed Sci 45:341–347

Watanabe K, Peloquin SJ (1991) The occurrence and frequency of 2n pollen in 2x, 4x, and 6x wild, tuber-bearing *Solanum* species from Mexico, and Central and South America. Theor Appl Genet 82:621–626

Werner JE, Peloquin SJ (1991) Occurrence and mechanisms of 2n egg formation in 2x potato. Genome 4:975–982

Yerk GL, Peloquin SJ (1989) Evaluation of tuber traits of 10, 2x(2EBN) wild species through haploid x wild species hybrids. Am Potato J 66:731–739

Yerk GL, Peloquin SJ (1990) Performance of haploid x wild species, 2x hybrids (involving five newly evaluated species) in 4x x 2x families. Am Potato J 67:405–417

Zhang LH, Mojtahedi H, Kuang H, Baker B, Brown CR (2007) Marker-assisted selection of Columbia root-knot nematode resistance introgressed from *Solanum bulbocastanum*. Crop Sci 47:2021–2026

Zimnoch-Guzowska E, Dziewonska M (1989) Breeding of potatoes at the diploid level. pp 163–171. In: Louwes K, Toussaint H, Dellaert L (eds) Parental line breeding and selection in potato breeding. Pudoc, Wageningen, the Netherlands

Zimnoch-Guzowska EEL (1993) Resistance to *Erwinia* spp. in diploid potato with a high starch content. Potato Res 36:177–182

Zimnoch-Guzowska E, Sieczka M (1999) Diploid and tetraploid parental line breeding focused on resistance to pathogens and quality traits. In: Jakuczun H, Domanski L (eds) Proceedings of the 14th triennial conference of the European association for potato research. Sorrento, Italy, pp 329–330

Perspectives of Advanced Genetics and Genomics Approaches to Exploit *Solanum* Wild Crop Relatives for Breeding

13

Massimo Iorizzo, Molla Fentie Mengist, and Nunzio D'Agostino

Abstract

Solanaceae crop breeders have abundant germplasm resources at their disposal in wild relatives and landraces. Over the past few decades, this germplasm has been used to transfer economically important traits into *Solanaceae* crops, mostly to overcome biotic stresses and improve quality. Extensive work has been done to understand interspecific crossability, identify large-scale differences in karyotypes, and establish phylogenetic relationships among and within wild and cultivated *Solanaceae* species. However, pre-breeding work to overcome linkage drag and sexual barriers is still largely based on phenotypic selection and traditional breeding schemes, which remain very time-consuming. This chapter provides an overview of the most recent breakthroughs approaches that can facilitate and expedite the effective use of wild *Solanaceae* species in breeding programs.

M. Iorizzo (✉)
Department of Horticultural Sciences, Plants for Human Health Institute, North Carolina State University, 600 Laureate Way, Kannapolis, NC 28081, USA
e-mail: miorizz@ncsu.edu

M. F. Mengist
Plants for Human Health Institute, North Carolina State University, 600 Laureate Way, Kannapolis, NC 28081, USA

N. D'Agostino
Department of Agricultural Sciences, University of Naples Federico II, via Università 133, 80055 Portici, Naples, Italy

13.1 Introduction

As the previous Chapters pointed out, *Solanaceae* crops have an exceptionally large pool of wild relatives that harbor a reservoir of alleles for traits related to agronomic and qualitative features. While the primary strategy for crop improvement remains the recurrent selection among elite modern varieties, scientists and breeders are increasingly looking at wild species as sources of novel material to widen the genetic background of *Solanaceae* crops and to develop a new generation of cultivars with improved yield and quality, resistance to pest and disease, and resilience to the challenges posed by climate change. This wild germplasm is highly accessible to breeders and allied scientists, promising increasing opportunity for genetic studies and, ultimately, crop improvement.

Until present, extensive taxonomic work has been carried out and clarified the phylogenetic relationships among and within wild and cultivated *Solanaceae* species (see Chaps. 1 and 4). Cytogenetic studies helped to uncover differences in karyotypes among cultivated and wild *Solanum* species, a diversity with a significant impact on introgression and pre-breeding

© Springer Nature Switzerland AG 2021

D. Carputo et al. (eds.), *The Wild Solanums Genomes*, Compendium of Plant Genomes,
https://doi.org/10.1007/978-3-030-30343-3_13

programs (see Chap. 2). Population genetic studies demonstrated that a significant reduction of genetic diversity occurred during the domestication process across all *Solanaceae* crops (see Chaps. 3, 6, 7, 12, and this chapter). In many cases, sexual barriers exist between wild and cultivated *Solanaceae* species. Linkage drag represent a major challenge since multiple cycles of backcrosses are necessary to break linkages between desirable and undesirable genes in introgressions of exotic germplasm. Despite these challenges, over the past few decades, multiple strategies have been used to transfer economically important traits from wild species into *Solanaceae* crops, mostly to overcome biotic stresses and improve quality (see Chaps. 3, 5–7, 12, and this chapter). Except for a few examples, marker-assisted selection (MAS) is not widely used in *Solanaceae* crop breeding programs and pre-breeding work is still largely based on phenotypic selection and traditional breeding schemes, which remains very time-consuming. Nonetheless, this extensive pre-breeding work laid solid foundations for implementing time and cost-effective strategies first to characterize and, then, use wild species in *Solanaceae* crop improvement. In some crops like potato, new efforts are underway to develop diploid breeding material with self-compatible germplasm, which can facilitate the use of wild germplasm. The surge of cost-effective genomic and biotechnology techniques holds promise to expedite modern breeding. These technologies include the application of omics-scale strategies for rapid gene discovery and advanced techniques to characterize gene functions. Perhaps within this context re-breeding efforts should focus more on: (i) the phenotypic characterization of wild accessions to be used as parents; (ii) on the establishment of suitable populations (diversity panels of large mapping populations) for large-scale genetic studies; and (iii) the expansion of the genetic and genomic studies aiming at the identification of genes controlling economically important traits. This will enable the use of MAS or/and directly transfer genes of interest from wild plant species to cultivated crops.

This chapter provides an overview of the genetic and genomic work that can facilitate and expedite the use of wild *Solanaceae* species in breeding programs. Successful achievement of these objectives would enrich the toolbox of breeders and allied scientists (e.g., geneticists, pathologists, physiologists, and taxonomists), who can, therefore, face challenges with renewed enthusiasm and achieve more easily their scientific and breeding objectives.

13.2 Capturing the Extent of Genomic Variation in Cultivated and Wild *Solanum* Species Through Pan-Genome Analysis

The availability of a high-quality reference genome sequence is of exceptional value for basic research and breeding purposes (see Chaps. 7–10). However, the genome of a single individual does not properly represent the broad diversity within a particular species or closely related lineage. Recently, a number of breakthroughs in high-throughput sequencing technologies and the drastic drop in sequencing costs have fostered the whole genome re-sequencing of additional individuals within a species or a clade. This has already allowed analyzing gene space and investigating the allele spectrum, thus providing unprecedented opportunities for genetic diversity assessment at population level. The availability of an ever-increasing number of sequenced genomes within a species prompted scientists to revisit the concept of "reference genome" (Hurgobin and Edwards 2017). Indeed, the use of a reference genome from a single individual has major implications on read mapping and evidently introduces bias in SNP discovery and in the detection of structural variations.

One of the most recent strategies to capture and investigate the extent and distribution of genetic diversity is the construction of a pan-genome (Li et al. 2014; Golicz et al. 2016; Zhao et al. 2018). The pan-genome, whose concept has been introduced by Tettelin et al. (2005), is the full complement of genes in a given species. It is made up of a 'core' genome, which includes genes shared among all individuals of the species, and a 'dispensable' genome (also referred to as 'accessory' or 'variable' genome), which comprises genes present in some but not all individuals (Zhao et al. 2018). The 'dispensable' genome plays a key role in the adaptation to diverse environments (Li et al. 2014). To cover the extent of genomic variation present within closely related species, the pan-genome concept can be expanded behind a single cultivated species.

The best strategy to develop a pan-genome is to sequence not only representative accessions of the cultivated species under study, which should be collected from different geographical locations and should represent phylogenetically distant lineages, but also a sizable number of wild relatives. This is necessary for capturing the greatest possible genetic variability within a lineage that may contribute to the detection of beneficial alleles, possibly associated with important agronomic traits, which can be introgressed into elite cultivars. It seems clear that it is more advisable to use a pan-genome as "reference" and that this guarantees plenty of advantages (Hurgobin and Edwards 2017). Firstly, it allows increasing the overall number of identified SNPs, quantifying and classifying SNPs contributed by each individual in the pan-genome, as well as discriminating SNPs from 'core' and 'dispensable' genome. The identification of the latter can be useful in characterizing molecular fingerprints to be exploited for population genetics studies. Most importantly, a pan-genome will enable the establishment of more efficient high-throughput genotyping platforms that avoid ascertainment biases toward the cultivated crops (Thomson 2014). Furthermore, the joint analysis of genomic sequences allows additional information to be captured in order to uncover genotype–phenotype relationships. In addition to SNP variation, which is the most common type of intraspecific genetic variation, the comparison between individual genomes could also result in the identification of structural variations (SVs; i.e., inversions and translocation) and copy number variations (CNVs). Structural variations are generally on a chromosomal scale, affect large fragments of DNA, and can lead to phenotypic changes. They are often associated with negative consequences, as they can hinder homologous recombination between sequences of the elite cultivar and its donor relatives, thus hampering the introgression of desirable traits into the recipient crop (Gaiero et al. 2016). On the other hand, it has been observed that SVs could be an important driver of genetic adaptation to new or changing environments (Stapley et al. 2010). CNV refers to the variability between individuals in the number of copies of genomic regions. Gene copy number variations affect gene dosage and can modify the regulation of gene expression. Gene CNVs have been associated with phenotypic variability and agronomic trait diversity in crops, which have not been captured by SNP-based studies (Bai et al. 2016).

Despite all the advantages mentioned above, the construction of a pan-genome can still be demanding in sequencing costs as well as challenging in computational time and storage capability, especially in the case of plant species with highly heterozygous, large, and complex genomes. A possible and advantageous alternative is to focus only on the gene-coding portion of the genome. Therefore, the construction of a pan-transcriptome via RNA sequencing seems to be an obligatory choice, as it represents a proxy for pan-genome (Hirsch et al. 2014; Jin et al. 2016). A pan-transcriptome is especially helpful to investigate at what extent gene expression variation contributes to phenotypic diversity. The growing interest and advantages to develop a pan-genome fostered the development of multiple computational tools to build and analyze the pan-genome (Contreras-Moreira et al. 2017; Tahir Ul Qamar et al. 2019). In the last years, several pan-genomes have been made publicly available in different crops (Hirsch et al. 2014; Li et al. 2014; Golicz et al. 2016; Zhao et al. 2018) with the aim of identifying the genetic basis that underlies complex traits and characterizing the adaptive genetic variation to different ecological niches.

In *Solanaceae,* a few genomes of wild relatives have been sequenced and used for comparative genomics (see Chaps. 7–11). Extensive genomic variations (including SNPs, SV, and CNV) have been detected (Qin et al. 2014; Bolger et al. 2014; Aversano et al. 2015; Razali et al. 2018), some of which affect important agronomic traits. Recently, the first tomato pan-genome, which includes some wild species, has been developed (Gao et al. 2019). This study remarked that the genomes of wild tomato accessions encode for a higher number of genes, thus suggesting a gradual loss of genes during domestication. Functional annotation and gene set enrichment analysis indicated that genes responsible for defense response were over-represented in wild genomes.

Expanding this study to other tomato wild species and other *Solanaceae* crops will be a key step to capture and characterize the gene load lost during domestication and to deliver a valuable resource breeders and allied scientists can exploit to identify candidate genes controlling economically important traits.

13.3 Marker-Trait Association Mapping in Wild *Solanaceae* Germplasm

Marker-trait association mapping methods, such as linkage-based quantitative trait *loci* (QTL) mapping and genome-wide association studies (GWAS), have been used to identify the genomic region(s) underlying important agronomic, nutritional quality, biotic and abiotic resistance traits in a wide range of crops (Dempewolf et al. 2017; Migicovsky and Myles 2017). Until present, marker-trait association studies in *Solanaceae* focused mostly on cultivated and economically important species (Dempewolf et al. 2017). Most of the genetic studies involving wild *Solanaceae* species were based on QTL mapping (see Chaps. 3, 5–7, and this chapter), with a relatively low marker density and populations with a small number of individuals, which limited mapping resolution. However, given the availability of reliable, cost-effective, and robust genotyping platforms, nowadays scientists can effectively exploit wild *Solanaceae* germplasm collections for marker-trait association mapping by applying different strategies. Expanding QTL/*loci* discovery and identification of markers tightly linked to economically important traits should be a primary focus of future research.

GWAS has some advantages over linkage mapping: (i) usually result in higher-resolution mapping; (ii) interrogate a broader genetic base and consequently are not restricted to alleles segregating from two parents only; (iii) require less time for population development (there is no need of mating design). On the other hand, GWAS has some limitations such as (i) the confounding effect of population substructure and relatedness can limit the detection of marker-trait associations; and (ii) it requires higher marker density (Stich et al. 2013; Crossa et al. 2017). Where a sizable (>150–200) number of wild accessions and a large number of phenotypic data points for the traits of interest are available, GWAS can be directly applied in wild species. Following identification of molecular markers associated with important traits, major-effect QTLs can be used to (i) develop a genotyping assay for marker-assisted selection (MAS); (ii) identify genes that underpin key traits and use them as potential targets for genome editing or *recombinant DNA technology* (Rao et al. 2015; Lin et al. 2019). GWAS has been applied to identify candidate *loci* associated with salt tolerance in a collection of tomato accessions, including the wild tomato species, *Solanum pimpinellifolium* (Rao et al. 2015). Lin and colleagues (2019) demonstrated that in *S. pimpinellifolium,* LD decay very rapidly (18-kb windows at $r^2 = 0.1$), indicating the potential of single-gene resolution in GWAS when marker density is not a limitation. The study also demonstrated that the use of the 8 K tomato SNP array for genotyping, whose design was based on the genome of the cultivated species, introduces bias into GWAS. Overall, this study highlighted the potential of GWAS for effective gene discovery. However, the number of publicly available GWAS that used wild *Solanaceae* species is low, partly because within each species, the number of wild accessions is often limited, and this affects the statistical power to detect marker-trait associations. Furthermore, most of the key agronomic traits are either lacking in wild species or are in strong linkage with undesirable traits (i.e., linkage drag). GWAS can be used in composite populations that include wild and cultivated species, where wild and domesticated accessions share segregating polymorphisms and are not fully differentiated at genetic level (Xu et al. 2013; Sauvage et al. 2014). However, this approach requires that the trait of interest segregate within the wild species/accessions, otherwise possible population structure signals will prevent the identification of marker-traits associations (Migicovsky and Myles 2017).

The alternative approach to GWAS is to develop bi-parental or multi-parental (e.g., multi-parent advanced generation inter-cross, MAGIC) mapping populations. By using this approach, wild and cultivated accessions are crossed and elite F_1 progenies are subjected to successive backcrosses with the recurrent parent. Depending on the crop, it may take several years to obtain advanced generations (e.g., from F_4 to F_9) that can be used for linkage-based QTL mapping (Dempewolf et al. 2017; Migicovsky and Myles 2017; Petit et al. 2017; Bali et al. 2018; Zhang et al. 2018; Brog et al. 2019). Even with the current abundance of genotyping platforms, this remains a time-consuming process, especially when the beneficial allele is in strong LD with other undesirable traits. Furthermore, crossing barriers such as cross-incompatibility and hybrid sterility also hampered these efforts (Dempewolf et al. 2017; Migicovsky and Myles 2017). Despite these challenges, this approach has been successfully used in *Solanaceae* species (Merk et al. 2012; Capel et al. 2015; Liu et al. 2015; Bali et al. 2018; Brog et al. 2019). A more efficient approach for linkage-based QTL mapping is to develop and use mapping populations derived from a cross between or within inter-fertile wild species. Wild accessions carrying or not the traits of interest are firstly selected and then used to develop bi-parental populations. Then, the construction of a genetic map and QTL mapping can be performed using the resulting progenies (Hein et al. 2009; Verzaux et al. 2011; Dempewolf et al. 2017; Yang et al. 2017). Depending on the species and the degree of divergence between the two parents, the mapping population can be F_1, F_2, backcross populations (BC_n), recombinant inbred lines (RILs), or near-isogenic lines (NILs) (Kuhl et al. 2001; Villamon et al. 2005; Gebhardt et al. 2006; Nakitandwe et al. 2007; Yang et al. 2017; Mengist et al. 2018; Nachtigall et al. 2018). As an example, this approach has been widely used in the identification of: (i) late blight resistance genes in potato species (Kuhl et al. 2001; Villamon et al. 2005; Yang et al. 2017; Nachtigall et al. 2018); (ii) cyst-nematode resistance in wild potato species (Bryan et al. 2002); (iii) bacterial speck disease *resistance genes in wild tomato species* (Zhao et al. 2013; Thapa et al. 2015); (iv) resistance to two-spotted spider mite in tomato species (Salinas et al. 2013); and (v) genes associated with fruit quality traits in wild tomatoes (Capel et al. 2015). In those cases where a bi-parental mapping approach is the only option for genetic studies, increasing population size or to develop advanced RIL populations will be required to increase recombination frequency and optimize mapping resolution. For traits that have a relatively non-complex inheritance, initial QTL mapping can still be performed using a relatively small number of progenies (N = 150–200). Following QTL mapping, a fine-mapping strategy should be used to narrow the genomic region(s) associated with the trait(s) of interest. This will enable identification of marker tightly linked to traits of interest and explore candidate genes to be used for functional characterization.

Given the relatively low cost of genotyping and sequencing the opportunity to expand marker-trait association studies in wild crop relatives exists. Identification of markers or genes associated with economically important traits directly in the wild species will enable to use MAS for selecting improved interspecific line and open opportunity to transfer the gene of interest directly through transformation techniques. This approach can dramatically reduce the time needed to overcome crossing barriers as well as to remove the "linkage drag" that comes from donor parents.

Overall, wild *Solanaceae* germplasm has retained important agronomic traits. As genotyping costs have continued to significantly drop with time, the opportunity to use wild accessions in plant breeding programs has grown. The identification of QTLs/genes via bi-parental, multi-parental, and GWAS would undoubtedly speed up breeding activities.

13.4 Genome Editing: A New Tool to Study Gene Function and Accelerate Gene Transfer

Genome editing (also referred to as gene editing) is a very effective tool for high-throughput functional genomics in crops. It is revolutionizing our ability to decipher gene functions and transcriptional regulation as well as to understand and model complex biological processes. The term genome editing groups all those technologies that can cause changes (insertion/deletion/substitution) at specific sites in the DNA sequence of a species. It relies on the use of engineered site-directed nucleases (SDNs, "molecular scissors") that generate site-specific double-strand DNA breaks (DSBs) within a target genomic region. Among SDNs are zinc finger nucleases (ZFNs), transcription activator-like effector nucleases (TALENs), and clustered regularly interspaced short palindromic repeats (CRISPR)-associated protein (CRISPR/Cas). Only with the spread of CRISPR/Cas, however, genome editing became widely used. DSBs can be repaired either by non-homologous end joining (NHEJ) or by homology-directed repair (HDR), being the latter the least frequent in plants. The NHEJ pathway is error-prone and generally is responsible for insertions/deletions (InDels) at the target site. This interesting feature can be exploited to knock-down or knock-out genes as well as to alter gene expression. In the case of HDR, a homologous sequence serves as a donor DNA to repair the DSBs, thus allowing precise gene modifications or replacement/insertion of a large DNA fragment at target *loci* (Cardi et al. 2017; Taranto et al. 2018).

An array of applications of genome editing technologies can be listed in functional genomics research. At present, the simplest and most widely used application is single-gene knock-out mediated by CRISPR/Cas. Knock-out through NHEJ-mediated disruption can typically result in small/medium-sized InDels and, thus, in complete loss of gene function. As this technique is very precise and with governable off-target effects, it is overcoming the method of silencing gene expression based on RNA interference (Guo et al. 2016). Furthermore, it turned out to be successful in editing all the alleles of polyploid species at the same time (Hilton and Gersbach 2015), showing its great potential for gene functional studies.

CRISPR/Cas can also be also applied for the simultaneous targeting of multiple genes (multiplex gene knock-out). For example, Zsögön et al. (2018), applied a genome engineering strategy based on a multiplex CRISPR/Cas9 approach to generate loss-of-function alleles for six *loci* associated with key domestication traits in tomato [general plant growth habit (*SELF-PRUNING*), fruit shape (*OVATE*) and size (*FASCIATED* and *FRUIT WEIGHT 2.2*), fruit number (*MULTIFLORA*), and nutritional quality (*LYCOPENE BETA CYCLASE*]. The authors proposed a reverse genetic approach for the de novo domestication of the wild *S. pimpinellifolium*. Indeed, the simultaneous editing of six genes resulted in the modification of fruit number, size, shape, nutrient content, and plant architecture of the engineered *S. pimpinellifolium*. As domestication traits have also been identified and characterized in many crops, multiplex gene editing could be broadly applied in molecular breeding programs aiming at the exploitation of the genetic diversity present in crop wild relatives (Zsögön et al. 2018).

Many key agronomic traits in crops are determined by variations in a single or a few nucleotides (i.e., point mutations) both in the gene-coding or regulatory region. The desired "edits" (i.e., precise base editing) can be introduced by the HDR mechanism. However, recently, CRISPR base editors have been developed to edit specific *loci* without the generation of a DSB (Komor et al. 2016; Yang et al. 2016; Eid et al. 2018). An inactive Cas9 (otherwise known as dead Cas9) is fused with adenine or cytidine deaminase offering many possibilities for precise gene editing.

The system based on dead Cas9 fused to a cytidine deaminase (responsible for C → T or G → A substitution) has been successfully used in tomato and rice to introduce multiple herbicide-resistance point mutations (Shimatani et al. 2017). Gene replacement (one-for-one substitution of a DNA fragment) or gene knock-in (the insertion of a new sequence in a target *locus*) are additional options of HDR-mediated gene targeting to be exploited for the development of new varieties with improved traits. CRISPR/Cas could also be used to remove medium/large sized DNA regions by generating two DSBs, thus introducing large deletions. This approach is particularly useful if the function of a gene cluster must be investigated or in case deleterious regions need to be removed (Hilton and Gersbach 2015). Finally, by combining the dead Cas9 with protein domains with different activities, it is possible to alter gene expression and investigate the mechanisms of transcriptional regulation (Murovec et al. 2017).

Overall, the availability of (i) the complete DNA sequence of many crops and their wild relatives, (ii) QTL and candidate genes controlling economically important traits, and (iii) the plethora of gene editing techniques is boosting basic researches that aim to elucidate gene functions, and it is revolutionizing crop improvement. Establishing effective transformation protocols is a key step needed to ultimately deliver superior genes and transfer them from the wild into *Solanaceae* crops.

References

Aversano R, Contaldi F, Ercolano MR, Grosso V, Iorizzo M, Tatino F, Xumerle L, Dal Molin A, Avanzato C, Ferrarini A, Delledonne M, Sanseverino W, Cigliano RA, Capella-Gutierrez S, Gabaldón T, Frusciante L, Bradeen JM, Carputo D (2015) The *Solanum commersonii* genome sequence provides insights into adaptation to stress conditions and genome evolution of wild potato relatives. Plant Cell 27:954–968. https://doi.org/10.1105/tpc.114.135954

Bryan GJ, McLean K, Bradshaw JE, De Jong WS, Phillips M, Castelli L, Waugh R (2002) Mapping QTLs for resistance to the cyst nematode globodera pallida derived from the wild potato species *Solanum vernei*. Theor Appl Genet 105:68–77. https://doi.org/10.1007/s00122-002-0873-9

Bolger A, Scossa F, Bolger ME, Lanz C, Maumus F, Tohge T, Quesneville H, Alseekh S, Lichtenstein G, Fich EA, Conte M, Keller H, Schneeberger K, Schwacke R, Ofner I, Vrebalov J, Xu Y, Osorio S, Aflitos SA, Schijlen E, Ryngajllo M, Kimura S, Kumar R, Koenig D, Headland LR, Maloof JN, Sinha N, Van HRCHJ, Mao L, Vogel A, Arsova B, Panstruga R, Fei Z, Rose JKC, Zamir D, Carrari F, Giovannoni JJ, Weigel D, Fernie AR (2014) The genome of the stress-tolerant wild tomato species. Nat Genet 46:1034–1039. https://doi.org/10.1038/ng.3046

Bai Z, Chen J, Liao Y, Wang M, Liu R, Ge S, Wing RA, Chen M (2016) The impact and origin of copy number variations in the Oryza species. BMC Genomics 17:1–12. https://doi.org/10.1186/s12864-016-2589-2

Bali S, Robinson BR, Sathuvalli V, Bamberg J, Goyer A (2018) Single nucleotide polymorphism (SNP) markers associated with high folate content in wild potato species. PLoS ONE 13:e0193415. https://doi.org/10.1371/journal.pone.0193415

Brog YM, Osorio S, Yichie Y, Alseekh S, Bensal E, Kochevenko A, Zamir D, Fernie AR (2019) A Solanum neorickii introgression population providing a powerful complement to the extensively characterized *Solanum pennellii* population. Plant J 97:391–403. https://doi.org/10.1111/tpj.14095

Capel C, Fernández del Carmen A, Alba JM, Lima-Silva V, Hernández-Gras F, Salinas M, Boronat A, Angosto T, Botella MA, Fernández-Muñoz R, Granell A, Capel J, Lozano R (2015) Wide-genome QTL mapping of fruit quality traits in a tomato RIL population derived from the wild-relative species *Solanum pimpinellifolium* L. Theor Appl Genet 128:2019–2035. https://doi.org/10.1007/s00122-015-2563-4

Cardi T, D'Agostino N, Tripodi P (2017) Genetic transformation and genomic resources for next-generation precise genome engineering in vegetable crops. Front Plant Sci 8:241. https://doi.org/10.3389/fpls.2017.00241

Contreras-Moreira B, Cantalapiedra CP, García-Pereira MJ, Gordon SP, Vogel JP, Igartua E, Casas AM, Vinuesa P (2017) Analysis of plant pan-genomes and transcriptomes with GET_HOMOLOGUES-EST, a clustering solution for sequences of the same species. Front Plant Sci 8:184. https://doi.org/10.3389/fpls.2017.00184

Crossa J, Pérez-Rodríguez P, Cuevas J, Montesinos-López O, Jarquín D, de los Campos G, Burgueño J, González-Camacho JM, Pérez-Elizalde S, Beyene Y, Dreisigacker S, Singh R, Zhang X, Gowda M,

Roorkiwal M, Rutkoski J, Varshney RK (2017) Genomic selection in plant breeding: methods, models, and perspectives. Trends Plant Sci 22:961–975. https://doi.org/10.1016/j.tplants.2017.08.011

Dempewolf H, Baute G, Anderson J, Kilian B, Smith C, Guarino L (2017) Past and future use of wild relatives in crop breeding. Crop Sci 57:1070–1082. https://doi.org/10.2135/cropsci2016.10.0885

Eid A, Alshareef S, Mahfouz MM (2018) CRISPR base editors: genome editing without double-stranded breaks. Biochem J 475:1955–1964. https://doi.org/10.1042/bcj20170793

Gebhardt C, Bellin D, Henselewski H, Lehmann W, Schwarzfischer J, Valkonen JPT (2006) Marker-assisted combination of major genes for pathogen resistance in potato. Theor Appl Genet 112:1458–1464. https://doi.org/10.1007/s00122-006-0248-8

Gaiero P, van de Belt J, Vilaró F, Schranz ME, Speranza P, de Jong H (2016) Collinearity between potato (*Solanum tuberosum* L.) and wild relatives assessed by comparative cytogenetic mapping. Genome 60:228–240. https://doi.org/10.1139/gen-2016-0150

Golicz AA, Bayer PE, Barker GC, Edger PP, Kim H, Martinez PA, Chan CKK, Severn-Ellis A, McCombie WR, Parkin IAP, Paterson AH, Pires JC, Sharpe AG, Tang H, Teakle GR, Town CD, Batley J, Edwards D (2016) The pangenome of an agronomically important crop plant *Brassica oleracea*. Nat Commun 7:13390

Guo Q, Liu Q, Smith NA, Liang G, Wang M-B (2016) RNA silencing in plants: mechanisms, technologies and applications in horticultural crops. Curr Genomics 17:476–489. https://doi.org/10.2174/1389202917666160520103117

Gao L, Gonda I, Sun H, Ma Q, Bao K, Tieman DM, Burzynski-Chang EA, Fish TL, Stromberg KA, Sacks GL, Thannhauser TW, Foolad MR, Diez MJ, Blanca J, Canizares J, Xu Y, van der Knaap E, Huang S, Klee HJ, Giovannoni JJ, Fei Z (2019) The tomato pan-genome uncovers new genes and a rare allele regulating fruit flavor. Nat Genet 1. https://doi.org/10.1038/s41588-019-0410-2

Hein I, Birch PRJ, Danan S, Lefebvre V, Odeny DA, Gebhardt C, Trognitz F, Bryan GJ (2009) Progress in mapping and cloning qualitative and quantitative resistance against *phytophthora infestans* in potato and its wild relatives. Potato Res 52:215–227. https://doi.org/10.1007/s11540-009-9129-2

Hirsch CN, Foerster JM, Johnson JM, Sekhon RS, Muttoni G, Vaillancourt B, Penagaricano F, Lindquist E, Pedraza MA, Barry K, de Leon N, Kaeppler SM, Buell CR (2014) Insights into the maize pan-genome and pan-transcriptome. Plant Cell 26:121–135. https://doi.org/10.1105/tpc.113.119982

Hilton IB, Gersbach CA (2015) Enabling functional genomics with genome engineering. Genome Res 25:1442–1455. https://doi.org/10.1101/gr.190124.115

Hurgobin B, Edwards D (2017) SNP discovery using a pangenome: has the single reference approach become obsolete? Biology 6:21. https://doi.org/10.3390/biology6010021

Jin M, Liu H, He C, Fu J, Xiao Y, Wang Y, Xie W, Wang G, Yan J (2016) Maize pan-transcriptome provides novel insights into genome complexity and quantitative trait variation. Sci Rep 6:18936. https://doi.org/10.1038/srep18936

Kuhl J, Hanneman R, Havey M (2001) Characterization and mapping of Rpi1, a late-blight resistance locus from diploid (1EBN) Mexican *Solanum pinnatisectum*. Mol Genet Genomics 265:977–985

Komor AC, Kim YB, Packer MS, Zuris JA, Liu DR (2016) Programmable editing of a target base in genomic DNA without double-stranded DNA cleavage. Nature 533:420–424. https://doi.org/10.1038/nature17946

Li Y, Zhou G, Ma J, Jiang W, Jin L, Zhang Z, Guo Y, Zhang J, Sui Y, Zheng L, Zhang S, Zuo Q, Shi X, Li Y, Zhang W, Hu Y, Kong G, Hong H, Tan B, Song J, Liu Z, Wang Y, Ruan H, Yeung CKL, Liu J, Wang H, Zhang L, Guan R, Wang K, Li W, Chen S, Chang R, Jiang Z, Jackson SA, Li R, Qiu L (2014) De novo assembly of soybean wild relatives for pangenome analysis of diversity and agronomic traits. Nat Biotechnol 32:1045

Liu J, Zheng Z, Zhou X, Feng C, Zhuang Y (2015) Improving the resistance of eggplant (*Solanum melongena*) to verticillium wilt using wild species *Solanum linnaeanum*. Euphytica 201:463–469. https://doi.org/10.1007/s10681-014-1234-x

Lin Y-P, Liu C-Y, Chen K-Y (2019) Assessment of genetic differentiation and linkage disequilibrium in solanum pimpinellifolium using genome-wide high-density SNP markers. G3 Genes Genomes Genetics 9:1497–1505. https://doi.org/10.1534/g3.118.200862

Merk HL, Ashrafi H, Foolad MR (2012) Selective genotyping to identify late blight resistance genes in an accession of the tomato wild species *Solanum pimpinellifolium*. Euphytica 187:63–75. https://doi.org/10.1007/s10681-012-0729-6

Migicovsky Z, Myles S (2017) Exploiting wild relatives for genomics-assisted breeding of perennial crops. Front Plant Sci 8:460. https://doi.org/10.3389/fpls.2017.00460

Murovec J, Pirc Ž, Yang B (2017) New variants of CRISPR RNA-guided genome editing enzymes. Plant Biotechnol J 15:917–926. https://doi.org/10.1111/pbi.12736

Mengist MF, Alves S, Griffin D, Creedon J, McLaughlin MJ, Jones PW, Milbourne D (2018) Genetic mapping of quantitative trait loci for tuber-cadmium and zinc concentration in potato reveals associations with maturity and both overlapping and independent components of genetic control. Theor Appl Genet 131:929–945. https://doi.org/10.1007/s00122-017-3048-4

Nakitandwe J, Trognitz FC, Trognitz BR (2007) Genetic mapping of solanum caripense, a wild relative of pepino dulce, tomato and potato, and a genetic resource for resistance to potato late blight. In: Acta horticulturae. pp 333–342

Nachtigall M, König J, Thieme R (2018) Mapping of a novel, major late blight resistance locus in the diploid (1EBN) Mexican *Solanum pinnatisectum* Dunal on chromosome VII. Plant Breeding 137:433–442. https://doi.org/10.1111/pbr.12580

Petit J, Bres C, Mauxion JP, Bakan B, Rothan C (2017) Breeding for cuticle-associated traits in crop species: traits, targets, and strategies. J Exp Bot 68:5369–5387. https://doi.org/10.1093/jxb/erx341

Qin C, Yu C, Shen Y, Fang X, Chen L, Min J, Cheng J, Zhao S, Xu M, Luo Y, Yang Y, Wu Z, Mao L, Wu H, Ling-Hu C, Zhou H, Lin H, Gonzalez-Morales S, Trejo-Saavedra DL, Tian H, Tang X, Zhao M, Huang Z, Zhou A, Yao X, Cui J, Li W, Chen Z, Feng Y, Niu Y, Bi S, Yang X, Li W, Cai H, Luo X, Montes-Hernandez S, Leyva-Gonzalez MA, Xiong Z, He X, Bai L, Tan S, Tang X, Liu D, Liu J, Zhang S, Chen M, Zhang L, Zhang L, Zhang Y, Liao W, Zhang Y, Wang M, Lv X, Wen B, Liu H, Luan H, Zhang Y, Yang S, Wang X, Xu J, Li X, Li S, Wang J, Palloix A, Bosland PW, Li Y, Krogh A, Rivera-Bustamante RF, Herrera-Estrella L, Yin Y, Yu J, Hu K, Zhang Z (2014) Whole-genome sequencing of cultivated and wild peppers provides insights into capsicum domestication and specialization. Proc Natl Acad Sci 111:5135–5140. https://doi.org/10.1073/pnas.1400975111

Rao ES, Kadirvel P, Symonds RC, Geethanjali S, Thontadarya RN, Ebert AW (2015) Variations in DREB1A and VP1.1 genes show association with salt tolerance traits in wild tomato (*Solanum pimpinellifolium*). PLoS ONE 10:e0132535. https://doi.org/10.1371/journal.pone.0132535

Razali R, Bougouffa S, Morton MJL, Lightfoot DJ, Alam I, Essack M, Arold ST, Kamau AA, Schmöckel SM, Pailles Y, Shahid M, Michell CT, Al-Babili S, Ho YS, Tester M, Bajic VB, Negrão S (2018) The genome sequence of the wild tomato *Solanum pimpinellifolium* provides insights into salinity tolerance. Front Plant Sci 9. https://doi.org/10.3389/fpls.2018.01402

Stapley J, Reger J, Feulner PGD, Smadja C, Galindo J, Ekblom R, Bennison C, Ball AD, Beckerman AP, Slate J (2010) Adaptation genomics: the next generation. Trends Ecol Evol 25:705–712

Salinas M, Capel C, Alba JM, Mora B, Cuartero J, Fernández-Muñoz R, Lozano R, Capel J (2013) Genetic mapping of two QTL from the wild tomato *Solanum pimpinellifolium* L. controlling resistance against two-spotted spider mite (*Tetranychus urticae* Koch). Theor Appl Genet 126:83–92. https://doi.org/10.1007/s00122-012-1961-0

Stich B, Urbany C, Hoffmann P, Gebhardt C (2013) Population structure and linkage disequilibrium in diploid and tetraploid potato revealed by genome-wide high-density genotyping using the SolCAP SNP array. Plant Breeding 132:718–724. https://doi.org/10.1111/pbr.12102

Sauvage C, Segura V, Bauchet G, Stevens R, Do PT, Nikoloski Z, Fernie AR, Causse M (2014) Genome-wide association in tomato reveals 44 candidate loci for fruit metabolic traits. Plant Physiol 165:1120–1132. https://doi.org/10.1104/pp.114.241521

Shimatani Z, Ishii H, Teramura H, Nishida K, Miura K, Kondo A, Komatsu H, Arazoe T, Ezura H, Takayama M, Yamamoto T, Kashojiya S, Terada R, Ariizumi T (2017) Targeted base editing in rice and tomato using a CRISPR-Cas9 cytidine deaminase fusion. Nat Biotechnol 35:441–443. https://doi.org/10.1038/nbt.3833

Tahir Ul Qamar M, Zhu X, Xing F, Chen L-L (2019) ppsPCP: a plant presence/absence variants scanner and pan-genome construction pipeline. Bioinformatics. https://doi.org/10.1093/bioinformatics/btz168

Taranto F, Nicolia A, Pavan S, De Vita P, D'Agostino N (2018) Biotechnological and digital revolution for climate-smart plant breeding. Agronomy 8:277. https://doi.org/10.3390/agronomy8120277

Tettelin H, Masignani V, Cieslewicz MJ, Donati C, Medini D, Ward NL, Angiuoli SV, Crabtree J, Jones AL, Durkin AS, DeBoy RT, Davidsen TM, Mora M, Scarselli M, Margarit y Ros I, Peterson JD, Hauser CR, Sundaram JP, Nelson WC, Madupu R, Brinkac LM, Dodson RJ, Rosovitz MJ, Sullivan SA, Daugherty SC, Haft DH, Selengut J, Gwinn ML, Zhou L, Zafar N, Khouri H, Radune D, Dimitrov G, Watkins K, O'Connor KJB, Smith S, Utterback TR, White O, Rubens CE, Grandi G, Madoff LC, Kasper DL, Telford JL, Wessels MR, Rappuoli R, Fraser CM (2005) Genome analysis of multiple pathogenic isolates of streptococcus agalactiae: implications for the microbial "pan-genome." Proc Nat Acad Sci 102:13950–13955. https://doi.org/10.1073/pnas.0506758102

Thomson MJ (2014) High-throughput SNP genotyping to accelerate crop improvement. Plant Breed Biotechnol 2:195–212. https://doi.org/10.9787/pbb.2014.2.3.195

Thapa SP, Miyao EM, Michael Davis R, Coaker G (2015) Identification of QTLs controlling resistance to *Pseudomonas syringae* pv. tomato race 1 strains from the wild tomato, *Solanum habrochaites* LA1777. Theor Appl Genet 128:681–692. https://doi.org/10.1007/s00122-015-2463-7

Villamon FG, Spooner DM, Orrillo M, Mihovilovich E, Perez W, Bonierbale M (2005) Late blight resistance linkages in a novel cross of the wild potato species *Solanum paucissectum* (series Piurana). Theor Appl Genet 111:1201–1214

Verzaux E, Budding D, de Vetten N, Niks RE, Vleeshouwers VGAA, van der Vossen EAG, Jacobsen E, Visser RGF (2011) High resolution mapping of a novel late blight resistance gene *Rpi-avl1*, from the wild bolivian species *Solanum avilesii*. Am J Potato Res 88:511–519. https://doi.org/10.1007/s12230-011-9218-z

Xu J, Ranc N, Muños S, Rolland S, Bouchet J-P, Desplat N, Le Paslier M-C, Liang Y, Brunel D, Causse M (2013) Phenotypic diversity and association

mapping for fruit quality traits in cultivated tomato and related species. Theor Appl Genet 126:567–581
Yang L, Briggs AW, Chew WL, Mali P, Guell M, Aach J, Goodman DB, Cox D, Kan Y, Lesha E, Soundararajan V, Zhang F, Church G (2016) Engineering and optimising deaminase fusions for genome editing. Nat Commun 7:13330
Yang L, Wang D, Xu Y, Zhao H, Wang L, Cao X, Chen Y, Chen Q (2017) A new resistance gene against potato late blight originating from *Solanum pinnatisectum* located on potato chromosome 7. Front Plant Sci 8:1729
Zhao Q, Zhao B, Zhang Q, Yu B, Cheng L, Jin R, Wang Y, Zhang J, Wang D, Zhang F (2013) Screening for chip-processing potato line from introgression of wild species germplasms with post-harvest storage and chip qualities. Am J Potato Res 90:425–439
Zhang S, Yu H, Wang K, Zheng Z, Liu L, Xu M, Jiao Z, Li R, Liu X, Li J, Cui X (2018) Detection of major loci associated with the variation of 18 important agronomic traits between *Solanum pimpinellifolium* and cultivated tomatoes. Plant J 95:312–323. https://doi.org/10.1111/tpj.13952
Zhao Q, Feng Q, Lu H, Li Y, Wang A, Tian Q, Zhan Q, Lu Y, Zhang L, Huang T, Wang Y, Fan D, Zhao Y, Wang Z, Zhou C, Chen J, Zhu C, Li W, Weng Q, Xu Q, Wang ZX, Wei X, Han B, Huang X (2018) Pan-genome analysis highlights the extent of genomic variation in cultivated and wild rice. Nat Genet 50:278–284. https://doi.org/10.1038/s41588-018-0041-z
Zsögön A, Čermák T, Naves ER, Notini MM, Edel KH, Weinl S, Freschi L, Voytas DF, Kudla J, Peres LEP (2018) De novo domestication of wild tomato using genome editing. Nature biotechnology

The manufacturer's authorised representative in the EU is Springer
Nature Customer Service Centre GmbH, Europaplatz 3, 69115 Heidelberg,
Germany. If you have any concerns regarding our products, please
contact ProductSafety@springernature.com

Printed and bound by CPI Group (UK) Ltd, Croydon, CR0 4YY
30/09/2026
02225691-0017